AF070967

CMS/CAIMS Books in Mathematics

Volume 15

Series Editors
Karl Dilcher, Department of Mathematics and Statistics, Dalhousie University, Halifax, Canada

Frithjof Lutscher, Department of Mathematics, University of Ottawa, Ottawa, Canada

Nilima Nigam, Department of Mathematics, Simon Fraser University, Burnaby, Canada

Keith Taylor, Department of Mathematics and Statistics, Dalhousie University, Halifax, Canada

Associate Editors
Ben Adcock, Department of Mathematics, Simon Fraser University, Burnaby, Canada

Martin Barlow, University of British Columbia, Vancouver, Canada

Heinz H. Bauschke, University of British Columbia, Kelowna, Canada

Matt Davison, Department of Statistical and Actuarial Science, Western University, London, UK

Leah Keshet, Department of Mathematics, University of British Columbia, Vancouver, Canada

Niky Kamran, Department of Mathematics and Statistics, McGill University, Montreal, Canada

Mikhail Kotchetov, Memorial University of Newfoundland, St. John's, Canada

Raymond J. Spiteri, Department of Computer Science, University of Saskatchewan, Saskatoon, Canada

CMS/CAIMS Books in Mathematics is a collection of monographs and graduate-level textbooks published in cooperation jointly with the Canadian Mathematical Society- Societé mathématique du Canada and the Canadian Applied and Industrial Mathematics Society-Societé Canadienne de Mathématiques Appliquées et Industrielles. This series offers authors the joint advantage of publishing with two major mathematical societies and with a leading academic publishing company. The series is edited by Karl Dilcher, Frithjof Lutscher, Nilima Nigam, and Keith Taylor. The series publishes high-impact works across the breadth of mathematics and its applications. Books in this series will appeal to all mathematicians, students and established researchers. The series replaces the CMS Books in Mathematics series that successfully published over 45 volumes in 20 years.

Raymond J. Spiteri · Kyle Klenk

Extreme-Scale Computing

A Practical Introduction with C++

Springer

Raymond J. Spiteri
Department of Computer Science
University of Saskatchewan
Saskatoon, SK, Canada

Kyle Klenk
Department of Computer Science
University of Saskatchewan
Saskatoon, SK, Canada

ISSN 2730-650X ISSN 2730-6518 (electronic)
CMS/CAIMS Books in Mathematics
ISBN 978-3-031-89032-1 ISBN 978-3-031-89033-8 (eBook)
https://doi.org/10.1007/978-3-031-89033-8

© The Editor(s) (if applicable) and The Author(s), under exclusive license to Springer Nature Switzerland AG 2025

This work is subject to copyright. All rights are solely and exclusively licensed by the Publisher, whether the whole or part of the material is concerned, specifically the rights of translation, reprinting, reuse of illustrations, recitation, broadcasting, reproduction on microfilms or in any other physical way, and transmission or information storage and retrieval, electronic adaptation, computer software, or by similar or dissimilar methodology now known or hereafter developed.
The use of general descriptive names, registered names, trademarks, service marks, etc. in this publication does not imply, even in the absence of a specific statement, that such names are exempt from the relevant protective laws and regulations and therefore free for general use.
The publisher, the authors and the editors are safe to assume that the advice and information in this book are believed to be true and accurate at the date of publication. Neither the publisher nor the authors or the editors give a warranty, expressed or implied, with respect to the material contained herein or for any errors or omissions that may have been made. The publisher remains neutral with regard to jurisdictional claims in published maps and institutional affiliations.

This Springer imprint is published by the registered company Springer Nature Switzerland AG
The registered company address is: Gewerbestrasse 11, 6330 Cham, Switzerland

If disposing of this product, please recycle the paper.

Preface

The year 2022 ushered in the era of exascale computing. Exascale computing refers to software applications and hardware routinely performing computations on the order of 10^{18} floating-point operations per second. Such computations have a profound impact on humanity, enabling simulations of weather, climate, and human and energy systems as well as artificial intelligence and big data processing.

There arc two main reasons to use distributed computing. The first is because your problem is too big to fit on one machine.[1] The second is because the computation you want to perform is taking too long and would benefit from utilizing more processes than are available on one machine.[1] In both cases and assuming algorithms are already optimal, the strategy is to throw more computational resources at the problem to reduce the bottlenecks. When you begin to push the envelope of what is possible in terms of computational resources that can thrown at a problem, you are entering the realm of *extreme-scale computing*.

Extreme-scale computing (ESC) is the term we use to describe the utilization of the highest class of available supercomputers, with particular emphasis on the practice of developing software for such systems. When designing software to run effectively at such extreme scales, it is often critically important to consider the hardware on which the program is to run. In other words, abstractions applied to the problem alone may not lead to effective execution on a given piece of hardware. With effectiveness in mind, it is advisable to look *first* to useful abstractions of the hardware and *then* to abstractions of the problem that cooperate harmoniously with the hardware abstractions. Those who ignore this perspective do so at their own risk!

ESC faces a myriad of challenges, including:

- heterogeneous compute hardware,
- constantly evolving and emerging supercomputing architectures, and
- network filesystems and input/output (I/O).

This book, assembled from a series of graduate-level lectures on ESC, aims to curate the wide array of topics involved in the performant programming of exascale-

[1] out of the machines to which you have access

class supercomputers. The material is presented in such a way as to expose a "non-expert" programmer (e.g., a domain scientist) to all facets of parallel programming with the following goals in mind:

1. to better interact with supercomputing resources and the technical staff associated with them and
2. to prepare for more in-depth and focused training programs such as the Argonne Training Program on Extreme-Scale Computing (ATPESC).[2]

The programming language of choice for this book is C++. The main reason for this choice is because C++ provides many abstractions that can map *performantly* to different types of hardware.

This book is organized as a course that can be delivered in approximately 36 lecture hours. There are eight parts divided into chapters that each represent a one-hour lecture. Although we have managed to cover all the material presented in one course, some basic material may be omitted or advanced material added at the discretion of the instructor, depending on the desired learning outcomes for the students. The detailed structure of the material is as follows.

Part I: ESC software management *(3 chapters)*. Much of the communication between humans and supercomputers is done via text. This part is intended to bring readers up to speed with standard tools required to build software on remote (shared) systems. It can safely be covered in less depth by the initiated. The topics covered include an introduction to the Linux command line (`bash`) and `ssh`, basic version control with `git`, and finally, building software and managing dependencies with `make` and `CMake`.

Part II: Programming patterns and modern C++ *(6 chapters)*. Although Fortran is unlikely to ever completely disappear, C++ seems to be the lingua franca of the ESC world at present. This part provides an introduction to *modern* C++, i.e., programming that emphasizes patterns (algorithms) and containers (iterators) as opposed to raw loops and pointers. The topics include serial algorithms à la the C++ Standard Template Library (STL), introduction to thread safety, and extensions to parallel programming patterns.

Part III: Hardware considerations *(2 chapters)*. In order to take full advantage of ESC hardware, it is essential to explore useful abstractions for understanding architectures from standalone x86 CPUs up to distributed networks (clusters) of GPUs. This part describes current and near-future hardware that make up the ESC landscape. Topics include components of modern computers, peripherals and accelerators, and cluster architectures.

Part IV: Distributed-memory programming *(6 chapters)*. At large scales, the existence of multiple independent memory spaces in hardware is a given, and so the ability to design (and understand) programs that work within such environments is of great importance. This part provides an introduction to MPI as a first approach to parallel programming because it requires program designs to work with disjoint memory spaces. Basic MPI usage is covered as well as a brief treatment of advanced topics in MPI-3.

---

[2] `https://extremecomputingtraining.anl.gov/`

Part V: Shared-memory and accelerator programming *(9 chapters).* It can be tricky business to program hardware where many compute resources (*threads*) have access to a single memory space. This part looks at how `OpenMP` and `CUDA` can be used to program multicore CPUs and GPUs, respectively, and how the `Kokkos` performance portability library provides the necessary abstraction layer to run performantly on either. Topics include an in-depth treatment of `OpenMP`, `CUDA`/GPU, and `Kokkos` programming.

Part VI: Parallel file systems and parallel input/output *(3 chapters).* Moving to binary file types that support parallel file systems and parallel I/O is necessary for dealing with extreme-scale data. This part gives a brief overview on using `HDF5` and `NetCDF4` for portable binary formats. Topics include the powerful and ubiquitous HDF5 and NetCDF parallel file system formats as well as an introduction to the Visualization Toolkit (VTK) for parallel I/O.

Part VII: Debugging and profiling ESC applications *(3 chapters).* Programming would not be programming without bugs, and debugging parallel code is much more involved than debugging serial code. This part gives an overview of tools that can be used for analyzing the behavior of shared and distributed parallel programs. The topics include coverage of profiling and debugging both CPU code and GPU code using `gdb`, `valgrind`, gprof, cuda-gdb, and *NVIDIA Nsight Compute.*

Part VIII: Numerical libraries for ESC *(4 chapters).* When it comes to solving extreme-scale scientific computing problems, best practice is to use existing software libraries geared toward the nature of the problem. As examples, this part considers generic software libraries for numerical linear algebra and the fast Fourier transform along with the PETSc and Trilinos software ecosystems. We conclude with a brief treatment of the actor model of concurrent computing as a potential paradigm shift in how ESC is performed.

The authors are indebted to the many colleagues and students who helped make this book better. In no particular order other than alphabetical, we wish to thank Cary Bernath, Alain Gervais, Olivier Fisette, Victoria Guenter, Greg Oster, and Reza Rafati Bonab. A special word of thanks goes to Kevin Green, who put a lot of effort into getting this work off the ground.

Saskatoon, Saskatchewan, Canada *Raymond J. Spiteri*
April, 2025 *Kyle Klenk*

Contents

Part I ESC software management

1 Linux command line 3
1.1 Two whys 3
1.2 Connecting to remote systems with SSH 5
1.2.1 Simplifying SSH connections 5
1.2.2 Managing SSH connections 6
1.2.3 SSH connections from Windows 8
1.3 Exploration of a Linux file system 8
1.3.1 File system structure 8
1.3.2 Directory navigation 9
1.3.3 Modifying files 11
1.3.4 File Permissions 12
1.3.5 Standard input, standard output, and standard error 14
1.4 Text editors and additional command-line tools 15
1.5 Shell scripting 17

2 Version control and repositories 19
2.1 Two whys 19
2.2 Repository structure 19
2.3 Setting up Git to contribute to remote repositories 21
2.4 Branches, commits, and pushing 23
2.5 Pull, fetch, merge, rebase 25
2.6 Tools for working with Git 26

3 Building software 27
3.1 Why? (Only one this time) 27
3.2 Compiling and linking 27
3.3 Makefiles 29
3.4 GNU Autotools 31
3.5 CMake 31

3.5.1 Using CMake ... 32
3.5.2 Writing CMake code ... 33
3.5.3 Variables and flow control ... 34
3.5.4 Functions and macros ... 35
3.5.5 Creating build targets ... 35
3.5.6 Installation rules ... 37
3.5.7 Tests ... 37
3.5.8 A simple CMake example ... 37
3.6 Managing external dependencies ... 38
3.6.1 Package managers ... 39
3.6.2 Modules ... 39
3.6.3 Finding existing dependencies from CMake ... 40
3.6.4 Build your own dependencies within CMake ... 40

Part II Programming patterns and modern C++

4 The C++ ecosystem ... 45
4.1 Inception and origin ... 45
4.2 Modern C++ standards ... 46
4.3 Compilers and standard implementations ... 47
4.3.1 Compilers ... 47
4.3.2 Common standard implementations ... 48

5 Primitive C++ ... 51
5.1 Basic program structure ... 51
5.2 Variables and types ... 52
5.3 Functions ... 53
5.4 Control flow ... 54
5.4.1 Iteration ... 55
5.4.2 Branching ... 56
5.5 Basic Input/Output ... 57

6 Advanced C++ ... 59
6.1 Memory, arrays, and pointers ... 59
6.2 Structs, classes, and objects ... 64
6.2.1 Inheritance and virtual functions ... 67
6.3 Templates ... 70
6.3.1 Type template parameters ... 70
6.3.2 Non-type template parameters ... 71

7 Modern C++ and guidelines ... 73
7.1 `auto` and type deduction ... 73
7.2 Memory ownership ... 74
7.2.1 Unique pointers ... 75
7.2.2 Shared pointers ... 76
7.2.3 Working with smart pointers ... 78

7.3 Functions, functions, functions 79
7.3.1 Function pointers 79
7.3.2 Function objects 80
7.3.3 Lambda functions 80
7.4 Header and implementation files 81

8 The standard template library 85
8.1 STL containers 85
8.1.1 Sequential containers 85
8.1.2 Associative containers 87
8.1.3 Adapters 88
8.2 STL iterators 88
8.3 STL Algorithms 90
8.3.1 Non-modifying sequence 91
8.3.2 Modifying sequence 91
8.3.3 Partitioning 92
8.3.4 Sorting 92
8.4 Other STL components 93

9 Parallel programming patterns 95
9.1 Loop dependence 95
9.1.1 Resolving loop dependencies 97
9.2 STL parallel algorithms 98
9.3 Parallel programming patterns primer 101

Part III Hardware considerations

10 Laptops, desktops, and workstations 109
10.1 A conceptual history of the CPU 109
10.1.1 The classical von Neumann machine 109
10.1.2 Extending von Neumann: caching 110
10.1.3 Extending von Neumann: MISD 111
10.1.4 Extending von Neumann: SIMD 112
10.1.5 Putting it all together 112
10.2 Components of a modern computer 113
10.2.1 Motherboards 113
10.2.2 CPUs 114
10.2.3 Random access memory 118
10.2.4 Persistent storage devices 118

11 Accelerators and cluster architectures 121
11.1 Peripheral connections and accelerators 121
11.1.1 PCIe 121
11.1.2 GPUs 123
11.2 Cluster architectures 125
11.2.1 Large shared-memory machines 125

11.2.2 Distributed clusters 125
11.2.3 Distributed file systems 128
11.3 Additional Resources 129

Part IV Distributed-memory programming

12 Introduction to MPI 135
12.1 A brief history 135
12.2 Practical considerations 136
12.2.1 Compiling MPI programs 137
12.2.2 Running MPI programs 137

13 MPI Basics 139
13.1 Basic Program Structure 139
13.2 MPI return codes 141
13.3 MPI Data types 142
13.4 Communicating Messages 143

14 Point-to-point communication 145
14.1 Blocking communication 145
14.2 Tags and the `MPI_Status` structure 148
14.3 Non-blocking point-to-point communication 149

15 Collective communication 153
15.1 The why 153
15.2 Broadcasting and Reducing 154
15.3 Scattering and gathering 156
15.3.1 Scattering operations 156
15.3.2 Gathering operations 159
15.4 Barriers and synchronization 161

16 Advanced MPI 163
16.1 Communicators 163
16.1.1 Group communicators 163
16.1.2 Splitting communicators 165
16.1.3 Topological communicators 166
16.2 Derived datatypes 168
16.2.1 Primitive datatypes 169
16.2.2 Contiguous datatype 169
16.2.3 Strided datatype 170
16.2.4 Indexed datatype 172
16.2.5 Struct datatype 173
16.2.6 C++ objects and deep copying 175

17 MPI scaling and recent advanced features 177
17.1 Scaling and efficiency 177
17.1.1 Strong scaling 178
17.1.2 Amdahl's law 179
17.1.3 Weak scaling 180
17.2 Advanced collectives 181
17.2.1 Non-blocking collective communication 181
17.2.2 Neighborhood collective communication 182
17.3 One-sided communication 183
17.3.1 Creating windows 184
17.3.2 Accessing remote windows 184
17.3.3 Accumulation of remote data 185
17.3.4 One-sided memory model 185

Part V Shared-memory and accelerator programming

18 OpenMP I 191
18.1 Compiling, linking, and running OpenMP programs 191
18.2 Parallel sections 192
18.3 Clauses 195
18.3.1 The `num_threads` clause 195
18.3.2 The `shared` and `private` clauses 195
18.3.3 The `reduction` clause 197

19 OpenMP II 199
19.1 Parallel for loops 199
19.2 The `schedule` clause 201
19.3 Single threads 202
19.4 Thread synchronization 203

20 OpenMP III 205
20.1 Race conditions 205
20.1.1 `atomic` clause 205
20.1.2 `critical` clause 206
20.1.3 Race condition examples 207
20.2 OpenMP Sections 210
20.3 OpenMP Tasks 211

21 GPU/CUDA programming I 213
21.1 Introduction and GPU refresher 213
21.2 CUDA basics 214
21.3 CUDA programming model 216
21.4 CUDA programming environment 217
21.4.1 CUDA runtime API 217
21.4.2 Monitoring device activity 219

22 CUDA/GPU programming II 221
22.1 Thread management 221
22.1.1 Indexing and dimensions 222
22.1.2 Thread synchronization 222
22.2 Memory management 223
22.2.1 CUDA memory basics 223
22.2.2 Shared memory 226
22.3 Error Handling 230

23 CUDA/GPU Programming III 233
23.1 Branching 233
23.2 Warp Shuffling 234
23.3 CUDA Streams 237
23.4 Multiple Devices 239
23.4.1 Unifying Memory 240

24 Kokkos I 243
24.1 Kokkos overview 243
24.1.1 What is Kokkos? 243
24.1.2 Obtaining and installing Kokkos 244
24.2 Basic data parallel patterns 244
24.2.1 Lambda functions 246
24.2.2 Equivalence to STL and OpenMP 247
24.3 Execution Spaces 247
24.4 Memory Spaces 248

25 Kokkos II 251
25.1 Views 251
25.1.1 Layouts 253
25.1.2 Memory Traits 253
25.2 Mirrors 255
25.2.1 Dual View 256
25.3 Subviews 257

26 Kokkos III 259
26.1 Hierarchical parallelism 259
26.1.1 Thread teams 260
26.1.2 `TeamThreadRange` 261
26.1.3 Vector-level parallelism 262
26.1.4 Scratch pads 263
26.2 Tasking 264

Part VI Parallel file systems and parallel input/output

27 HDF5 . . . 273
27.1 Introduction . . . 273
27.2 Data Model . . . 273
27.2.1 Groups and Datasets . . . 273
27.2.2 Dataspaces . . . 275
27.2.3 Datatypes . . . 275
27.2.4 Properties and Attributes . . . 277
27.3 Working with HDF5 . . . 278
27.4 Reference material . . . 280

28 NetCDF . . . 281
28.1 NetCDF data model . . . 281
28.1.1 Classic Data Model . . . 281
28.1.2 Enhanced Data Model . . . 284
28.2 Subsetting . . . 287
28.3 NetCDF Command-Line Tools . . . 288
28.4 Parallel access . . . 289
28.5 Additional reference material . . . 289

29 VTK . . . 291
29.1 Introduction . . . 291
29.2 Using VTK . . . 291
29.3 VTK Objects and the Visualization Pipeline . . . 293
29.3.1 Data Objects . . . 293
29.3.2 Datasets . . . 293
29.3.3 Process Objects . . . 295
29.4 VTK Reading and Writing . . . 296
29.4.1 VTK File Formats . . . 296
29.4.2 Reader and Writer Objects . . . 297
29.4.3 Example from `deal.II` . . . 297
29.4.4 VTK Rendering Engines . . . 298
29.5 Additional reference material . . . 298

Part VII Debugging and profiling ESC applications

30 Debugging with `GDB` . . . 303
30.1 `GDB` . . . 303
30.1.1 Configuring `GDB` . . . 303
30.1.2 The `GDB` Environment . . . 304
30.1.3 Common Commands . . . 304
30.1.4 Catchpoints and Watchpoints . . . 306
30.2 `Valgrind` . . . 307
30.2.1 Setup and Usage . . . 307
30.2.2 `Memcheck` . . . 307
30.2.3 `Memcheck` with MPI . . . 309
30.2.4 DRD . . . 309

30.3 MPI debugging 310
30.3.1 Using GDB and Valgrind 311
30.3.2 Debugging large problems 311
30.4 Additional resources 312

31 Profiling serial and shared-memory code 313
31.1 gprof 313
31.1.1 Usage 313
31.1.2 Profiling summaries 314
31.1.3 Run-time estimation 316
31.2 Cachegrind 317
31.2.1 Cache Misses 317
31.2.2 Branch Prediction 320
31.2.3 Cachegrind Behavior 321
31.2.4 Beyond Cachegrind 321
31.3 Codes for profiling demonstrations 321
31.4 Additional resources 323

32 Debugging and profiling compute kernels 325
32.1 CUDA-GDB 325
32.1.1 Compiling CUDA Code for Debugging 325
32.1.2 Breakpoints 326
32.1.3 Focus 326
32.1.4 Program Execution 327
32.1.5 Inspecting State 328
32.2 Profiling 328
32.3 NVIDIA Nsight Compute 328
32.3.1 Metrics 329
32.3.2 Replays 329
32.3.3 Supplementary Information 330
32.4 Additional resources 330

Part VIII Numerical libraries for ESC

33 Linear algebra and FFTW 335
33.1 BLAS 335
33.1.1 BLAS example codes 336
33.2 Multi-precision GPU computation 339
33.3 LAPACK 343
33.3.1 LAPACKE examples 345
33.4 FFTW 347
33.4.1 Using FFTW 348
33.4.2 Parallel FFTW 349
33.5 Additional Resources 350

34 PETSc . . . 351
34.1 PETSc Setup . . . 351
34.2 PETSc Vectors and Matrices . . . 352
34.3 PETSc Solvers . . . 354
34.3.1 Linear Solvers . . . 354
34.3.2 Nonlinear Solvers . . . 356
34.3.3 Time-Steppers . . . 358
34.4 Parallelism . . . 358
34.5 PDE Example . . . 359
34.5.1 Direct vs. Iterative Linear System Solvers . . . 361
34.6 Additional Resources . . . 363
34.7 Addendum . . . 363

35 The Trilinos ecosystem . . . 365
35.1 Obtaining Trilinos . . . 365
35.2 Teuchos . . . 366
35.2.1 Teuchos Core . . . 366
35.2.2 Teuchos ParameterList class . . . 367
35.2.3 Teuchos Communication . . . 368
35.2.4 Teuchos Numerics . . . 369
35.3 Tpetra . . . 370
35.3.1 Using Tpetra . . . 370
35.4 Additional Resources . . . 372

36 The actor model of concurrent computing . . . 373
36.1 Motivation . . . 373
36.2 The Actor Model . . . 374
36.3 Using the Actor Model . . . 376
36.4 Fault Tolerance . . . 378
36.5 Distributed Computing . . . 380
36.6 Advanced Actor Model Features . . . 381
36.7 Addtional Resources . . . 382

Index . . . 383

Part I
ESC software management

This first part of the book is intended to bring readers up to speed on three key aspects of working with scientific software:

1. the Linux environment including `ssh` (Chapter 1),
2. version control with `git` (Chapter 2), and
3. building software and managing dependencies (Chapter 3).

The goal of this part is to impart a basic comfort in managing and exploring an ESC environment from a Linux command-line interface.

Chapter 1
Linux command line

1.1 Two whys

Why use Linux? Not all operating systems are created equal, and some operating systems are more amenable to ESC than others. Although Windows and macOS are both popular, we believe any serious attempt at ESC requires Linux. Currently, *all* the TOP500[1] supercomputers are running some flavor of Linux, while Windows has not appeared on the list since June 2015.

The usual way to interact with a Linux operating system is through a command-line interface (CLI), or simply *the command line*. For the beginning user, the CLI is often quite unintuitive. When confronted with an unfamiliar CLI tool, our advice is to consult its `man` pages. Searching Google or asking an artificial intelligence (AI) language model like ChatGPT can also prove just as useful; however, the `man` pages are the most readily available source of information from the command line, so we take the time to introduce them here.

The `man` pages for a command can be accessed with

```
$ man <command>
```

where <`command`> is the name of the command for which you want to see the manual page. The `man` command itself has a manual page, which can be accessed with

```
$ man man
```

This command opens up the manual pages in a pager such as `more` (Section 1.4). The pager allows you to scroll through the manual page with the arrow keys, and you can exit the pager with the `q` key. The manual pages for `man` are shown in Figure 1.1

From Figure 1.1, we can see that the basics of the `man` pages are introduced. The `NAME` section gives the name of the command and a brief description of its purpose. The `SYNOPSIS` section shows the basic usage of the command, and the `DESCRIPTION` section provides a more detailed explanation of the command's functionality. We encourage you to explore the `man` pages further because Figure 1.1 only shows a

[1] `https://www.top500.org/`

© The Author(s), under exclusive license to Springer Nature Switzerland AG 2025
R. J. Spiteri and K. Klenk, *Extreme-Scale Computing*, CMS/CAIMS Books in Mathematics 15, https://doi.org/10.1007/978-3-031-89033-8_1

```
    MAN(1)

NAME
       man - an interface to the system reference manuals

SYNOPSIS
       man [man options] [[section] page ...] ...
       man -k [apropos options] regexp ...
       man -K [man options] [section] term ...
       man -f [whatis options] page ...
       man -l [man options] file ...
       man -w|-W [man options] page ...

DESCRIPTION
       man  is  the  system's manual pager.  Each page argument given to man is
       normally the name of a program, utility or function.  The manual page
       associated with each of these arguments is then found and displayed.
       A section, if provided, will direct man to look only in that section of
       the manual.  The default action is to search in all of the available
       sections following a pre-defined order (see DEFAULTS), and to show only
       the first page found, even if page exists in several sections.

       The table below shows the section numbers of the manual followed by the
       types of pages they contain.

       1   Executable programs or shell commands
       2   System calls (functions provided by the kernel)
       3   Library calls (functions within program libraries)
       4   Special files (usually found in /dev)
       5   File formats and conventions, e.g. /etc/passwd
       6   Games
       7   Miscellaneous (including macro packages and conventions), e.g.
           man(7), groff(7), man-pages(7)
       8   System administration commands (usually only for root)
       9   Kernel routines [Non standard]

       A manual page consists of several sections.

       Conventional section names include NAME, SYNOPSIS, CONFIGURATION,
       DESCRIPTION, OPTIONS, EXIT STATUS, RETURN VALUE, ERRORS, ENVIRONMENT,
       FILES, VERSIONS, CONFORMING TO, NOTES, BUGS, EXAMPLE, AUTHORS, and SEE
       ALSO.

       The following conventions apply to the SYNOPSIS section and can be used
       as a guide in other sections.

       bold text          type exactly as shown.
       italic text        replace with appropriate argument.
       [-abc]             any or all arguments within [ ] are optional.
       -a|-b              options delimited by | cannot be used together.
       argument ...       argument is repeatable.
       [expression] ...   entire expression within [ ] is repeatable.

       Exact rendering may vary depending on the output device.  For instance,
       man will usually not be able to render italics when running in a
       terminal, and will typically use underlined or colored text instead.

       The command or function illustration is a pattern that should match all
       possible invocations.  In some cases it is advisable to illustrate
       several exclusive invocations as is shown in the SYNOPSIS section of
       this manual page.
... (more options and other headings)
```

Fig. 1.1 Man page entry for man

small portion of the full output. More information about `man` pages can be found at https://www.kernel.org/doc/man-pages/.

CLI programs not distributed with Linux may or may not have man pages associated with them. If not, you might consider trying to execute the program with a `-h` or `--help` option. If the above fails, you can try to find documentation online, or as stated above, turn to Google or ChatGPT for additional help.

Why use a CLI for interacting with Linux? When using shared computing resources, a CLI is the primary means for interacting with these systems. GUI (Graphical User Interface) tools are available, but using them effectively on shared resources necessitates knowledge of CLI functionality. Additionally, the commands used in the CLI can be scripted, allowing for repeated actions to be saved and executed within a *shell script*.

For example, imagine you need to perform a similar computation on many different data sets, each with slight variations in algorithmic parameters. A shell script can be constructed to loop over the data sets and configuration files, running all desired configurations, and organizing their respective outputs. Moreover, shell scripts are used to submit jobs to a job scheduler on a supercomputer. Details of shell scripting can be found in Section 1.5. We also advocate the use of actors as a more efficient and robust way to perform repeated tasks; a brief description of actors is given in Chapter 36.

1.2 Connecting to remote systems with SSH

ESC relies on resources that are not sitting on your desk, meaning local login is not possible. The principal method for connecting with remote resources is the secure shell (SSH) protocol. SSH is the most widely used protocol for secure remote connections, even across unsecured networks.

To SSH into a remote system, SSH needs to be enabled on the target machine, and a user account you can login to needs to exist. If the above conditions are met, you can SSH into the target machine using the command

```
$ ssh username@machine.address
```

where `username` is your username on the target machine and `machine.address` is either the Internet Protocol (IP) address or the hostname of the target machine.

1.2.1 Simplifying SSH connections

Trying to remember SSH targets can be cumbersome, especially when the target machine has a long address. Luckily, simplifying SSH connection commands can be done through a configuration file that SSH reads. This file is located within the home directory on your local machine at `~/.ssh/config`. The config file is usually created the first time you connect to a remote machine, but if it does not exist, it can be created manually.

To simplify your SSH connection commands, add or modify the entry for the desired target machine like so:

```
Host machine
    HostName machine.address
    User username
```

This allows you to access the target machine `machine` as

```
$ ssh machine
```

One nuance of using SSH (that quickly becomes irritating) is you must enter your password with each connection. This is both inefficient and insecure. Thankfully, we can avoid password authentication in favor of *key-based authentication*, which uses a public/private key pair created through the `ssh-keygen` command to authenticate you on the remote system automatically. The authentication works by storing the public key on the remote system and using the private key to verify your identity. The public key can be shared with anyone without compromising security, whereas the private key should *never* be shared.

Running the `ssh-keygen` command and accepting the default options should produce the files `~/.ssh/id_rsa` and `~/.ssh/id_rsa.pub`, which are the private and public keys, respectively. During the key-generation process, you are asked to enter a passphrase. Leaving this field blank is acceptable, but some organizations may require a passphrase that encrypts your private key for extra security. When a passphrase is used, it must be entered each time the private key is used. The contents of `~/.ssh/id_rsa.pub` can be appended to the file `~/.ssh/authorized_keys` on the remote system. This copying can be performed automatically with the command:

```
$ ssh-copy-id machine
```

At this point, you should be prompted from the target machine for your password one last time. Now, provided the remote system is configured to permit automatic authentication (most commonly they are), a password will no longer be required upon each login. To check whether the process was successful, try to `ssh` to the machine and verify that it no longer prompts for a password. Additional SSH key pairs can be generated and used for different purposes; for example, one key pair can be used to sign in to remote machines, and another can be used to interact with a Git repository. The `-i` option followed by the path to the private key can be used to specify which key (sometimes called the identity file) to use when connecting to a remote machine.

1.2.2 Managing SSH connections

To both reduce the performance impact and decrease re-authenticating with multiple SSH connections, we can use *SSH multiplexing*. SSH multiplexing enables a given machine to reuse existing connections by correctly setting the `ControlMaster` and

ControlPath variables. The easiest way to enable this feature for all remote targets is to set the first two lines in your ~/.ssh/config file to

```
ControlMaster auto
ControlPath ~/.ssh/control-%h-%p-%r
```

This ensures that only one SSH *tunnel* to a given machine is ever opened up, and any further activity that uses the SSH protocol will go through that tunnel. The benefit of ContorlMaster is that it reduces the overhead of opening new SSH connections because there is no need to re-authenticate each time. The auto keyword for ControlMaster means that SSH will try to use a master connection if one exists, or it will open a new one if no such connection exists. ControlPath is set to a file name, with %h evaluated to the remote host name, %p evaluated to the port, and %r evaluated to the remote username. With the above settings, a temporary file is saved inside the local ~/.ssh/ directory when a master connection exists. For example,

```
$ ls ~/.ssh/
control-machine.address-22-username
```

triggers the existing SSH tunnel to be used whenever username tries to connect to machine.address through the (default) port 22. When all SSH connections to that machine have been terminated, the file is deleted, and the SSH tunnel is closed. The tunnel can also be closed manually by deleting the ControlPath file associated with it.

All actions associated with ControlMaster need not be set in the ~/.ssh/config file, for example, if you do not wish for all your connections to behave in the same way. Command-line options can be supplied to set and use these properties for any given connection; for further information, we refer to the ssh man page

```
$ man ssh
```

A more in-depth guide that explores using further options available with SSH can be found at https://www.digitalocean.com/community/tutorials/ssh-essentials-working-with-ssh-servers-clients-and-keys as well.

To put everything together, a basic SSH config file example is given in Listing 1.1.

Listing 1.1 Example SSH config file

```
# Comments in this config file start with #, like in bash

# Control can be set to apply to all connections
ControlMaster auto
ControlPath ~/.ssh/control-%h-%p-%r

# ServerAliveInterval sends a keepalive packet every 15 seconds
# ServerAliveCountMax will close the connection after 3 missed keepalives
Host machine1
    HostName machine1.domain.ca
    User myusername
    ServerAliveInterval 15
    ServerAliveCountMax 3

# You can set up multiple machines
Host machine2
    HostName machine2.domain.ca
    User myusername

# Sometimes you need to access one machine through another.
# This is done through a ProxyCommand.
# The ProxyCommand can use existing Host definitions
Host machine3
    ProxyCommand ssh machine1 %h %p
    User myusername
```

For more information about the SSH configuration file, we refer to its man page:

```
$ man ssh_config
```

1.2.3 SSH connections from Windows

Before Windows 10, SSH was not a native feature available to Windows users. Instead, third-party software such as PuTTY was required to establish SSH connections. However, with the most recent releases of Windows, SSH is now a native feature of the operating system. Accordingly, the information presented above on SSH is also applicable to Windows users running Windows 10 or later. Some details are different, such as the application used to launch the Windows CLI, but the general functionality (and command names) should remain the same.

For users who do not have SSH enabled natively, PuTTY is the alternative we recommend. The program along with its documentation can be found at `https://putty.org/`

1.3 Exploration of a Linux file system

Now that we have learned how to connect to a Linux system remotely, we can turn towards navigating and exploring a Linux file system from the command line.

1.3.1 File system structure

To first understand how to navigate the file system, we must begin with describing its structure. Within Linux, everything is considered a file, including directories and connected devices. Directories are special files that store lists of the files they contain. Hardware devices such as hard drives, USB drives, mice, and keyboards will also appear as files within the file system.

Fundamentally, there are two categories of files: *text* files and *binary* files. Text files are files that are encoded in text-based formats, such as ASCII or UTF-8. These files can easily be read by humans via text editors. Binary files are non-text files that exist in various formats and can be accessed by specialized programs or tools. Examples of binary files include executables, compiled libraries, and binary data files. Executables and libraries require compilation to work with a given system, while binary data files typically follow a particular schema and require specific tools to interface with them.

The Linux file system itself is organized into a tree structure, with the root directory designating the beginning. The root directory is denoted by `/`, and the entire file system is contained within it. Some common directories that make up the Linux file system along with brief descriptions are given in Table 1.1.

Directory	Purpose
/	The *root* (highest-level) directory of the system
/bin/	Binary files (executables) available to all users
/dev/	Device files for all hardware devices
/etc/	System-level configuration files for programs
/home/	The user home directories
/lib/, /lib64/	Libraries for linking programs
/opt/	External software, separated into subdirectories
/usr/	User software
/usr/bin/	Executables for user software (managed by system package manager)
/usr/include/	Header files for user software (managed by system package manager)
/usr/lib/	Libraries for user software (managed by system package manager)
/usr/local/	A local version of /usr/ for manually installed software

Table 1.1 Directories in Linux systems and their expected contents. This is not an exhaustive list but does represent the basic directories with which you should be familiar. We note that some directories are managed by a system package manager, which is a tool that automates the process of installing, upgrading, configuring, and removing software packages. Package mangers are more formally introduced in Section 3.6.1.

Whenever you connect to a machine, you begin life inside your home directory by default. This directory usually contains the path `/home/username`, where `username` is your username on the system. A typical user does not need to interact with most of the directories in Table 1.1. The goal here is to just give a sense of how things look and where things are. Usually on shared resources such as supercomputers or high-end workstations, users will see additional *mounted* file systems. Mounted file systems are additional file systems, like an additional hard drive, that get attached to the main tree structure of the file system through the `mount` command. Such file systems usually are available for a specific intended purpose such as storing large amounts of data or providing a parallel file system for speedy I/O. Every resource is generally different, so you should always consult the documentation specific to the compute resources being used.

1.3.2 Directory navigation

Having introduced the Linux file system structure, we shift focus to navigating, exploring, and interacting with it from the command line. Some key commands for navigating and modifying the file system are:

- `ls` - list the contents of a directory
- `pwd` - print working directory
- `cd` - change directory
- `mkdir` - make directory

- cp - copy file
- mv - move/rename file

Mastering these to the point of not having to look them up every time is essential for efficient navigation. The following paragraphs provide more detail to familiarize us with each command.

Beginning with ls, simply running

```
$ ls
```

at the command line shows us the files within our current directory. It is helpful to think of this command as *listing* the contents of a directory. For example, basic usage of ls might look like

```
$ ls
Desktop/      Pictures/   Videos/
Documents/    Public/
```

By default, just the file names are shown, but most commands have *options* that can be used to adjust their behavior. Options are often specified by a dash followed by a single letter (in which case they are also referred to as *flags*). With the ls command for instance, the -l (for 'long format') option can be added to retrieve more information about a directories contents, as such

```
$ ls -l
drwxr-xr-x  4 username student      4096 Apr 24 10:49  Desktop/
drwxr-xr-x  6 username student      4096 Apr 24 10:50  Documents/
drwxr-xr-x  4 username student      4096 Apr 24 10:49  Pictures/
drwxr-x---  2 username student      4096 Sep 12  2018  Public/
drwxr-xr-x  4 username student      4096 Apr 24 10:49  Videos/
```

As we can see, there is much more information provided, and the output is now ordered by column including, file permissions, number of links, owning user, owning group, file size in bytes, last edited date, and file name. We provide more information on these details in Section 1.3.4.

The ls command has introduced us to showing the contents of a directory, but how do we know which directory we are in? And how do we navigate to other directories? The pwd command prints the path of the current directory, and can easily be remembered as 'print working directory'. Example usage of pwd might look like

```
$ pwd
/home/username
```

which would be the home directory for username. To change our current working directory, we use the cd (change directory) command like so:

```
$ cd <directory>
```

The `<directory>` argument can either be an *absolute* or *relative* path to the desired directory. An absolute path starts from the root directory with /, whereas a relative path is relative to our current working directory. Some additional arguments that can be used with cd are .., ~, and -. The .. argument moves us up one directory, ~ moves us to our home directory (as does cd with no argument), and - moves us to the last directory we were in. Putting all these together, we might go through a shell session like

```
$ ls
Desktop/     Pictures/   Videos/
Documents/   Public/
$ cd Documents/
$ pwd
/home/username/Documents
$ cd /usr/local/  # navigates to /usr/local/
$ pwd
/usr/local
$ cd -            # navigates to /home/username/Documents
$ cd ..           # navigates to /home/username
$ pwd
/home/username
```

The next two commands have more to do with modifying files than navigating directories, and so we introduce them along with additional details for modifying files in Section 1.3.3.

1.3.3 Modifying files

Navigating through the file system is great, but the majority of our time is generally spent modifying it. The first command we introduce is mkdir, which is used to make a new directory and is crucial for keeping our files organized. Example usage looks like

```
$ mkdir <directory>
```

where `<directory>` is the name of the new directory, which can be an arbitrary string of characters with no spaces. Both an absolute and relative path can be used to specify the location of the new directory.

Likewise, we can also remove an *empty* directory with the rmdir command like so

```
$ rmdir <directory>
```

If the directory is not empty, we get an error. This is helpful to ensure we do not delete directories that contain data we have not moved elsewhere. It is always an excellent idea to double-check the contents of a directory before removing it because

deleting things from the Linux command line is permanent; i.e., there is no trash folder or other general way to recover removed files.

Alternatively, we can remove a directory and all of its contents with the `rm` command and the `-r` option as follows

```
$ rm -r <directory>
```

This command deletes all files within a directory recursively. Some Linux configurations ask for confirmation before deleting each file; others do not. To enable this behavior, we can use the `-i` option, then `rm` will prompt us with y(es)/n(o) for every item being deleted. Additionally, to circumvent the confirmation prompt, we can use the `-f` (force) option. This option forces the removal of all that it is asked. We strongly advise care with using `rm` because it cannot be stated often enough that **deleting things via the Linux command line is permanent**.

We have two commands at our disposal to organize files and other directories into our newly created directories. The commands are `cp` and `mv` for copy and move, respectively. Copying files and directories is done differently. For a file, we use

```
$ cp filename new_filename
```

and for directories

```
$ cp -r directoryname new_directoryname
```

noting again the `-r` option to indicate that the operation is to be performed recursively.

The `mv` command is used both to move and rename files and directories. For example,

```
$ mv file_or_directory_name new_name
```

can have different behaviors depending on the state of `file_or_directory_name` and `new_name`.

- If `new_name` does not exist, then a simple rename of `file_or_directory_name` is performed, whether it is a file or directory.
- If both files exist, `file_or_directory_name` is renamed to `new_name`, overwriting its contents, and `file_or_directory_name` will no longer exist.
- If `new_name` is an existing directory, then `file_or_directory_name` is moved inside the `new_name` directory.

1.3.4 File Permissions

Recalling the output of `ls -l` from Section 1.3.2, we see important information about which files we and other users can interact with and how. All files and directories have a set of permissions associated with them in addition to an owning user and group.

The permissions of a file or directory are presented in the first column of the output from `ls -l` and generally consist of a ten-character string that tells us about who can access the file and in which ways. An additional + may be present at the end of the permissions string to indicate the presence of *Access Control Lists* [2] on the file. The first character is the directory flag, and the remaining nine characters are organized into groups of three: read, write, and execute. The first group (characters 2-4) corresponds to the permissions of the file owner, the second group (5-7) corresponds to the group owner, and the last group (8-10) corresponds to all other users. With all characters, a full specification looks like

```
drwxrwxrwx
```

where the characters represent: **d**irectory, **r**ead, **w**rite, and e**x**ecute. If a certain permission is not set, the character is replaced with a dash, `-`. As an example, the permission specification for a text file that has read and write access for its owner and group and read access for all other users looks like

```
-rw-rw-r--
```

This specification means that the owner and any member of the owning group can both see and edit the contents of the file, whereas any other user on the system can only read the contents. To determine the owning user and group of a file, we can reference the third and fourth columns of the `ls -l` output.

The permissions of a file or directory can be changed on the command line using the `chmod` (for 'change file mode') command. A frequent task is to add execute permissions to files (in the case of a script, for example), which can be achieved with

```
$ chmod +x filename
```

noting that execute permissions are given at each level, i.e., the above `-rw-rw-r--` would become `-rwxrwxr-x`. Similarly, permissions can be removed at all levels using - instead of +, and multiple permissions can be changed at once. For example, running

```
$ chmod -rw filename
```

would change `-rwxrwxr-x` to `---x--x--x`. This style of modifying permissions can be narrowed specifically to `u`–owner, `g`–group, or `o`–other by prefixing the `+/-`. For example,

```
$ chmod g-w filename
```

would convert `-rwxrwxr-x` to `-rwxr-xr-x`.

Up to this point, we have only used the `chmod` command to modify permissions by adding or removing them. However, it is also possible to set the permissions to a specific state using a numerical shorthand for `chmod`. The shorthand looks like

```
$ chmod xyz filename
```

[2] Although we do not cover Access Control Lists explicitly, they are a way to provide more fine-grained control over user access to files and directories. See `https://en.wikipedia.org/wiki/Access-control_list` for more information.

where x represents the state of the owner permissions, y the state of the group permissions, and z the state of the other permissions. Each value of the state is a number 0–7 that sets the level's permissions. Perhaps better thought of as a 3-bit binary (octal) number, where the numeric integers correspond to the values in Table 1.2. This shorthand allows explicit setting of all permissions of a file in a

Value	Binary	Permissions
0	000	---
1	001	--x
2	010	-w-
3	011	-wx
4	100	r--
5	101	r-x
6	110	rw-
7	111	rwx

Table 1.2 Numeric shorthand values for chmod permissions.

concise way. For example,

```
$ chmod 755 filename
```

sets the permissions to -rwxr-xr-x, which is often a desirable permissions state for an executable file.

Finally, to use chmod to change permissions on a file or directory, a user must either be the owner of the file or in the *group* of users associated with the file.

1.3.5 Standard input, standard output, and standard error

Whenever a command provides output, this output is written to a *stream*. Typically, the output of a command gets written to the standard output stream, stdout. The other two streams are called standard input, stdin, and standard error, stderr, and are used for command input and error output, respectively. These streams are generally used to give the results of a command and can conveniently be redirected to other streams. For example, one can take the stdout from one command and redirect it to a file or even to the stdin of another command.

The stdout stream can easily be redirected to a file in two different ways. The first, >, overwrites the file, e.g.,

```
$ ls -l > file_list.txt
```

replaces the contents of file_list.txt with a listing of the current directory's files. The second, >>, appends to the end of the file, e.g.,

```
$ ls -l / >> file_list.txt
```

appends a listing of the root directory to the end of file file_list.txt.

The `stderr` stream operates in a slightly different fashion. For example, if you try to remove a non-empty directory, you find that the output message cannot be redirected in this way:

```
  $ rmdir tmp_dir > rmdir.out
rmdir: failed to remove 'tmp_dir': Directory not empty
```

because the error message gets written to the `stderr` stream (even though it ends up appearing on the screen) — there is no output stream to be redirected to `rmdir.out`. We note here that any attempt to redirect output to a file creates the desired file regardless of whether anything comes out of the stream or not; i.e., the previous example produces an empty `rmdir.out`. To redirect `stderr`, one uses 2> or 2>> as in, for example,

```
$ rmdir tmp_dir 2> rmdir.out
$ rmdir tmp_dir 2>> rmdir.out
```

which produces a file `rmdir.out` that contains two copies of the error message.

A common act is to redirect both `stdout` and `stderr` to the same place; this is achieved with &>

```
$ rmdir tmp_dir &> rmdir.out
```

However, this command cannot be used to append both streams to a file. To do that, we must consider the more general operation of combining streams with 2>&1 which can be thought of as "take stream 2 and put it in the same place as stream 1", noting that stream 1 is `stdout`. Stream redirections are always read from left to right, and so to append both `stdout` and `stderr` to a file, it must be specified as

```
$ rmdir tmp_dir >> rmdir.out 2>&1
```

The act of redirecting `stdout` from one command to be used as `stdin` for another command is called *piping*. It uses the syntax | to perform the redirection. The ability to construct chains of operations such as

```
$ cmd1 | cmd2 2>&1 | cmd3 | ...
```

Combined with text files and text-processing tools, such chains make for a versatile component in processing hoards of text.

1.4 Text editors and additional command-line tools

Text files are the most common type of files interacted with from the command line. Because text files are so common, it is a great idea to familiarize yourself with how to efficiently use a text editor for dealing with them. Some common text editors include `emacs`, `vim`, `nano`, `sublime`, and `vscode`. We refrain from taking up arms in the religious wars over text editors, but we do strongly suggest that one get familiar with both `emacs` and `vim` due to their power and ubiquity—`emacs` being quite useful as a generic interface to text and `vim` for quick (and sometimes dirty) edits on any remote Linux system.

Comprehensive reference cards for `emacs` and `vim` can be found at `https://www.gnu.org/software/emacs/refcards/pdf/refcard.pdf` and `https://michaelgoerz.net/refcards/vimqrc.pdf`, respectively.

In Linux, the extension of a file really only serves the purpose of informing users of its potential contents. File extensions are not needed on any file, but, in the spirit of good communication, it is recommended to follow naming conventions when dealing with source code for any particular language or data files of specific data formats. For example, it is useful to have C++ source files with extension `.cpp`, C++ header files with `.hpp`, shell scripts with `.sh`, and HDF5 files with `.h5`. Using consistent extensions also helps software (e.g., text editors or analysis software) filter files and determine special modes under which to open a particular file.

At the command line, there are some additional basic and useful operations on text files:

- Write contents of a file to `stdout`:

  ```
  $ cat <filename>
  ```

  The `cat` command also produces output for binary files, but it is likely to be unintelligible. For files that are long, it may be preferable to view them in a *pager* like `more` or `less`. We encourage you to explore such tools via their `man` pages.
- Search a text file using `grep`. There are two common ways to do this. First, by using a `filename` directly in the `grep` command,

  ```
  $ grep <pattern> <filename>
  ```

  Alternatively, `grep` can read from `stdin` as well, allowing its use in pipes. For example, the following produces the same effect of the `grep` call above

  ```
  $ cat <filename> | grep <pattern>
  ```

  but will run more slowly.
- Cutting a specific value/column from a line of text. This operation comes up frequently in post-processing output log files. As an example, consider the output of `ls -l`:

  ```
    $ ls -l tmp_file
  -rw-r--r-- 1 username username 0 Sep 10 09:17 tmp_file
    $ ls -l tmp_file | cut -d " " -f 8
  09:17
  ```

  Piping through the above `cut` command splits with the **d**elimiter " " and takes the eighth **f**ield, in this case being the last modified time.

Any program that is called from the command line must either exist in your `PATH` environment variable or be specified by a relative or absolute path. You can inspect your `PATH` variable by running

```
  $ echo $PATH
/usr/local/sbin:/usr/local/bin:/usr/sbin:/usr/bin:/sbin:/bin
```

The way to interpret this is as an ordered list, separated by colons, of locations where the shell looks for executables. For the locations in your PATH, you do not need to explicitly specify a path to the executable. To determine which executable is being used, invoke the which command

```
  $ which ls
/bin/ls
```

If you ever see output that is unexpected for a given program, a useful sanity check is to ensure you are running the correct program!

To execute programs in your current directory, you need to prepend the executable name with ./, but you can also run programs by relative or absolute paths:

```
$ ./program  # run program in the current working directory
$ subdirectory/program # run program located in subdirectory
$ /home/username/bin/program # run program by absolute path
```

1.5 Shell scripting

To finish our discussion of the Linux command line, we introduce shell scripting. Shell scripts are files containing sequences of shell commands that are intended to run in a specified order. An example of a shell script is given in Listing 1.2.

The details of what the script actually does are not important at the moment. As a rule, one should be cautious about running scripts from the internet or unknown sources; however, this script is safe to run and will not make any changes to your system. We also encourage you to return to this script and analyze it later when you have more experience with scripting.

Shell scripts are sometimes referred to as bash scripts and carry the file extension .sh. This file extension is not unique to bash scripts and is commonly used for shell scripts in general. Bash is generally the default shell (CLI) on Linux systems. Because of how Linux deals with files, the file extension may not accurately reflect the interpreter of the script. Therefore, for a text file to be executable, it must contain a *shebang* (#!) of the form:

```
#!<interpreter> [optional arguments]
```

This tells the Linux system to use the specified <interpreter> and its optional arguments to run the contents of the file.

For shell scripts (bash), common shebangs are

```
#!/usr/bin/bash
```

```
#!/usr/bin/env bash
```

The first one specifies precisely to use the version of bash present at /usr/bin/bash. The second one is more generic, telling the env program to use the bash environment. The difference really only comes in to play when virtual environments[3] are being used or if multiple versions of bash are present. A script with either version of shebang can be executed with

```
$ ./script.sh
```

Listing 1.2 Example bash script.

```
#!/usr/bin/env bash

# Create a temporary working dir
TMPDIR=.tmp
STARTDIR=pwd

mkdir $TMPDIR
cd $TMPDIR

# Do something basic
for i in {0..10}
do
  if [ $((i%2)) -eq 0 ]
  then
    echo $i >> even.txt
  else
    echo $i >> odd.txt
  fi
done

for f in *.txt
do
  echo $f
  cat $f
done

echo "Second-last odd number:"
tail -2 odd.txt | head -1

# Go back to starting directory and clean up
cd $STARTDIR
rm -rf $TMPDIR
```

Shell scripts do not necessarily need a shebang in them. If they do not, then they cannot simply be executed; rather, the script needs to be passed as an argument to the bash (or any other script interpreter) command:

```
$ bash script_with_no_shebang.sh
```

An introductory tutorial to using bash can be found at https://linuxconfig.org/bash-scripting-tutorial-for-beginners. The most useful elements of that tutorial are

- variable declaration and usage,
- flow control (if/then/else), and
- looping constructs.

[3] So for example, this is more of a concern when dealing with Python scripts.

Chapter 2
Version control and repositories

2.1 Two whys

Why version control? Version control (VC) allows us to easily track changes to text-based files. Although VC can also track changes to binary files, it is primarily designed for human-readable files, such as source code. VC provides a comprehensive history of all changes to files that have been placed under it. This history allows us to inspect the differences line-by-line between any two versions of a file, a feature that is often necessary for developers to understand software updates.

Why Git? Git is known as a *decentralized VC system*. This means that every copy of a Git repository is itself a complete repository. In contrast, Subversion (svn) is a *centralized VC system*, which represents an older model of VC, where all actions center around a main repository that all users "check out" from and "check in" to. Both systems have their advantages and disadvantages and are implemented within various different software packages apart from Git and svn. However, Git is by far the most popular VC system at present and offers the following benefits:

- Nearly every operation is a local operation (no synchronization with the main *remote* repository is necessary after every change/commit because every repository is a complete repository).
- Git scales up to more developers modifying code simultaneously (because of its decentralized nature and simplified branch management).

2.2 Repository structure

A Git repository is a collection of files and directories that are tracked by Git. There are typically two types of Git repositories: *remote* and *local*. The remote repository is intended to be the core repository for a project. These repositories are often hosted

© The Author(s), under exclusive license to Springer Nature Switzerland AG 2025

R. J. Spiteri and K. Klenk, *Extreme-Scale Computing*, CMS/CAIMS Books in Mathematics 15, https://doi.org/10.1007/978-3-031-89033-8_2

with hosting services such as GitHub[1], which provide many additional features, like a web interface, for managing community contributions to hosted projects. A local repository is typically a *clone*—a copy of a remote repository—made available through a directory that is created on your local machine. This local repository is where you make changes and save them through Git *commits*. A commit is a snapshot of the repository at a given point and is usually accompanied by a message describing the changes made since the previous commit. Once your changes are ready to be applied to the remote repository, the *push* operation is used to send your local commits to the remote repository for others to use.

For Git projects, the remote repository is usually considered the definitive version of the project. For example, the definitive version of the Trilinos project is the publicly available version on GitHub at `https://github.com/trilinos/Trilinos`.

It is possible to, however, have many remote repositories for a single project. This is achieved by *forking* a repository, a process that creates a new remote repository that is a copy of the original. Forking is not a Git operation but rather a feature of GitHub (and other Git hosting services). Therefore, it is important that projects dictate a definitive version of their project's repository. GitHub makes this easy by automatically tracking forks of a repository. For example, the Trilinos repository has over 500 forks at the time of this writing, each potentially exhibiting a range of differences from the original. This may seem like overkill, but the logic is straightforward: If you fork a repository to your GitHub account, you can `clone` your forked repository to your local machine and make changes. The benefit of this is the `Trilinos` account does not need to manage your permissions on their original repository, and you can freely change things on your own. If you complete some potentially useful changes, you can offer that the changes from your forked repository be merged into their main repository via a *Pull Request*. Pull Requests are another feature of GitHub and other Git hosting services: they allow users to request that changes from one repository be merged into another, granting the community a chance to review and discuss the changes before they are merged.

When you clone a repository, it is automatically set up to track the remote repository from which it was cloned. This remote repository is usually called `origin`. You can see all the remote repositories that your local repository is tracking with the command

```
$ git remote
```

When you perform a *push*, your local changes will, by default, update the `origin` remote repository. Likewise, when you perform a *pull*, your local repository will, by default, update from `origin`. However, you can add as many remote repositories as you like for your local repository to track. If for example, you have forked the Trilinos repository and cloned your fork to your local machine, you can add the original Trilinos repository as a remote repository with the command

```
$ git remote add upstream https://github.com/trilinos/Trilinos.git
```

[1] `https://github.com`

The name `upstream` is common for the original repository from which a fork was created, but any name can be used. The reason for adding the original repository as a remote repository is because you may have obtained write access to the original Trilinos repository and can now push changes directly to it. Another more common scenario for using remote repositories is to push and pull changes between two Git projects that come from the same code base. In this scenario, you can add another remote repository to your local repository and merge the changes between them. These repositories do not necessarily have to have been forked from each other or even be hosted on the same hosting service. All that is required is that their git history originates from some common commit. Setting up remote repositories is recommended for collaborating with other developers and obtaining changes from multiple repositories of the same originating project.

2.3 Setting up Git to contribute to remote repositories

The easiest way to set up Git to interact with remote repositories is to set up SSH keys with the remote repository hosting service. GitHub, for example, uses the SSH protocol to enable automatic authentication once you have set up your SSH keys with your GitHub account. The instructions for setting up SSH keys with GitHub can be found at `https://docs.github.com/en/github/authenticating-to-github/adding-a-new-ssh-key-to-your-github-account`. We highly recommend that you do this for the automatic authentication.

In order to push changes to a repository, you need to have *push access*. In general, for repositories that you do not own, you can approach this in one of three ways: get the repository owner to grant you push access, create a fork (introduced in Section 2.2) of the repository to some location where you have such access, or clone the repository to your local machine and push the entire repository to an empty remote repository that you own. The second method is handled in GitHub simply by using the `Fork` button. When a repository is forked, GitHub simply creates a copy of the repository under your username. Forking can be viewed as the streamlined version of cloning a repository and pushing it to a new remote repository, while maintaining a connection to the original repository automatically through GitHub.

The typical workflow for contributing to a repository that you do not own is:

- Fork the repository to your account.
- Clone the forked repository to your local machine.
- Make your changes and commit to branches in the cloned repository.
- Push your changes to your forked repository.
- Submit a *Pull request* for your changes to the developers of the main repository.

Cloning a repository for use is simple. GitHub offers two protocols for interacting with repositories hosted on their site: HTTPS and SSH. Again, we recommend the use of the SSH protocol when interacting with your repositories; see Figure 2.1. When you click on the green `Code` button, you can then select the URL for the

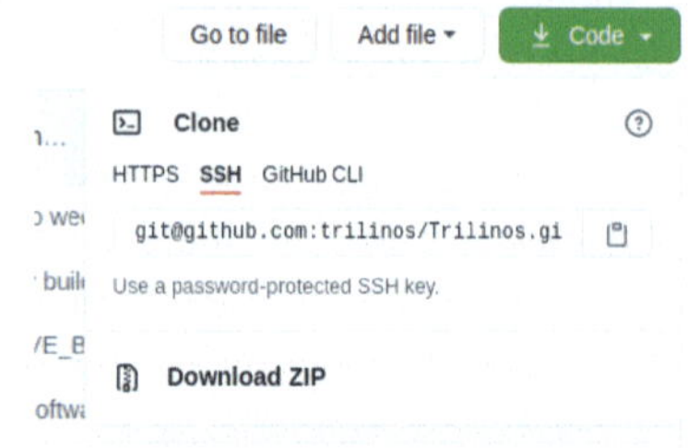

Fig. 2.1 Obtaining code through the SSH protocol on GitHub.

desired protocol you wish to use. Assuming this protocol is SSH, you can then simply clone the repository to your local machine with something like

```
$ git clone git@github.com:<username>/<reponame>
```

If your SSH setup has been done correctly, this should not prompt you for a password.

On your local machine, you should take the time to set up your Git configuration. Git tries to populate the necessary fields of your Git user information based on your local system account, but depending on how your system account is set up, Git may not have the information you wish to tie to your Git user accounts. This most commonly occurs when you are using the HTTPS protocol to interact with the remote repository.

You can adjust global settings for Git via

```
$ git config --global user.name <Full name>
$ git config --global user.email <email address>
```

Global settings are stored in a configuration file in the user's home directory: `~/.gitconfig` in Linux, for example. Global settings are the default for newly cloned repositories.

Any particular repository can have local Git settings set within it. *After* you have cloned a repository, you can enter any subdirectory in the clone and issue commands such as, for example,

```
$ git config user.name <Full name>
```

to adjust how your name appears in commits made to that repository. These local configuration settings get stored in the file `.git/config` within your local repository.

Warning: DO NOT manually modify the contents of anything in a repository's .git/ subdirectory. Editing items manually in the `.git/` subdirectory of a repository can cause the internals of Git to become out of sync in that repository. Using Git commands is the safest way to interact with your repositories.

2.4 Branches, commits, and pushing

Git branches are essentially named pointers to a specific commit. Commits in Git are stored as SHA-1 hashes and are *snapshots* of the repository. The primary branch of a Git repository is usually referred to as the `main` branch. We can inspect what branches are available in the local repository by using

```
git branch
```

Your local repository also *tracks* remote branches, and you can see all the remote branches using

```
git branch -r
```

Remote branches display as `<remote name>/<branch name>`, e.g., `origin/main`.

We can create a new local branch using

```
$ git branch <new branch name>
```

that creates a new named branch pointing to the current commit.

Git keeps a special pointer called `HEAD` that points to where you are in the repository. Usually, `HEAD` points to a named branch (which in turn points to a specific commit hash in the repo), but it is possible to point `HEAD` to any commit by simply checking out that commit by its SHA-1 hash.

We can checkout a specific branch using the `checkout` command,

```
$ git checkout <branch-name or commit-hash>
```

When checking out a commit hash, it is not necessary to use the full 20-character SHA-1 hash but rather only sufficiently many of the first characters until a commit can be uniquely identified. If `HEAD` is not pointing to a branch, it is considered to be in a *detached* state (kind of like a Git-Dullahan). This is sometimes referred to as being an *unnamed* or an *anonymous* branch and is denoted in some software with `@` as its name. Commits can be made while in this state, but **we highly recommend you avoid doing this** because such commits cannot be pushed to a remote repository until they are added to a branch—the committed code may end up buried somewhere in your local repository. **Always make your commits to a named branch.**

Once you are on a branch and have made some changes to your files, you can inspect these changes by checking the status of the repository via

```
$ git status
```

which provides a summary of the modifications that occurred within the repository, files in the *staging area*, and untracked files. It is the files in the staging area that are added to the next commit. Files must be *added* to the staging area before they can be committed. This is done with the command

```
$ git add <filename>
```

which adds a single file or a list of files to the staging area. All modified files can be added to the staging area with

```
$ git add .
```

however, this command should be preceded with `git status` to ensure that only the files you wish to add are added.

To see specifically what has changed in the files, you can run

```
$ git diff [path/to/file]
```

where an optional argument with the path to a specific file can be supplied. The default behavior is to list the `diffs` of all modified files.

Once you have added all the changes you wish to include in a given commit, you can issue the commit with

```
$ git commit -m "Commit message describing the changes"
```

At this point, we mention that you can have your command-line Git set up to call an editor on certain commands that could possibly benefit from it. The command `git commit` is one of these commands. You can set your *core* editor as a global config option, for example,

```
$ git config --global core.editor "emacs -nw"
```

that runs the quoted command (in this case, `emacs` in terminal mode) whenever an editor is needed. This is useful for when you want to have more detailed information on what the commit is doing, using multiple lines.

The decentralized nature of Git means that you do not need any synchronization with remote repositories as you make your commits. Commits are made in your local repository, and multiple commits can be made before you wish to *push* them to the remote. The updated local state of a branch can be pushed to a remote with the command

```
$ git push <remote> <branch>
```

that pushes branch <`branch`> to remote <`remote`>. For example,

```
$ git push origin main
```

pushes the local state of your `main` branch to the remote `origin`. Similarly, if you have local changes to multiple branches, you can push all of them to the remote repository with

```
$ git push <remote> --all
```

Trying to push a branch that has changed in the remote repository results in an error. The remedy for this is to figure out how to handle potential conflicts that exist in your new local commits and the updated remote commits. This is done via a *merge* or *rebase* of your commits, discussed in the next section.

2.5 Pull, fetch, merge, rebase

To obtain the most recent remote version of your current branch, we *fetch* the commits using the command

```
$ git fetch <remote>
```

In addition to this, we can fetch the updated versions of all branches using

```
$ git fetch --all
```

noting that this fetches all branches of all registered remotes.

Once the remote repository has been updated, it may be that the local commits you have made are no longer synchronized with the remote branch in question. We can try to merge our local branch with the updated state of the remote with

```
$ git merge <remote>/
```

If there are no conflicts between the changes in the local branch and the updated repository version, then Git can perform what is called a *fast forward*, where it just moves the branch pointer forward, automatically merging the states in effect. If there are files with conflicts, Git presents these files to you and indicates that they have to be merged manually. Inside files with conflicts, Git inserts both versions of the changes, using the markers <<<<<<<, =======, and >>>>>>> to separate them, for example

```
text not affected by the conflict
<<<<<<<
conflicted text from one version
=======
conflicted text from other version
>>>>>>>
```

At this point, *you* must decide how to deal with the conflict. You can manually edit the text files or refer to Section 2.6 for a few pointers on setting up your environment with tools to ease dealing with conflicts. Once you have fixed your changes, you summarize them in a commit, often called a *merge commit*.

The `git pull` operation is essentially a combination of the previous two commands: a `fetch` followed by a `merge`:

```
$ git pull <remote>
```

fetches the changes of your current branch and performs the merge command in an attempt to update your current branch. Sometimes this is what you want, usually when either your local or remote branches have not changed much, but other times you may want to inspect the commits of the remote branch before deciding what to do for a merge. In this latter case, the two-stage process of fetch and then merge/rebase is the better choice.

Rebasing a branch is slightly different from merging. In a rebase of a branch, Git attempts to rewrite *each* commit as if it were off of the new point being rebased.

This is different from the merge, where all the changes from a merge attempt get summarized in a merge commit. In the case where a branch is fast-forwardable, merging and rebasing have the same effect, essentially just moving the branch pointer.

Rebasing is generally used to maintain a clean, linear project history. The most common use-case we have found for using `rebase` over `merge` is when working on a branch long enough that the main branch (or whatever branch it has derived from) has changed. When this happens, it is better to `rebase` your branch off of the new state of the `main` branch before continuing to work on your branch. This has the effect of your branch appearing to be a linear history off of `main`.

The alternative would be to merge the new state of `main` to your branch and then continue working. Your branch now involves synchronization with commits that are unrelated to the changes your branch is introducing. Such a state of affairs makes history tracking, and thus bug tracking, more difficult for future developers.

When you run

```
$ git rebase <location>
```

Git tries to rebase your branch starting from `<location>`, which is any kind of commit reference. It marches through each commit, notifying you whenever a commit has a conflict. You then have to address the conflicts on a per-commit basis.

You should **only** use `rebase` on private/isolated scope branches of your repositories. Because it essentially rewrites history, rebasing of main branches (say `main` or `dev`) of main repositories should be avoided. The intent should always be to keep feature/fix branches as linear as possible; they can then be merged into the more mainline branches as large, linear chunks of commits.

2.6 Tools for working with Git

Using Git from the command line is all well and good, but there are numerous tools that can be used to help visualize and manipulate what is going on with your branches and remote repositories. If you have a favorite editor or development environment, it is a good idea to look into what plugins it has for dealing with Git repositories. For example, `emacs` has the `magit` package (`https://github.com/magit/magit`) that gives exceptionally smooth integration of Git functionality within the editor.

There are also standalone tools that can help with viewing the branching structure and dealing with merges. Examples of these include:

- git repository dtools,
- diff tools, and
- merge tools.

Chapter 3
Building software

3.1 Why? (Only one this time)

Another fact of life in extreme-scale computing is software must be built for the machine that executes it. Not only is this a general necessity, but when it comes to ESC in particular, optimized builds for a given machine are often desired because they can add a significant boost to performance.

The process of building software often starts on your local machine. It is only after it is developed and tested that software is migrated to a supercomputer for production runs. In today's ESC environment, it is often the case that the software is built with many dependencies. Automating the build process is crucial to ensure that the software is built correctly and efficiently.

Most supercomputers typically have many optimized libraries already installed as *modules*; see Section 3.6.2. Using the module system from the supercomputer that you plan to use is one way to help streamline the migration process and avoid unnecessary duplication and other maintenance errors. These module systems can be implemented on your local machine, allowing you to use the same libraries and compilers *with the same paths* as on the supercomputer. Most supercomputer system administrators can help you configure and use the module system on both your local machine and the supercomputer.

3.2 Compiling and linking

At the time of writing, the primary languages used to write ESC software are C, C++, and Fortran. These languages are known as *compiled* languages, meaning that their source code needs to be converted to machine-readable code by a separate program called a *compiler*. In contrast, *interpreted* languages such as Python can be run directly by the *interpreter* without any compilation step. Interpreted languages, however, are generally much slower than compiled languages and thus are not the

© The Author(s), under exclusive license to Springer Nature Switzerland AG 2025

R. J. Spiteri and K. Klenk, *Extreme-Scale Computing*, CMS/CAIMS Books in Mathematics 15, https://doi.org/10.1007/978-3-031-89033-8_3

focus for creating ESC software. Therefore, we only discuss the C, C++, and Fortran programming languages in this chapter.

A compiler is a special program that converts source code, which is written in a high-level programming language, into executable code, which can be understood and run by a computer. There are a few common families of compilers: GCC, LLVM, and Intel, with their command-line compiler names given in Table 3.1. Included in the LLVM family is the Clang compiler, and we note that the "Clang" and "AppleClang" compilers are considered different compilers, with Clang being the C++ compiler built on the LLVM[1] technology stack and AppleClang being a version of Clang used by macOS that has branched off of the mainline Clang. Having software that compiles across all compilers and system types is a non-trivial task, and a great deal of effort is required to accomplish such a feat. In this book, we generally just consider the GCC compilers and deal mostly with C++.

Compiler	C	C++	Fortran
GCC	gcc	g++	gfortran
LLVM	clang	clang++	flang
Intel	icc	icpc	ifort

Table 3.1 Common compilers usable from command line

Building C, C++, and Fortran software is generally broken down into three stages:

1. Preprocessing (expanding) macros and #includes.
2. Compiling object code from source files.
3. Linking object code together into an executable or library.

In the preprocessing stage, all #<preprocessor directive>s and macro definitions are expanded with their actual values. For example, #include directives paste the entirety of the included file at that line; #if, #ifdef, and #ifndef blocks conditionally leave sections of code based on their evaluation.

The compilation stage takes a source code file (a *compilation unit*) and converts it to the machine-readable object code. To successfully compile C++ code into object code, a compilation unit needs to know how the declarations look for all functions and classes it uses, but it does not need to know their precise definitions. This is where the separation of header and implementation files comes in to play and is further discussed in Part II.

Linking object code into libraries allows those libraries themselves to be linked into other executables. Libraries can either be static (.a in Linux and MacOS) or dynamic (.so in Linux or .dylib in MacOS). Static libraries are groupings of object code (sometimes referred to as *archives*) that must be inserted into any executable or other library that wishes to use them. Dynamic libraries are groupings of object code that do not get inserted into calling executables or code. Rather, a reference to the shared library gets inserted into calling code at link time.

[1] https://clang.llvm.org/

3.3 Makefiles

Makefiles are a quick-and-dirty approach to automate the compile-and-link process for mildly complicated projects. They provide a simple way to set rules for *building* different project elements depending on what *dependencies* of the elements have changed. This is done by specifying rules for *targets* and listing the files (or other targets) on which they depend.

Makefiles are generally just given the default name Makefile and can simply be invoked with the command

```
$  make
```

in the directory with the makefile. This runs the first target in the makefile. As usual, we can inspect various other command-line arguments and usage using

```
$ man make
```

In particular, some useful invocations are

```
$ make <target>  # builds specific target
$ make -k        # keep going if a target fails
$ make -f <file> # speicfy a file other than Makefile
```

The general form for a rule in a makefile is

```
<target>: <list of prerequisites>
    <command 1>
    <command 2>
    ...
```

where <target> specifies the name of a build target, <list of prerequisites> is a space-delimited list of other targets (or files) that <target> requires, and <command X> is a list of commands that the target needs to execute. The list of commands **must be indented by tab characters**. This is not us trying to incite the age-old debate of "tabs vs spaces" in source code, but rather it is a hard fact of the requirements of a makefile—they do not parse correctly unless the indents are tabs.

There are some useful constructs to help condense the complexity of makefile rules. First, we look at automatic variables. These are variables that get populated based on the rule being executed. For example, the most common automatic variables you see are

- $@ - name of the target rule,
- $< - name of first prerequisite,
- $^ - names of all prerequisites.

For example, the rule

```
hello: hello.cpp
    g++ -o hello hello.cpp
```

is equivalent to

```
hello: hello.cpp
	g++ -o $@ $<
```

which in this case (because there is only a single prerequisite) is equivalent to

```
hello: hello.cpp
	g++ -o $@ $^
```

Next, we consider pattern rules. These are rules that apply to all files of a specific pattern (or, say, a specific type). Pattern rules use exactly one % character to define the target pattern in the rule. The % character shows how the prerequisites relate to the target name. For example,

```
%.o: %.cpp
	g++ -c $<
```

defines a pattern rule that can be run on *any* .cpp file to create the associated object file .o.

Makefiles are language-agnostic, so they can be used for any sort of *build* procedure where you need to execute certain command-line operations depending on changed files. However, as complexity of a project increases, it can become difficult to keep your build structure organized using only makefiles, particularly if you need cross-platform builds or if there are many external dependencies. For this, you may need to start looking up to *meta*-build managers, as we discuss in Section 3.4 and Section 3.5.

Makefiles can have variables defined inside them. The definition and usage of variables are similar to those in bash, i.e., with an equals sign *immediately* after the variable name, but they can also be defined with the := assignment operator.

```
VAR1 = <value1>
VAR2 := <value2>
```

When using variables, we typically use the dollar-brace notation, and to help keep things understandable, we generally use round brackets for makefile variables and the usual curly braces for environment variables. For example, we might see

```
hello.exe: $(MAKEFILE_VARIABLE)
	gcc -o $@ $^ ${ENVIRONMENT_VARIABLE}
```

This is not a general convention, but we have found that distinguishing between them can assist other developers in debugging build problems more efficiently.

An example makefile demonstrating some of the above is found in Listing 3.1.

Listing 3.1 Example makefile file

```
# Example makefile
CXX=g++
CXXFLAGS=
LDFLAGS=

# "phony" targets that aren't named after actual files
.PHONY: all clean

# default target is the first one
all: hello

# build an executable dependent on a source file
```

```
hello: hello.cpp
	$(CXX) -o $@ $(CXXFLAGS) $< $(LDFLAGS)

# generally want make to help us clean up
clean:
	rm hello
```

The comprehensive documentation on GNU make can be found at `https://www.gnu.org/software/make/manual/html_node/index.html`.

3.4 GNU Autotools

As mentioned above, manually specifying makefiles can become cumbersome as project complexity increases. In this case, we turn to meta-build systems, starting with GNU Autotools. We do not go into detail for how to set up a project with Autotools because CMake is more widely used and preferred for new projects. However, we do want to provide a brief overview of Autotools because you may encounter projects that use it.

The usual usage of Autotools is a three-stage procedure:

```
$ ./configure  # set up makefiles that work on your system
$ make         # build the libraries/executables
$ make install # install the software in desired location
```

Generally, any input needed for an Autotools-based build comes at the `./configure` stage. The most common option used with this is likely *prefix*, which tells the installation where the software should be installed. This looks like

```
$ ./configure --prefix=/path/to/installation/dir
```

Outside of that, a given project typically provides its own set of options for configuration. These options can get quite complex depending on what other libraries the software makes use of. Thus, the best general advice we can provide is to *read the installation documentation* of whatever is being built and explore the output of

```
$ ./configure --help
```

You can try configuring many times without risk of hurting your system. Generally, if the configuration stage fails, suggestions as to some possibilities for how to remedy the problem are offered.

3.5 CMake

CMake is now the most widely used build system for C++ projects. It functions like GNU Autotools, using a configure, build, and install procedure, but provides a simpler project build language. CMake can be considered a programming language itself, one designed specifically to reason about a project's build structure. It was initially designed for C++ projects, but it has been extended to include both C and Fortran, in addition to other languages.

We focus on *modern* CMake, introduced in version 3.0, which significantly simplified the build of complicated projects through the use of *build targets* and *properties*. Specifically, we discuss how to use target commands like

- `add_executable()`
- `add_library()`

and work with their target properties over the older approach that used commands like `include_directories()` and `link_libraries()`.

General documentation for CMake can be found at `https://cmake.org/documentation/`.

3.5.1 Using CMake

The CMake build specification for a project can be found in files with the name `CMakeLists.txt`. There should be at least one of these files in the root directory of the project.

The best approach to building software using CMake is to create a new directory for the build and run the CMake commands from that directory. In practice, this goes something like

```
$ mkdir path/to/build/dir
$ cd path/to/build/dir
$ cmake [options] path/to/source/dir
$ make
$ make install
```

All configuration for the project should occur by specifying command-line options when running `cmake`. It is useful to think of two different types of command-line options. First, there are the options that specify *how* CMake should be run. These are options specified with the double dash (`--`). Some useful options include

- `--version` – print version of CMake being invoked and quit,
- `--graphviz=[file]` – generate a graph of the dependencies in the project,
- `--help` – print the documentation for command-line usage.

Second, there are the configuration variables, which are used to specify how the current project should be configured or built on the system. For example, compiler warning levels can be specified using the `-W` prefix:

- `-Wdev` – enable developer warnings,
- `-Wno-dev` – suppress developer warnings,
- `-Wdeprecated` – enable deprecation warnings,
- `-Wno-deprecated` – suppress deprecation warnings.

Configuration variables can also be specified in the cmake line and prefixed by -D. Some useful variables are

- -DCMAKE_INSTALL_PREFIX – where to install the build,
- -DCMAKE_BUILD_TYPE=<Debug|Release|RelWithDebInfo|MinSizeRel> – what type of build to perform,
- -DCMAKE_CXX_COMPILER=<path/to/compiler> – manually specify the C++ compiler to be used,
- -DCMAKE_BUILD_SHARED_LIBS=<ON|OFF> – build libraries as shared or not.

Specific CMake variables available to a given project can be explored using

- ccmake (curses CMake) – a text user interface using the curses library,
- cmake-gui – a full graphical user interface.

These applications are quite useful when it comes to seeing what kind of options are available for any given project, and they allow you to generate a CMake configuration from within them. However, we recommend that once you find the options you need for compiling a project in a desired way, you should create a custom command-line invocation for CMake using the options. This allows for more reproducible builds and enables scripted builds for different configurations of the project.

3.5.2 Writing CMake code

When writing CMake code, it is important to think of the CMake code itself as production-quality code. This keeps the build system clean and organized. From the root level CMakeLists.txt file, the instructions for how to build all the source files can be specified using relative paths to each file, but a better approach is to provide a CMakeLists.txt file in each directory that specifies the instructions locally. In this approach, subdirectories to be added to the project are specified in the CMakeLists.txt file of each directory.

Every project needs to start by setting a cmake minimum version, project name, and project version number:

```
cmake_minimum_required(VERSION 3.28)
project(<name> VERSION <number>)
```

Because CMake is evolving quickly, there are various elements of functionality and behavior that are only available in newer versions. We suggest that you use a recent version as your minimum version because it can easily be obtained on any system that we would likely use. This number can be subsequently increased to include functionality that is only available in newer versions. Generally, you want to set the minimum required version to the oldest version that has all the features you want (ensure you test your build against that version of CMake!).

From the root level CMakeLists.txt, subdirectories can be added by using the function

```
add_subdirectory(<directory_name>)
```

In the subdirectory `<directory_name>`, a `CMakeLists.txt` specifies what is done in that directory.

3.5.3 Variables and flow control

Variables in CMake come in various forms. First, there are *local variables*, which can be assigned using the `set(<variable> <val1> <val2> ...)` function and can be used with a similar dollar-sign evaluation as in bash. For example,

```
set(MY_VALUE 42)
set(MY_STRING "Some number:")
string(APPEND MY_STRING " ${MY_VALUE}")
```

noting also a brief demonstration of the `string(APPEND ...)` function. More information about string manipulation can be found at `https://cmake.org/cmake/help/latest/command/string.html`.

Second, there are *cache variables*, which are variables that can be set from the command line. CMake creates a cache of variables, stored as a file `CMakeCache.txt` in the build directory. You can specify variables to be added to the cache using

```
set(<variable> <value>... CACHE <type> <docstring>)
```

where the valid `<type>`s include:

- `BOOL` – Boolean `ON/OFF` value.
- `FILEPATH` – Path to a file.
- `PATH` – Path to a directory.
- `STRING` – A line of text.

The cached variables are what the CMake visualization tools (like `ccmake`) display for a given project.

Conditional blocks of CMake code can be created using the `if`, `elseif`, `else`, and `endif` statements. For example,

```
if(variable)
  # If variable is 'ON', 'YES', 'TRUE', 'Y', or non zero number
else()
  # If variable is '0', 'OFF', 'NO', 'FALSE', 'N', 'IGNORE',
  #   'NOTFOUND', '""', or ends in '-NOTFOUND'
endif()
```

3.5.4 Functions and macros

Functions and macros can be defined in CMake to allow re-use of patterns in a build. The main difference between a function and a macro is the scope of the variables used within them. Variables used inside functions remain local to the function, unless otherwise specified by setting the variable using PARENT_SCOPE. For example,

```
set(<variable> <value> PARENT_SCOPE)
```

sets <variable>, which persists beyond the function call. Any variables defined and used inside a macro, on the other hand, persist beyond the macro by default.

Functions can be defined in CMake to be re-used with different arguments. The general form for defining a function looks like

```
function(<name> [<arg1> ...])
  <command1>
  <command2>
  ...
endfunction()
```

So for example, a function defined as

```
function(my_function IN_STRING)
  message(STATUS "my_append argument: ${IN_STRING}, followed by ${ARGV}")
endfunction()
```

(noting usage of the ARGV variable as mentioned in https://cmake.org/cmake/help/latest/command/function.html) can be called as

```
my_function("This is an input string")
my_function("We can pass in more arguments than declared"
            "like this"
            "and this")
```

CMake macros can be defined and used similarly to functions, with the mentioned caveat of persisting variables by default. Further documentation on macros can be found at https://cmake.org/cmake/help/latest/command/macro.html.

3.5.5 Creating build targets

Build targets can either be *executables* or *libraries*. Executables are the actual programs that are run on a computer, and libraries are organized groupings of tools that can be linked into and used by executable programs. Either of these types of targets can be specified with

```
add_executable(<name> [source1] [source2] ...)
add_library(<name> [STATIC|SHARED|MODULE|OBJECT]
            [source1] [source2] ...)
```

The inclusion of source files when specifying the build targets is optional. Source files for a target can be specified later by modifying the *properties* of the target. In general, all objects inside the build have properties associated with them. These start out at a global level and then get passed through to more local levels, such as directories, individual targets, and even individual files. By default, targets inherit the properties of the current directory, and the properties of a target can then be modified for that particular target's needs.

Expanding on the *type* of library, we have:

- `STATIC` – static libraries,
- `SHARED` – shared libraries,
- `MODULE` – groupings of code that do not need to be linked into other targets but may be loaded dynamically at runtime,
- `OBJECT` – groupings of source code that get compiled to object code but are not archived into a library.

A common design decision for the build system is to organize the build tree into sub-libraries that then either get linked into a main target library or a main executable for usage. This design allows the build to be represented as smaller targets with more functional coherence. When the sub-libraries are not intended to be distributed (exported or used by external programs), it makes sense for them to be `OBJECT` libraries.

Once targets are specified, their properties can be set. For example, a target in question may need to specify the required *include directories* to properly find header files used in the source codes of that target. This requirement is accomplished with

```
target_include_directories(<name> <INTERFACE|PRIVATE|PUBLIC>
                                 [dir1] [dir2] ...)
```

We note the three possible choices for how the link dependencies work:

- `INTERFACE` – target is made part of the link interface but is not linked to,
- `PRIVATE` – target is linked to and not made part of the link interface,
- `PUBLIC` – target is linked to and made part of the link interface.

These three choices are used in all CMake functions that start with `target_` and specify things to do with a defined target:

- `target_compile_definitions(...)`
- `target_compile_features(...)`
- `target_compile_options(...)`
- `target_include_directories(...)`
- `target_link_directories(...)`
- `target_link_libraries(...)`
- `target_link_options(...)`

3.5.6 Installation rules

Files and targets can be marked for installation with the `install()` command. This process can look as simple as

```
install(TARGETS <target_name> DESTINATION <location>)
install(FILES <file_name> DESTINATION <location>)
```

We note that `<location>` is relative to the variable `CMAKE_INSTALL_PREFIX`.

With install rules specified, `make install` moves the appropriate components to their desired location. If you are using CMake 3.15+, you may also install with the command

```
$ cmake --install .
```

from the build directory.

3.5.7 Tests

Tests in CMake allow you to specify a particular command, `ctest`, to run your built executable(s). You can enable testing with the `enable_testing()` function and add tests with the `add_test(...)` function https://cmake.org/cmake/help/latest/command/add_test.html. To specify the conditions for passing a test, you can set properties of tests via the `set_tests_properties(...)` function. A list of available properties that can be set on tests can be found at https://cmake.org/cmake/help/latest/manual/cmake-properties.7.html.

Putting these together, setting up a test in a CMakeLists.txt file comes down to the following:

```
# Turn on testing
enable_testing()
# Add a test
add_test(NAME <test_name> COMMAND <test_command_line>)
# Give the test a passing condition
set_tests_properties(<test_name> PROPERTIES <property_type> <property_value>)
```

3.5.8 A simple CMake example

Finally, we go through a simple CMake example to demonstrate what a simple but complete `CMakeLists.txt` file looks like. The following example compiles a program from two files and demonstrates how to link an external library. The library being linked to is MPI. The directory structure is assumed to look like the following:

```
| -- CMakeLists.txt
```

```
| -- build/
| -- source/
    | -- helloWorld.cpp
    | -- mpiFunction.cpp
    | -- mpiFunction.h
```

At this point, we are not concerned with what the .cpp files contain but rather only about the `CMakeLists.txt` file, which can be found in Listing 3.2.

Listing 3.2 Example `CMakeLists.txt` file

```
cmake_minimum_required(VERSION 3.12)

# Set up the project name
project(Example)

# We can use set to assign a variable to specific subdirectory paths
set(SOURCE_DIR ${CMAKE_CURRENT_SOURCE_DIR}/source)

# We can then assemble the files into one variable for a cleaner syntax
set(SOURCE
    ${SOURCE_DIR}/helloWorld.cpp
    ${SOURCE_DIR}/mpiFunction.cpp)

# We can use the find_package command to get CMAKE
# to automatically look for the library we wish to use
find_package(MPI)

# We can then include the necessary files we need
# CMake can help us with that by using libraryName_INCLUDE_PATH
include_directories(Example ${MPI_INCLUDE_PATH})

# We then create our executable
add_executable(Example ${SOURCE})

# Lastly, we link the library
target_link_libraries(Example PUBLIC
    ${MPI_LIBRARIES})
```

To build the project, we first need to ensure that our current work directory is the `build` directory. From here, we can use the following commands:

```
cmake ..
make
```

On success, we should have compiled a binary (that is properly linked to an appropriate MPI installation) named `Example` that is located in the `build` directory.

3.6 Managing external dependencies

When building software, you will almost undoubtedly need to use external software within your project. Often, external software comes in the form of a library (.so, .a, or .dylib files) that gets linked to your code. There are various ways to manage these dependencies. However, if none of the methods described in this section are available to you, you can always consider building the library yourself from the project's source code. This is often the most time-consuming method, but scripts can be written to automate the process. The benefit to building the library yourself is that you ensure you get the correct version and configuration of the library you need. The next sections introduce the most common methods to install external dependencies and how to manage them within CMake.

3.6.1 Package managers

Each Linux distribution usually comes bundled with a package manager that can be used to install, upgrade, and remove software. Common package managers include `apt` for Debian-based systems like Ubuntu, `yum` for Red Hat-based systems like CentOS, and `pacman` for Arch Linux. Package managers are a convenient method for installing software, but they are not always up-to-date with the latest versions of software. Furthermore, when using a shared system like a supercomputer, you typically do not have the privileges to interact with the system package manager.

In addition to system package managers, there are also package managers specific to certain languages or domains that have been emerging to assist with managing the *dependency hell* of modern software. For example, conan (`https://docs.conan.io/en/latest/`) is aimed at being a package manager for C/C++libraries. In the domain of scientific computing, the US-led E4S project (`https://e4s-project.github.io/`) has been developing the Spack package manager (`https://spack.readthedocs.io/en/latest/`) to provide pre-built libraries for many configurations of numerous high-performance libraries (in a variety of programming languages) across a wide range of computing architectures.

3.6.2 Modules

Supercomputers generally use a *module* system for including libraries that are compiled in an optimized way. We strongly recommend the use of available modules whenever possible. If you are starting a new project and are targeting a specific system on which to run it, you may want to limit your dependencies to those already available as modules on that system.

The basics of using modules can usually be found in the documentation for the system on which you are working. A popular piece of software for managing modules is the Lmod[2] tool developed at the Texas Advanced Computing Center (TACC). With this tool, modules can be loaded using the `module load <module_name>` command. Conveniently, loading modules automatically sets the appropriate environment variables for the module being loaded. Using the `module avail` command allows you to see what modules are available on the system. There are many commands available for manipulating your environment with modules, and we refer to the documentation specific to the system being used for the most relevant information.

Additionally, modules can also be installed locally on your own system. This allows you to mimic the same module setup as the supercomputer you are targeting. This is commonly done with CernVM-FS (CVMFS).[3] CVMFS enables you to mount a remote file system that contains the same software stack as the supercomputer you are targeting. Again, we refrain from going into details here because how this is done

---

[2] `https://www.tacc.utexas.edu/research/tacc-research/lmod/`

[3] `https://cvmfs.readthedocs.io/en/stable/`

or whether it is available depends on the organization managing the supercomputers you are targeting. However, it is worth noting so that you are aware of such options.

3.6.3 Finding existing dependencies from CMake

Due to CMake's popularity, many libraries have a `FindXXX.cmake` file that CMake can search for as part of its `find_package(...)` function `https://cmake.org/cmake/help/latest/command/find_package.html`. The `FindXXX.cmake` files are usually installed with the desired library, if it was built with CMake. To use this function, you call the function with the desired package name as an argument. For example,

```
find_package(XXX)
```

tries to find the `FindXXX.cmake` file for your project. Additional options can be supplied to `find_package(...)`, such as `REQUIRED` to ensure that if the package is not found, the build fails. Additionally, the `find_package(...)` function can specify a minimum version a library must have to be used in the build.

3.6.4 Build your own dependencies within CMake

If you depend on specific versions or configurations of certain libraries, you may want to consider having your CMake project download and build them itself. These are what CMake calls *external projects* and are handled by the `ExternalProject_Add(...)` function `https://cmake.org/cmake/help/latest/module/ExternalProject.html`. This function allows you to specify the stages of obtaining and building the desired library. These stages include

1. downloading the source using
    - direct url for a zipped archive, or
    - various version control protocols (Git, Subversion, etc);
2. configuring the build using
    - CMake or
    - general/custom configuration scripts, including GNU Autotools `configure` as discussed in Section 3.4;
3. building the dependency;
4. testing the built dependency;
5. installing the built dependency.

Using this procedure allows you to install your (customized) dependencies in a subdirectory of your project's build, allowing you to more easily deal with linking your project to them.

Part II
Programming patterns and modern C++

Although Fortran [4] is still widely used for ESC and can definitely meet the demands of extreme scales, C++ seems to be the standard language of choice for modern HPC applications. Even C++ is starting to show its age, as newer languages like Rust [5] and Julia [6] are starting to gain traction within the HPC community. At some level, a programming language is simply a tool to solve a problem, and over time, you will encounter (and perhaps have to learn) many programming languages. We focus on C++ in this book because of its current wide use and its exposure of many low-level details that are critical for performance and yet are often abstracted away in many newer languages.

This part introduces C++ programming in a manner that emphasizes writing *expressive* code while keeping in mind its performance characteristics. Chapter 4 opens with a brief history of the language, following the evolution of its standards and ending with a look at the various compilers. Chapter 5 explores *primitive* C++, which includes the fundamental components inherited from its C roots and are necessary for understanding performance implications. Building on this, Chapter 6 introduces the more advanced features of C++ that are essential for achieving our goal of writing expressive code. Chapter 7 moves into the modern age of C++, presenting a selection of its most recent features. Chapter 8 introduces the Standard Template Library (STL) and how it can be used to write more *expressible* code while keeping performance front of mind. Finally, Chapter 9 looks at common parallel programming patterns and how the evolution of C++ is aligning itself to incorporate these patterns into the STL.

Resources for digging deeper into the specifics of C++ constructs include

- `https://cplusplus.com` – Great source of information, tutorials, and references.
- `https://learncpp.com` – A website with free tutorials from beginner to expert levels.
- `https://cppreference.com` – Overall reference of *everything* in C++, listing even variations among the various standards.

---

[4] `https://fortran-lang.org/`

[5] `https://www.rust-lang.org/`

[6] `https://julialang.org/`

Chapter 4
The C++ ecosystem

In this chapter, we take a brief look at the *what* of C++: examining its inception, origin, and the landscape of the language as it appears at the time of writing.

4.1 Inception and origin

C++ is an *imperative* (i.e., non-assembly) programming language initially developed by Bjarne Stroustrup in 1979. The road to C++ started from Fortran, the progenitor of imperative languages. Algol evolved from Fortran in the sense that it added emphasis on structured programming through the introduction of code blocks. From Algol, the programming language landscape quickly bifurcated in terms of abstractions that either 1) allowed programmers to easily reason about the problem they are solving or 2) easily model the underlying hardware so that programs could be human-readable yet still run efficiently. C++ reunites these two branches, bringing the best of both worlds and, if you are not careful, also the worst.

Having experimented with many low-level languages, Stroustrup eventually came to the Simula programming language, which, in his opinion, was the first programming language that allowed programmers to modularize programs easily. Additionally, Simula came with a strong type system, increasing the ways that programmers could structure the organization of their programs. A programmer could define a *type*, and a type could use or *inherit* from other types. This idea of building programs around types is what we now refer to as *object-oriented* programming, and it influenced the initial design of C++ greatly.

Parallel to the existence of Simula, with its ease of abstracting program components to objects suitable to the problem, developments were made in simplifying how programmers can view the model of computer hardware. These developments, which began with BCPL (the Basic Combined Programming Language), culminated in the C language in 1972. In comparison to Simula programs, C programs could be compiled to machine code that made much better use of the available hardware.

© The Author(s), under exclusive license to Springer Nature Switzerland AG 2025

R. J. Spiteri and K. Klenk, *Extreme-Scale Computing*, CMS/CAIMS Books in Mathematics 15, https://doi.org/10.1007/978-3-031-89033-8_4

Stroustrup took the type extensibility of Simula and looked towards implementing it on top of C while retaining compatibility. This resulted in the birth of C++ in 1979, an object-oriented programming language with efficiency placed first and foremost in its design criteria. After many years of demonstrated ability, the first standard for C++ came in 1998 – C++98.

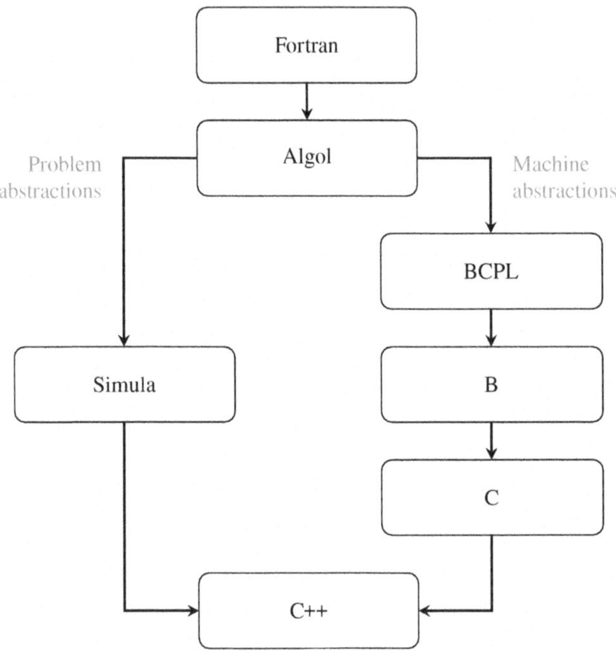

Fig. 4.1 Progression/influence of non-assembly programming languages. From Algol, there was a bifurcation in terms of what code should do: represent the problem well (Simula) or represent the hardware well (BCPL). C++ aims to bring the fruits of these disparate branches together.

4.2 Modern C++ standards

C++ was first standardized under the ISO[1] in 1998. This is the C++98 standard, with subsequent standards adopting the format of C++XX, where XX represents the last two digits of the calendar year. The first C++ standard was around an 8-year effort between Stroustrup, various industry partners, and the ISO.

The next standard took even longer, not being released until 2011 – C++11. This lengthy process was considered a failure of their standardization procedure – it took

[1] ISO – The International Organization for Standardization [sic]

so long that their initial name for the standard was C++0x[2], implying that it should have been completed at some point in the 2000s. The delay was largely due to their attempted 10-year cycle, which led to feature creep and the late discovery of some half-baked solutions to problems, such as *Concepts*. By the time all the loose ends surrounding the standard's features were tied up, it became a much larger standard than the original C++98, and a further three years had passed from the 2008 goal. From this failing, the ISO C++ committee aimed to release a new standard every three years to stay on track with concrete plans for the future standards. The committee has not missed a standard release since that time, now having released C++14, C++17, C++20, and C++23, with no sign of interruption ahead.

The *modern* standards of C++ are considered to be from C++11 onwards due to the large number of changes that had gone into that standard. Prior to C++11, syntax and semantic limitations of the C++ language made the usage of the standard template library quite verbose. Key language features to arrive in C++11 that admitted more expressive code were: lambda functions, uniform initialization syntax, and automatic (compile-time) type deduction. These features really streamlined the syntax required for programs to have access to the highly efficient *templated* components of the standard library.

What goes into a standard? C++ consists of two pieces: the *core language*, which defines the syntax and semantics of the language, and the *standard library*, which defines a lot of useful building blocks for designing applications. Most of the language's high-level usability comes from features in the standard library. The reason for this separated view is so that different standard library implementations can be used. This facility is quite useful in the realm of alternative architectures and embedded programming, which often have more limitations surrounding possible implementations. The C++ standard library contains specifications for containers, algorithms, input/output, and many other utilities. Further information on the Standard C++ Foundation can be found at `https://isocpp.org/`.

4.3 Compilers and standard implementations

4.3.1 Compilers

For C++, there are different compilers that are capable of parsing the core language, and they may or may not implement their own standard. The four most common compilers today come from:

- GCC – The GNU Compiler Collection, with its own history beginning with C. Cross platform.

[2] "I think of the x as hexadecimal" – Stroustrup

- Clang – A compiler frontend leveraging LLVM[3] in the backend for C, and C++. Cross platform.
- AppleClang – The Clang compiler that ships with Apple's macOS, branched and modified from older Clang variants. macOS only.
- MSVC – Microsoft Visual Studio Compiler. Windows only.

There are other compilers for C++ in existence, some retired such as the Borland C++ compiler, and others in specialized use such as Intel's C++ compiler, `icpc`. When it comes to ESC, it is most common to see the GCC and Intel C++ compilers – GCC because many high-performance libraries compile with GCC (or at least, compile *most easily* with GCC) and Intel because a lot of hardware used in supercomputers uses Intel components. However, Clang support is steadily growing due to the extensive toolchain[4] components that LLVM provides.

In this book, we focus most of our efforts on ensuring compatibility with the GCC `g++` compiler, occasionally noting particular sticking points of `g++` or particularly useful features of others.

An extensive list of which compilers are compliant with which aspects of the core language and standard library can be found at `https://en.cppreference.com/w/cpp/compiler_support`

4.3.2 Common standard implementations

As mentioned above, a C++ standard is split into two components, the core language and the standard library. All the existing C++ compilers work with their own implementations of the C++ standards. For example, GCC uses what is called the GNU Standard C++ Library, `libstdc++`, Clang uses the LLVM implementation of the standard library, `libc++`, and MSVC uses the Universal C Runtime Library, `ucrtbase.dll`. A summary of the compilers and their associated standard library implementations is given in Table 4.1.

Toolchain	Compiler	Library
GCC	`g++`	`libstdc++`
Intel	`icpc`	`libstdc++`
Clang	`clang++`	`libc++`
AppleClang	`clang++`	`libc++`
MSVC	`cl`	`ucrtbase.dll`

Table 4.1 Summary of compilers with their standard libraries.

[3] Originally an initialism for "Low-level virtual machine", LLVM is now the accepted terminology; i.e., the words behind the initialism have been dropped. LLVM provides many modular components on which software toolchains for many languages can be developed.

[4] A *toolchain* is a collection of programming tools that can be used together.

A key goal of a standard library implementation is its Application Binary Interface (ABI). The ABI is what defines how different binary pieces of software interact with each other. These pieces of software must *speak the same language* at the binary level—those that do are said to be ABI compatible.

ABI compatibility ensures that code that has been compiled with older versions of a standard library works with code compiled with newer versions of the standard library. This is generally useful because it allows libraries, which are potentially designed around an older standard, to be combined with newer libraries that take advantage of features in the newer C++ standards.

Perhaps a more relatable idea is that of the Application Programming Interface (API). An API defines how programmers can interact with a given library. If an API changes for a particular library, say by changing the number or type of arguments to a specific function, a programmer should not expect to be able to call the new function based on its old signature! The ABI is essentially the same idea but baked into the lower level of the binary interactions between libraries.

The standard libraries of the GCC and Clang compilers are ABI compatible for all modern C++ standards. There has been recent discussion in the standards committee about breaking ABI compatibility in an upcoming standard. This is because ensuring ABI compatibility can be quite restrictive to *how* certain components in the standard get implemented, locking them in to implementations that are consistent with how they were originally defined and implemented. This means, for instance, that better ways of handling some standard components may need to be foregone in the face of maintaining this backwards compatibility of ABI.

An ABI-breaking change would mean that older libraries would at least have to be recompiled with a new standard after an ABI break. In the realm of ESC, this is not such a grand issue—most software almost necessarily has to be compiled/recompiled for its target architecture for performance concerns—but it is more of a general concern for how the language itself should be defined and how it should progress.

Chapter 5
Primitive C++

This chapter introduces the basic components of C++ programming, including strong typing for variables and functions, controlling the flow of execution, and reading and writing from standard input and output.

5.1 Basic program structure

As with many things, an effective way to learn can be to cast aside fear of making mistakes and jump right in. We believe this is definitely true when it comes to learning a new programming language: jump in and start writing code. In this spirit, we begin with a simple, albeit unexciting, C++ program in Listing 5.1 to get accustomed to the syntax.

Listing 5.1 A basic C++ program file

```
/*
 * funktion_one: Function that just returns the value 1
 */
int funktion_one(void) {

  return 1;
}
/*
 *  Main program: Assigns output of function call to a variable.
 */
int main(int argc, char* argv[]) {

  // immediately assign to a variable at its declaration
  int number = funktion_one();

  return 0;
}
```

This program does not do much, but it does demonstrate a few points about syntax:

- Comments in C++ can be written in blocks (between `/*...*/`) or by line (`// ...`).
- Statements must end with a semicolon.
- Variables must have a type associated with them: `type variable_name`.
- Variables can be assigned at their declaration.

© The Author(s), under exclusive license to Springer Nature Switzerland AG 2025

R. J. Spiteri and K. Klenk, *Extreme-Scale Computing*, CMS/CAIMS Books in Mathematics 15, https://doi.org/10.1007/978-3-031-89033-8_5

- The `main` program has a specific signature: input arguments `int, char* []`[1] and return type `int`.
- Functions must be declared *before* they are used in any calling code.

5.2 Variables and types

Variables store data that can be manipulated programmatically. Every variable in C++ must be associated with a type. In Listing 5.1, `number` is a variable of type `int`. There are five fundamental types a variable can have: `void`, `bool`, `char`, `int`, and `float`. The `void` type is a special type, meaning *no* type, and it usually appears in function signatures. For example, in Listing 5.1, `void` is used in `funktion_one`'s signature to indicate that the function takes no arguments. The `bool` type is a boolean type that can be either `true` or `false`.

The remaining types, `char`, `int`, and `float`, all involve many subtypes. The exact size of these types, specifically how much memory is used to store them, are implementation-dependent. The standard (see Section 4.3.2 for details) only prescribes *minimum* size guarantees for these types. An example program that displays the sizes in bytes (1 byte = 8 bits) for the machine on which it is compiled and run on is given in Listing 5.2. The rest of this section discusses the `char`, `int`, and `float` types in more detail.

Character types. The most common character type seen is likely `char`, which is a 1-byte data type. We generally do not worry about alternatives to this basic character type, but it is worth noting that there are other character types that are larger, smaller, or aligned differently.

Integer types. The basic integer type is `int`. It is guaranteed by the standard to be at least 16 bits, but it is usually 32 bits. It is fairly safe to assume that when you see `int`, you are dealing with 32-bit integers.

Integer types can have *modifiers* applied to them. Modifiers have two categories:

1. Signedness: `signed` integers are the default and have a *sign bit*, allowing them to represent both positive and negative numbers; `unsigned` integers have no sign bit and can only represent non-negative numbers.
2. Size: `short` or `long` can be used to decrease or increase the number of bits used to represent an `int`.

As an example, one can have integer types like

```
short int          // at least 16 bit
unsigned long int // at least 32 bit, nonnegative
```

The range an integer value takes depends on how many bytes there are in the implementation of its type. To get the exact size of a data type on a given system, one can use the `sizeof(type)` function, again demonstrated in Listing 5.2. For output, it uses the `<iostream>` library, which we elaborate on in Section 5.5.

[1] We discuss the meaning of `*` and `[]` in the next chapter.

Floating-point types. Generally, we stick to `float` and `double`, which are the IEEE-754[2] standard types for 32- and 64-bit floating-point representations, respectively. The IEEE standard representation of floating-point numbers splits the bits into three categories: a single *sign bit*, the *exponent*, and the *mantissa*. A `float` has 8 exponent bits and 23 mantissa bits, whereas a `double` has 11 exponent bits and 52 mantissa bits. For further details on floating-point representations, we refer to `https://en.wikipedia.org/wiki/Floating-point_arithmetic#Internal_representation` and references therein[3].

Listing 5.2 A C++ program that prints sizes of some fundamental types

```
#include <iostream>

int main() {
    std::cout << "Size of bool: " << sizeof(bool) << " byte" << std::endl;
    std::cout << "Size of char: " << sizeof(char) << " byte" << std::endl;
    std::cout << "Size of int: " << sizeof(int) << " bytes" << std::endl;
    std::cout << "Size of short unsigned int: "<< sizeof(short unsigned int) << " bytes" << std::endl;
    std::cout << "Size of long long int: " << sizeof(long long int)  << " bytes" << std::endl;
    std::cout << "Size of float: " << sizeof(float) << " bytes" << std::endl;
    std::cout << "Size of double: " << sizeof(double) << " bytes" << std::endl;
    return 0;
}
```

5.3 Functions

Functions are the basic building blocks for organizing code into modular components that can be re-used. As seen in Listing 5.1, a function has a certain *signature* associated with it, that is, a list of arguments that must be supplied to the function as input and a return type for the function. The return type is all that gets returned from the function. However, *side effects* can also occur within the function's execution when it interacts with memory, which we introduce along with pointers in the next chapter.

The general structure of a function looks like

```
return_type function_name(in_type_1 var_1, in_type_2 var_2, ...) {
  ... function_body ...
}
```

The *body* of the function occurs between the curly braces and can span multiple lines. The previous example is a function that is both *declared* and *defined* in the same statement. The declaration of a function specifies which types go into the input arguments and its return type. The definition (or *implementation*) of a function defines the body of what the function does. They can be split up as

```
return_type function_name(in_type_1, in_type_2, ...);
// Some other code (functions that use function_name)
return_type function_name(in_type_1 var_1,
```

[2] The Institute of Electrical and Electronics Engineers `https://www.ieee.org/`

[3] The entirety of the Wikipedia article gives a good overview of the quirks in dealing with floating-point representation and arithmetic.

```
                        in_type_2 var_2, ...) {
    // Body of function_name ...
}
```

The reason for this separation is because a function must be declared in a compilation unit[4] before it is used so that a compiler can know the signature of the function in order to prepare the machine code to move variables of the correct sizes into the correct places for input and return. There *can only be one* implementation of a function, which may be at a later point in this file or in a different file altogether (as long as the compiler knows about the other file).

In C++, functions can be *overloaded*. That is, functions can have the same names provided they have different input arguments in their signatures. For example, the code in Listing 5.3 has three different definitions of the function `overloaded` depending on whether the input is `void`, an `int`, or a `double` variable.

Listing 5.3 A C++ program with an overloaded function. Also demonstrates the separation of declaration and implementation of functions.

```
#include <iostream>

int overloaded(void);
int overloaded(int);
// Declarations can optionally include variable names
double overloaded(double in_arg);

int main(int argc, char* argv[]) {

  int number1 = overloaded();
  int number2 = overloaded(number1+20);
  // cast number2 to double for next call
  double number3 = overloaded((double)number2);
  std::cout << "number1: " << number1 << std::endl;
  std::cout << "number2: " << number2 << std::endl;
  std::cout << "number3: " << number3 << std::endl;
  return 0;
}

int overloaded(void) {
  return 1;
}
int overloaded(int in_arg) {
  return in_arg;
}
double overloaded(double in_arg) {
  return 2*in_arg;
}
```

5.4 Control flow

Controlling the flow of a program has two main aspects: iteration and branching. Iteration is the act of running the same piece of code repeatedly (hopefully with different variable values!), and branching is the act of conditionally executing different pieces of code.

[4] which can simply be thought of as a *file* for this discussion

5.4.1 Iteration

There are three mechanisms for iteration in C++: for loops, while loops, and do-while loops.

For loops take three statements for defining their iteration:

- a *pre-loop* statement that is executed before the body. Generally, it is a good idea to initialize the iteration variable here.
- a *continue condition* that is checked at the end of each body evaluation and before ...
- a *post-iterate* statement that should modify the iteration variable before the next iterate starts.

Accordingly, for loops are useful when it is known exactly how many iterations are to be performed. The form of the for loop is

```
for(pre loop; continue condition; post iterate) {
  // loop body
}
```

The while loop is characterized by a single termination criterion and takes the form

```
while(continue condition) {
  // loop body
}
```

The continue condition is checked at the beginning of each iteration, and if it evaluates to `false`, the iteration terminates. Accordingly, while loops are useful when the number of iterations to be performed is unknown ahead of time. When programming while loops, you need to be careful about what variables are contributing to the *continue condition* to guard against infinite loops.

The difference with a do-while loop is where the continue condition is evaluated. A do-while loop takes the form

```
do {
  \\ loop body
} while (continue condition)
```

and so the *continue condition* is checked at the end of each iterate. Accordingly, a do-while loop is always guaranteed to execute at least once. Similar to the while loop, care needs to be taken to ensure the loop terminates for a given problem.

There are two common scenarios in which you may wish to alter the above loops. You may want to jump to the start of the next iteration to avoid doing some useless computation *on a given iterate*, or you may wish to terminate the loop prematurely, eschewing *all* further iterates. The keywords that do this are `continue` and `break`, respectively. They can be inserted into a loop and are generally used with a branching operation to check whether such behavior should be performed. An example of this might be to check some condition at the beginning of a loop to see whether this iterate should be performed or whether we can safely continue to the next iterate:

```
for(pre loop; continue condition; post iterate) {
  if (some other condition) { continue; }
  // loop body
}
```

5.4.2 Branching

Branching in C++ can be achieved with `if` statements. Yes, there is also a `goto` statement, but its use is generally discouraged because it can easily make code difficult to follow (so-called *spaghetti* code). Perhaps the only good use of `goto` in C++ is to efficiently *break* from inside deeply nested loops.

The most general form of the `if` statement looks like:

```
if (condition) {
  // execute if condition true
} else if (condition2) {
  // execute if condition is false, but condition2 is true
} else {
  // execute if neither of condition or condition2 is true
}
```

With a few points of usage:

- The `else if` and `else` portions are optional.
- There can be an arbitrary number of `else if` blocks.

If you find that you are using a lot of `else if` conditions testing equality in your code, it may be the case (no foreshadowed pun intended) that a `switch` statement may make the code more understandable. A `switch` statement takes the form:

```
switch(variable) {
  case val_1:
    // do something
    break;
  case val_2:
    // do something else
    break;
  case val_3:
  .
  .
  .
  default:
    // perform default behavior
}
```

The `variable` gets evaluated once on entry to the `switch` statement. Then the value of each `case` is checked sequentially. If a `case` matches, then the code in its block is executed. If none of the cases match, the `switch` statement *falls through* to the `default` case, which must be the last block in the statement. We note that all of the execution blocks in a `case` (except for `default`) need to include a `break` call to exit the `switch` statement.

5.5 Basic Input/Output

We have already seen the previous examples (Listings 5.2 and 5.3) contain the following components:

```
#include <iostream>
std::cout << ... << std::endl;
```

These commands are required to produce output from the program. The `#include` statement is a preprocessor[5] directive that tells the program to include the standard input/output streams library so that data can be read in through `stdin` or written out through `stdout` or `stderr`. The `std::cout` is read as: "The `cout` function in the `std` namespace," and its role is to direct the subsequent *stream* to `stdout`. Streams can be composed via the sequential double arrow (or *insertion*) operator <<. The `std::endl` inserts a new line character into the stream and *flushes* the buffer (printing it to screen immediately). Because of this flush, we generally recommend using a newline character `\n` instead of `std::endl` because excessive flushing of the output streams can cause lag in programs with heavy output. Output streams are flushed automatically when they reach a certain size or when the program receives a signal, so there is no risk of output being lost if it is not immediately flushed.

An example of reading from and writing to the `std` streams and various other pieces from this chapter are found in Listing 5.4.

Listing 5.4 Hello, all of the worlds. Unfortunately, the astronomers who discovered these worlds (likely exoplanets) lacked the creativity to give them normal names and instead identified them with unique non-negative integers. Fortunately, this leads to a simple integer loop to address them all.

```cpp
#include <iostream>
using namespace std;

// Print hello world a number of times
void print_hello_worlds(int n_times) {
  for(int i=0; i<n_times; ++i) {
    cout << "Hello, world number " << i << ".\n";
  }
}

int main(int argc, char* argv[]) {
  // Get number of worlds from input
  int n_worlds;
  cout << "How many worlds do you have? ";
  cin >> n_worlds;
  cout << "\n";

  // Greet all of the worlds
  print_hello_worlds(n_worlds);

  return 0;
}
```

[5] `#include <lib_name>` and `#include "lib_name"` are the standard ways of including libraries in C++ code.

This streaming output behavior is quite different from *formatted print statements* from other languages, including C. The logic is that the *insertion operator* << can be overloaded with different types, including more complicated classes as discussed in the next chapter, and allow simple code to combine standard output with other types. If formatted print statements are desired, one can use `#include <cstdio>` and the `printf(...)` functions, but we recommend using the standard input/output streams for C++ programs.

Chapter 6
Advanced C++

This chapter dives into more advanced C++ topics, including memory management, data structures, and template programming.

6.1 Memory, arrays, and pointers

C++ proivdes complete control over memory management. This is in stark contrast to languages like Python, where memory management is largely automatic. In this section, we discuss how memory is managed in C++, beginning with the two memory regions to which we have access. These two regions are the *stack* and the *heap*. The **stack** is a last-in, first-out data structure that is carefully managed by the CPU. Whenever a function defines a local variable, it gets added (pushed) to the stack. When a function returns, its local variables get deleted (popped) from the stack. Stack space can be quite small relative to the total size of a modern computer's RAM, leading us to the realm of heap memory.

The **heap** is a memory space that the programmer has to manage explicitly. This is different from the stack, where the CPU manages memory automatically. Memory on the heap must be manually allocated and deallocated and then accessed through a *pointer*. Heap memory is much larger than stack memory, but it is also generally slower to access. This is because allocation times are higher and there is less locality of data, resulting in more cache misses when using heap memory. Sometimes, you may see `malloc` (or `calloc`) and `free`, which are the older C-style functions for heap memory management that are still valid in C++. However, we recommend using `new` and `delete` because they have a cleaner syntax.

Arrays of data. Arrays are a common way to store multiple values within a single variable. An array can have one or more dimensions, and if the dimensions are known at compile time, they can be defined like so:

```
int od[10];
int td[2][3][4];
```

© The Author(s), under exclusive license to Springer Nature Switzerland AG 2025

R. J. Spiteri and K. Klenk, *Extreme-Scale Computing*, CMS/CAIMS Books in Mathematics 15, https://doi.org/10.1007/978-3-031-89033-8_6

Each dimension is *hard-coded* within its respective square brackets, and accessing the elements of an array is also done using square brackets, i.e., `od[i]` or `td[i][j][k]`. Defining arrays in this way stores all of their elements on the stack. Given the stack's limited size, it is recommended to be judicious in defining variables this way, especially if they require a lot of memory.

It is important to know that, in C++, arrays are stored linearly in *row-major* order. That is to say the right-most index varies most quickly as you traverse memory linearly. This is different from, say, Fortran, where arrays are stored in column-major order. See Figure 6.1 for a visualization of how the ordering looks for multidimensional arrays.

A_{000}	A_{001}	A_{010}	A_{011}	A_{020}	A_{021}	A_{100}	A_{101}	A_{110}	A_{111}	A_{120}	A_{121}

Fig. 6.1 Row-major memory layout of multidimensional arrays in C++. Example of how C++ indices translate to linear memory location for an array defined as `A[2][3][2]`.

This may seem like a trivial or inconsequential difference in theory, and in theory there is no difference between theory and practice. However, in practice, there often is. Knowing how data are laid out in memory is critical when it comes to efficiently using those data. A simple example shows this by iterating over a two-dimensional array,

```
double A[N][N];
// row major iteration - fast
for (int i = 0; i < N; ++i) {
  for (int j = 0; j < N; ++j) {
    A[i][j] = i * j;
  }
}
// column major iteration - slow
for (int j = 0; j < N; ++j) {
  for (int i = 0; i < N; ++i){
    A[i][j] = i * j;
  }
}
```

With g++, the previous two iteration attempts produce substantially different execution times regardless of the optimization level applied.[1] A simple code that demonstrates this can be found in Listing 6.1.

[1] In fact, the higher the optimization level, the more their performance diverges.

Listing 6.1 Row-major iteration over row-major array. Compiling this with `g++ -O3 traverse_array_fast.cpp` and running using the shell `time` command produces markedly different results if the `ii` and `jj` `for` loops are swapped.

```
#include <iostream>

#define N 1000

int main(int argc, char* argv[]) {
  double A[N][N];
  // traverse the whole array N times
  for (int iter = 0; iter < N; ++iter) {
    for (int i = 0; i < N; ++i) {
      for (int j = 0; j<N; ++j) {
              A[i][j] = i*j;
      }
    }
  }
  // Use a value of A so it doesn't get optimized out
  std::cout << A[N-1][N-1] << std::endl;
}
```

This performance difference is because of how modern CPUs *fetch* data from memory. CPUs can only read a full *cache line* at a time. On a typical Intel CPU, a cache line is 64 bytes (8 double precision variables). Thus, when iterating sequentially through the memory, the CPU will already have the necessary data (most of the time) and not need to fetch anything. This operation is relatively fast compared to the case where each new iteration deals with a location in memory that is far away and needs to be fetched. We speak more to this effect in Part III when we discuss hardware.

Pointers are variables that point to different locations in memory. They are defined by prepending an asterisk to a variable name, as in

```
int *p;
```

Pointers can be stored on both the stack and the heap, and their size depends on the OS architecture. For example, on a 32-bit system, pointers are 4 bytes, and on a 64-bit system, they are 8 bytes. This means pointers themselves take up little memory.

Pointers do not necessarily have to point to anything, and a pointer that is pointing to nothing is called a *null* pointer. Hence, the options for pointers are that they can point to some location on the stack, to some location in the heap, or to nothing at all. A way to get a pointer to point to something on the stack can be done by:

```
double x  = 5.0;
double *p = nullptr;
p = &x;
```

The above introduces the & unary operator. The & operator means "memory address of", so that final line is saying, "Point `p` at the memory address of `x`". On the other hand, on the previous line, the value of `p` is `nullptr` – a C++ keyword that explicitly states this is a *null* pointer. We note that assigning a newly declared pointer to `nullptr` is not mandatory; the statement `double *p;` compiles fine, for instance, but it is considered bad practice to do so.[2] When working with pointers, it is good practice to check that the pointer is actually pointing to something,

[2] There is nothing inherently *wrong* with a null pointer, but problems arise when you try to dereference one; i.e., you try to find out what it was pointing to!

```
if (p) {
  // Do something if p != nullptr
} else {
  // Do something else if p == nullptr
  // (print warning, cleanup,
  //  skip operating on *p,
  //  gracefully exit, etc)
}
```

because *dereferencing* a `nullptr` is undefined behavior. Dereferencing a pointer is the act of finding out what a pointer is pointing to and is done with the * unary operator. We can demonstrate some seemingly quirky behavior with the following code snippet:

```
double x  = 5.0;
double *p = nullptr;
p = &x;
cout << "The memory address p is pointing to is " << p << "\n";
cout << "p points to memory containing the value " << *p << "\n";
x = 42.0;
cout << "The memory address p is pointing to is " << p << "\n";
cout << "p points to memory containing the value " << *p << "\n";
*p = 2*x;
cout << "The value of x is: " << x << "\n";
```

See if you can predict what each output statement will produce before running it or reading on. The actual output is:

```
The memory address p is pointing to is 0x7fff9f2e72b0
p points to memory containing the value 5
The memory address p is pointing to is 0x7fff9f2e72b0
p points to memory containing the value 42
The value of x is: 84
```

noting that memory addresses are hexadecimal numbers that begin with `0x`, but their actual values change every time the program runs.

Similar to how * is used to both denote pointer variables and dereference them, & is used for not only obtaining the address of a variable, but it can also be used to define a *reference variable*. A reference variable differs only from a pointer variable in that it *must* point to a memory address. Continuing on with the previous snippets, we can introduce references like:

```
double &r = x;
cout << "The memory address r is referencing is " << &r << "\n";
cout << "r points to memory containing the value " << r << "\n";
r = 43;
cout << "The value of x is " << x << "\n";
```

Again, try to predict the output before executing the code. The output of this snippet (when included after the previous snippet) is:

```
The memory address r is referencing is 0x7fff9f2e72b0
r points to memory containing the value 84
The value of x is 43
```

A simple program demonstrating some things to expect from pointers and references is found in Listing 6.2.

Listing 6.2 Basic usage of pointers and references.

```
#include <iostream>
using namespace std;

int main(int argc, char* argv[]) {

  // a normal double-precision variable
  double x = 5.0;

  cout << "The memory address of x is                " << &x << "\n";

  // Pointers should initially be assigned to nullptr
  double *p = nullptr;
  // but can be reassigned to other memory addresses
  p = &x;

  // Display the pointer's memory address and contents
  cout << "The memory address p is pointing to is " << p << "\n";
  cout << "p points to memory containing the value " << *p << "\n";

  // modify by assignment to variable
  x = 42.0;
  cout << "The memory address to which p is pointing is " << p << "\n";
  cout << "p points to memory containing the value " << *p << "\n";

  // modify by assignment to dereferenced pointer variable
  *p = 2*x;
  cout << "The value of x is: " << x << "\n";

  // reference variables MUST be assigned at declaration
  double &r = x;
  cout << "The memory address r is referencing is " << &r << "\n";
  cout << "r points to memory conaining the value " << r << "\n";

  // modify by assignment to reference variable
  r = 43;
  cout << "The value of x is " << x << "\n";

  return 0;
}
```

Allocating memory with pointers. The primary use of pointers is to refer to memory that is allocated on the heap. To manage variables on the heap, we now take a closer look at the `new` and `delete` keywords.

Allocating, using, and cleaning up a variable on the heap can easily be done with

```
int *var = new int;
*var = 5;
delete var;
```

noting that using the variable requires dereferencing of the pointer.

Similarly, chunks of memory can be allocated on the heap and accessed via array indexing:

```
int n    = 20;
int *arr = new int[n];
arr[15]  = 10.2;
delete[] arr;
```

We note the usage of the `delete[]` keyword to indicate that a block of memory should be cleaned up. Memory that is allocated in the manner shown above is not guaranteed to be initialized to any particular value. This is useful because it saves time when overwriting the elements with values. Sometimes, however, you will want an array of *zero-initialized* data. In such a case, we can make use of the default initialization below:

```
int *var = new int();
int *arr = new int[n]();
```

The pointer `*var` and all `n` elements of `*arr` are *default initialized*, which for type `int` means initialized to 0.

An example program that demonstrates basic usage of variables and arrays on the heap is given in Listing 6.3.

Listing 6.3 Allocating memory on the heap with pointers

```
#include <iostream>
using namespace std;

int main(int argc, char* argv[]) {

  // new allocates a double in the heap, and p points to it
  double *p = new double;
  cout << "p points to memory address " << p << "\n";
  cout << "p contains the value " << *p << "\n";

  // clean up allocated memory when you're done with it
  delete p;

  // We use pointers to allocate arrays with sizes that are
  // not known at compile time.
  size_t size;
  cout << "How big of an array should we allocate (choose n > 5)? ";
  cin >> size;

  int *pp = new int[size]();
  cout << "Allocated int array with " << size
       << " elements on the heap.\n";

  cout << "The 5th value of the allocated array is " << pp[4] << "\n\n";

  cout << "i   address\n";
  // pp points to the address of the first element
  cout << "0   " << pp << " == " << &pp[0] << "\n";

  // We can look at the memory addresses of all of the allocated elements
  for(int i=1;i<5;++i){
    cout << i << "   " << pp+i << " == " << &pp[i] << "\n";
  }
  cout << "... etc ...\n";

  // However, we have to be careful with indexing dynamically allocated arrays
  // - accessing indices beyond size-1 is undefined behavior
  cout << "Value " << size+3 << " \"in the array\" is " << pp[size+3] << "\n";

  // clean up allocated array memory
  delete[] pp;

  return 0;
}
```

6.2 Structs, classes, and objects

With the basics of memory management and pointers introduced, we can now begin the fun of defining our own aggregate types with structures and classes. First mentioned in Section 4.1, C++ offers features that support object-oriented programming.

The philosophy behind object-oriented programming is to define data structures that encapsulate both *data* and *methods* that operate on that data, which are instantiated as *objects*. C++ provides two keywords for defining these data structures: `struct` and `class`.

Both structures and classes serve the same purpose, but they differ in the default privacy of their contents. Both can store their contents as any combination of `private`, `public`, or `protected` *members*; these three keywords are referred to as *access specifiers*. Structures default to `public`, whereas classes default to `private`. We note that for the rest of this section, we use the term *class* to refer to both structures and classes.

The general syntax for defining a structure and class is

```
struct MyStruct {
  // Private member variables and methods
public:
  // Public member variables and methods
protected:
  // Protected member variables and methods
};

class MyClass {
  // Private member variables and methods
public:
  // Public member variables and methods
protected:
  // Protected member variables and methods
};
```

The order in which the access specifiers are defined is not restricted, and they can each be used multiple times, as shown below:

```
class MyClass {
    // Private member variables and methods
  public:
    // Public member variables and methods
  private:
    // More private member variables and methods
  public:
    // More public member variables and methods
};
```

The access specifiers dictate how members are exposed to the surrounding elements of code and are defined as follows:

- `private` – Members are *only* accessible from within the class.
- `public` – Members can be accessed from outside the class.
- `protected` – Members can be accessed within a class (like `private`), as well as within classes that *inherit* from the *parent* class.

In contrast to a programming language like Python, where everything in a class is public and accessibility is managed by naming conventions, C++ provides explicit statements of what is visible to outside code. This feature allows for exposure of a limited public interface (which you often need to think *very* carefully about) for using classes. Behind the scenes of `public` methods, a class can do whatever it likes with `private` and `protected` members. Armed with this discussion of `protected`, the topic of inheritance is addressed in Section 6.2.1.

As mentioned, *members* of a class can be variables or methods (functions specific to a class). When an object is instantiated, it calls the *constructor* of its class to set up the state of its member variables. Constructors have the same name of the class, no return type, and there can be multiple constructors with different signatures. When an object is manually deleted or goes out of scope (when not created using `new`), the *destructor* is called. The destructor is always defined with the class name preceded by a tilde: `~ClassName`. Within the destructor, any manually allocated memory in the object should be freed at this point to avoid *memory leaks* (memory that is allocated but cannot be freed). A simple class with a constructor and destructor is

```
class MultiArray {

  int m_size;
  int *m_iarr;
  double *m_darr;

  public:
    MultiArray(int size) : m_size(size) {
      m_iarr = new int[size];
      m_darr = new double[size];
      for(int i=0; i<size; ++i) {
        // ...
      }
    }

    void print_arrays(void) {
      // ...
    }

    ~MultiArray() {
      delete[] m_iarr;
      delete[] m_darr;
    }
};
```

We note that the constructor, destructor, and `print_arrays` are all `public`, whereas the member variables are `private`. Before the constructor's body, an optional comma-delimited list can be used to initialize member variables in the form

```
: member_var1(input_var1), member_var2(input_var2), ...
```

This is the recommended way of initializing member variables, and it should be used whenever possible. The reason comes down to guaranteeing information about assignment helping compilers better optimize our code. A simple example using the above class is given in Listing 6.4.

Listing 6.4 Using objects allocated on the stack and on the heap

```
#include <iostream>
using namespace std;

class MultiArray {

  int m_size;
  int *m_iarr;
  double *m_darr;

public:
  MultiArray(int size) : m_size(size) {
    m_iarr = new int[size];
    m_darr = new double[size];
    for(int i=0; i<size; ++i) {
      m_iarr[i] = i;
      m_darr[i] = 1.0/i;
    }
  }

  void print_arrays(void) {
    cout << "i iarr darr\n";
      for(int i=0; i<m_size; ++i) {
        cout << i << " " << m_iarr[i] << " " << m_darr[i] << "\n";
      }
  }

  ~MultiArray() {
    delete[] m_iarr;
    delete[] m_darr;
  }
};

int main(int argc, char* argv []) {

  int size;
  cout << "How big the arrays be? ";
  cin >> size;

  cout << "A MultiArray on the stack:\n";
  MultiArray ma(size);
  ma.print_arrays();

  cout << "\nA MultiArray on the heap\n";
  MultiArray *ma_heap = new MultiArray(size);
  ma_heap->print_arrays();

  return 0;
}
```

6.2.1 Inheritance and virtual functions

Classes can be designed to inherit members of some *base class*. This is useful when, for example, you would like to have different object types and use them via the same *interface*. You can specify the common interface via the base class and the specifics of the types in different *derived* classes. The syntax for having a class inherit from a `BaseClass` is:

```
class DerivedClass : public BaseClass {
  // ...
};
```

Recalling the member access identifiers, the `DerivedClass` can access both `public` and `protected` members of the `BaseClass` but not the `private` members. An example of this in use is given in Listing 6.5

Listing 6.5 Simple inheritance of members of a base class

```
#include <iostream>

class BaseClass {
private:
  int a;

protected:
  int b = 10;

public:
  BaseClass(int in_a) : a(in_a) {}

  void print_a() {
    std::cout << " a = " << a << "\n";
  }
};

class DerivedClass : public BaseClass {
public:
  DerivedClass(int in_a) : BaseClass(in_a) {}

  void print_b() {
    std::cout << " b = " << b << "\n";
  }

  // This will not work as a is private in BaseClass
  // void print_a_subclass() {
  //   std::cout << " a = " << a << "\n";
  // }
};


int main(int argc, char* argv[]) {

  BaseClass base_class(4);
  DerivedClass derived_class(5);

  base_class.print_a();

  derived_class.print_a();
  derived_class.print_b();

  return 0;
}
```

The output of Listing 6.5 is:

```
a = 4;
a = 5;
b = 10;
```

In this example, the `BaseClass` has a `private` member `a`, a `protected` member `b`, a `public` method `print_a`, and a constructor that sets `a`. The `DerivedClass` inherits from `BaseClass` and uses the constrictor of `BaseClass` in its own constructor to set the value of `a`. The `DerivedClass` has access to `print_a` but cannot access `a` otherwise because it is `private`. The `DerivedClass` can access `b`, however, because it is `protected`, and we show this through the `print_b` method of `DerivedClass`.

A base class can also define its methods as *virtual* by preceding them with the `virtual` keyword. Virtual methods can be, but do not always have to be, overridden by derived classes. For example,

```
struct AbstractBase {
  virtual void fn();
};
```

declares a purely virtual class if `AbstractBase::fn` is never defined. Classes with only virtual methods are called *abstract classes* and are useful for defining interfaces that derived classes must implement. This means that classes that derive from `AbstractBase` *must* override `fn`, or the compiler will throw an error. Overriding a virtual method is done by defining using the `override` keyword, as shown below:

```
struct DerivedClass2 : public AbstractBase {
  void fn() override
  {
    // Body of DerivedClass2::fn
  }
};
```

If `AbstractBase::fn` included a definition, then `DerivedClass2` would not be *required* to override it and could simply inherit the `AbstractBase` implementation.

When virtual methods exist in classes, C++ makes use of *virtual tables* (vtables) to keep track of which functions the methods of a given object are actually referring to. More specifically, an object uses the vtable to determine which definition of a virtual method to call, ensuring the most derived version of the method is called. Calls to functions that are referenced by a vtable thus require an extra level of indirection *at runtime* to call the correct function. This may or may not be an issue for any given code,[3] but it can generally be avoided with good design.

To assist compilers in their attempts to resolve virtual function calls, C++ has the `final` keyword. When a class declaration is annotated with `final`, such as

```
class DerivedClass3 final : public AbstractBase {
  // ...
};
```

the compiler understands that nothing else derives from `DerivedClass3`. If some piece of code tries to derive from `DerivedClass3`, a compile-time error occurs. When the `final` keyword is used, compilers are able to reason that the member functions of the class can be determined statically; i.e., it can be found at compile time *which* function is being referenced from an overridden virtual function, and no vtable is needed for objects of type `DerivedClass3`. The `final` keyword can also be applied to individual member functions in a derived class.

A common design trap that can plague object-oriented systems is having *too much* inheritance. Long chains of inheritance often make it more difficult to reason about a given program, as well as make it more difficult to re-use subcomponents. Therefore, we advise that, when using objects, try to keep your inheritance hierarchy as shallow as possible. For example, having a purely virtual base class to define a desired interface and having all classes that derive from it marked `final` makes it easy to avoid any possibility of your inheritance hierarchy blowing up.

---

[3] vtable lookups can be effectively branch-predicted by compliers under many circumstances.

6.3 Templates

Now that we have the capacity to define our own types, we can start thinking of writing code that can function for many types. In C++, this style of *generic* programming is accomplished by the use of *templates*. Templates are type placeholders in the code that get evaluated (*instantiated*) at compile time. This approach is useful because it allows many different functions with similar signatures to exist without duplicating code and without incurring any runtime overhead.

6.3.1 Type template parameters

Templating a function is done by providing `typenames` for the desired types that should be templated. The `typenames` are provided as a comma-delimited list inside the angle brackets of `template<...>` before the function declaration. For example, a function that has one templated parameter might look like

```
template<typename T>
T function_name(T first_arg, T second_arg) {
  // ...
}
```

Here, every instance of `T` must be the same type when calling the function. This allows us to provide code that calls the function as

```
int i1, i2;
function_name(i1,i2);
double d1, d2;
function_name(d1,d2);
```

In doing this, the compiler can determine that it needs to compile two versions of `function_name`, one with `T=int` and another with `T=double`. In this example, because the return type is the same as both of the argument types, its type is well defined. We note that trying to use something like `function_name(i1,d1)` results in a compilation error because the type of *both* arguments do not match. To call something like `function_name(i1,d1)`, we require the function to be specified as something (possibly) like

```
template<typename T1, typename T2, typename T3=T1>
T3 function_name(T1 first_arg, T2 second_arg){ /* ... */ }
```

Here we now have separate template parameters for each argument, and we have also defined a third parameter for the return type that defaults to the type of `T1`. Possible ways to call this function include:

```
function_name(i1,i2); // same types, return type is int
function_name(d1,i2); // different types, return type is double
function_name<int,double,float>(i1,d2); // return type is float
```

A simple example demonstrating the usefulness of templates for numeric types is given in Listing 6.6.

Listing 6.6 Simple templated function that works for multiple types

```
#include<iostream>
using namespace std;

template<typename T1, typename T2>
auto add_vars(T1 first, T2 second) {
  return first + second;
}

int main(int argc, char* argv[]) {

  int i1=1, i2=2;
  long l1=10, l2=20;
  float  f1=100.1, f2=200.2;
  double d1=1000.1, d2=2000.2;
  string s1("Hello, "), s2=("World!");

  // Could easily be handled by single template parameter
  cout << "Adding integers:  " << add_vars(i1,i2) << endl;
  cout << "Adding long ints: " << add_vars(l1,l2) << endl;
  cout << "Adding floats:    " << add_vars(f1,f2) << endl;
  cout << "Adding doubles:   " << add_vars(d1,d2) << endl;
  cout << "Adding strings:   " << add_vars(s1,s2) << endl;
  cout << endl;

  // Mixing types requires more template parameters
  cout << "Integer+double:  " << add_vars(i1,d2) << endl;
  cout << "Double+long int: " << add_vars(d1,l2) << endl;
  // ERROR: no + operator defined for string + float
  //    cout << "String+float: " << add_vars(s1,f2) << endl;

}
```

It is important to note that when using templated functions, all operations within the templated functions must be well-defined for the types being used. Any operation that is not defined for a type results in a compiler error.

6.3.2 Non-type template parameters

In addition to deferring type specification until compile time, C++ also allows non-type template parameters, which must have a value rather than a type at compile time. With a class defined as

```
template<int n> class MyClass {
  double contents[n];
  ...
}
```

an object declared as

```
MyClass<20> object;
```

will have `n=20` defined inside. A simple example of how this can be used in practice is shown in Listing 6.7.

Listing 6.7 Usage of a non-type template parameter

```
#include<iostream>
using namespace std;

// Non-type template parameter of type int
template<int n>
class Compound {
  double contents[n];
public:
  // Compute the values at object construction
  Compound(double starting_val) {
    contents[0] = starting_val;
    for (int i=1; i < n; ++i) {
      contents[i] = contents[i-1]*starting_val;
    }
  }
  void print() {
    cout << contents[0] << " compounded " << n
         << " times: " << contents[n-1] << endl;
  }
};

int main(int argc, char* argv[]) {
  Compound<20> object(1.01);
  object.print();
}
```

Chapter 7
Modern C++ and guidelines

This chapter covers some guidelines and best practices when working with modern C++. The guidelines include discussion on 1) the `auto` keyword and type deduction, 2) how to think about memory ownership and how to use *smart pointers* (covering both *unique pointers* and *shared pointers*), 3) different mechanisms for handling functions, and 4) how to organize code into *header files* (.hpp or .h extension) and *implementation files* (.cpp extension). After reading this chapter, one should feel more confident in developing safer and more transparent C++ code.

7.1 `auto` and type deduction

As we define more involved templated classes or libraries of templated functions, declaring variables can become cumbersome. This is where the `auto` keyword comes into play. For statements where declaration coincides with initialization of a variable, `auto` can be used to automatically deduce the variable's type at compile time. Brief examples with numeric types include

```
auto a = 1+2;        // a will be int
auto b = 1.0+2.0;    // b will be double
auto c = 1.0f+2.0f; // c will be float
auto d(42.0);        // d will be double (constructor init)
```

For variables declared with `auto`, if information about the type is needed further in the code, the `decltype()` function can be used to determine its type. For example,

```
decltype(d);        // gets the type from a defined variable
decltype(a+d*c-2); // gets the type from an expression
```

Furthermore, `auto` can be used for function pointers as well. For example,

```
auto (*f)() -> double;
```

This declares `f` as a pointer to a function that returns a double. We also note the usage of the -> syntax to explicitly set the return type of the function. This usage

© The Author(s), under exclusive license to Springer Nature Switzerland AG 2025
R. J. Spiteri and K. Klenk, *Extreme-Scale Computing*, CMS/CAIMS Books in Mathematics 15, https://doi.org/10.1007/978-3-031-89033-8_7

of -> comes in handy when using lambda functions (Section 7.3.3) and templated functions (Section 6.3).

7.2 Memory ownership

We had a quick introduction to memory allocation and pointers in Section 6.1. Although the ability to work with memory addresses through pointers is powerful, it is also dangerous. To reduce the danger, we now introduce the idea of *smart pointers*, which contain additional information associated with a given pointer. In particular, we consider std::unique_ptr<T>—a pointer that is not copyable—and std::shared_ptr<T>—a *reference-counted pointer*, in which the number of copies of the pointer are maintained alongside the pointer. To access these two types of pointers, a C++ program needs the preprocessor directive #include<memory> from the standard library.

The main advantage of smart pointers is that they automatically call the destructors of the objects they point to when they go out of scope. Scope can be thought of as the lifetime of a piece of data. A piece of data *exists* when in scope but does not when it falls out of scope. For example, if we have two functions, one that is contained within the other, the inner function has access to the outer function's variables, but the outer function does not have access to the inner function's variables. This happens because when we enter the inner function, the outer function has not returned yet, so its variables are still on the stack and hence in scope. However, when the inner function returns, its variables are popped off the stack and are out of scope to the outer function. This perhaps shines light on *why* smart pointers are considered useful: when a pointer points to memory allocated on the heap and then goes out of scope and gets deleted, the memory allocated on the heap is no longer accessible. This makes it impossible to call the destructor of the object if allocated. On the other hand, smart pointers call the destructor automatically when they go out of scope.

Scope can also be actively managed inside a function as well by using curly braces. For example,

```
void function(void *ew) {
  // scope of *ew is this function
  double x; // scope of x is this function
  for (int i=0; i < 6; ++i) {
    // scope of i is this loop
  }
  {
    double y; // scope of y is between these braces
  }
  return;
}
```

Garbage collection is a phrase you may have heard, and (in computing anyway) it refers to the automatic cleanup of unreferenced memory. Some languages have garbage collection built into them. C++ does not have garbage collection because garbage collection can cause program performance inconsistencies. Instead, C++ offers smart pointers to automatically clean up resources (i.e., automatically call destructors) when they go out of scope, putting the onus on the programmer to manage the scope of such pointers.

7.2.1 Unique pointers

A *unique pointer* is a pointer that cannot be copied. In addition to this, it also ensures that the object it is pointing to gets deleted when the pointer goes out of scope. Example usage of a unique pointer in C++ is

```
unique_ptr<double> d1(new double(42.0));
cout << "Address: " << d1.get() << " value: " << *d1 << endl;
```

This code snippet creates a unique pointer of type double and initializes its value to 42.0. Now d1 can be treated just like a standard pointer. It can be dereferenced with *d1, and a standard pointer can be obtained with d1.get().

Trying to copy a unique_ptr<T> results in a compile-time error. This result is the best-case scenario because it tells the programmer that there is some flaw in their memory ownership assumptions before the code is even run.

Ownership of a unique_ptr<T> can be *moved* to another unique_ptr<T> using the std::move() function, for example:

```
unique_ptr<double> d2;
d2 = std::move(d1);
```

Now, d2 has ownership of the object at the end of its pointer, and its destructor will be called when d2 goes out of scope.

The constructor for a unique_ptr<T> may not sit well with some: it uses new to allocate memory and point to it, immediately passing that off to the unique_ptr<T> constructor. In general, one should opt to use std::make_unique<T>() for constructing a new unique_ptr<T> from scratch. We can replace the above code with

```
auto d1 = make_unique<double>(42.0);
```

We note the use of auto, which was introduced in Section 7.1. Sometimes you may wish to use an existing standard pointer (perhaps some data structure supplied by a library) to construct a unique_ptr<T>, in which case you *cannot* use make_unique<T>. In general, using explicit new and delete statements should be avoided whenever possible. However, using these keywords (as well as owning standard pointers) is sometimes unavoidable in lower-level data structures.

Example code demonstrating usage of unique_ptr<T> is found in Listing 7.1.

Listing 7.1 Unique pointers

```
#include<iostream>
#include<memory>
using namespace std;

int main(int argc, char* argv[]) {

  double *dp;

  // Inside a block, we set a unique_ptr to point to
  // memory that contains the double precision number 42.0
  {
    // unique_ptr<double> d1 = make_unique(new double(42.0));
    unique_ptr<double> d1 = make_unique<double>(42.0);
    // Raw pointers can be used to point to a unique_ptr,
    // but they are not responsible for managing its memory
    dp = d1.get();
    cout << "Inside block.\n";
    cout << "  Address: " << dp << " value: " << *dp << endl;
  }

  // When the unique_ptr goes out of scope,
  // the destructor for its type gets called automatically.
  cout << "After block.\n";
  cout << "  Address: " << dp << " value: " << *dp << endl;

  return 0;
}
```

The output for Listing 7.1 is

```
Inside block.
  Address: 0x612ba545ceb0 value: 42
After block.
  Address: 0x612ba545ceb0 value: 1.28872e-313
```

Notice that the value of the pointer is not 42 after the block because the value was deleted when the pointer went out of scope.

7.2.2 Shared pointers

A shared pointer shared_ptr<T> should be used when you want to have multiple owners of a raw pointer. The shared_ptr<T> keeps track of the *reference count* (number of owners) alongside the raw pointer, and it will not delete and clean up the raw pointer until the reference count falls to zero. Similar to unique pointers, we also have a std::make_shared<T>() function that can be used for creating a new shared_ptr<T>. This is shown by the example in Listing 7.2.

Listing 7.2 Reference counting with shared pointers.

```
#include<iostream>
#include<memory>
using namespace std;

class Number {
  private:
    double value;
    double *p;
  public:
    Number(double v) : value(v) {
      p = new double(v);
      std::cout << "Number(" << value << ") created.\n";
    }
    ~Number() {
      delete p;
      std::cout << "Number(" << value << ") destroyed.\n";
    }
    double get_value() const {
      return value;
    }
};

void print_address_ref_count(shared_ptr<Number> p) {
  // Note the reference counts reported inside this function
  std::cout << "  Address: " << p.get()
            << ", value: "   << p->get_value()
            << ", copies: "  << p.use_count() << endl;
}

int main(int argc, char* argv[]) {

  Number *raw;

  {
    // shared_ptr<double> original(new double(41.99));
    shared_ptr<Number> original = make_shared<Number>(41.99);
    raw = original.get();
    std::cout << "Originally.\n";
    print_address_ref_count(original);
    {
      shared_ptr<Number> copy_constructed(original);
      std::cout << "Plus one copy.\n";
      print_address_ref_count(copy_constructed);

      {
        shared_ptr<Number> assigned;
        // assignment copy constructs the shared_ptr
        assigned = copy_constructed;
        std::cout << "Plus another copy.\n";
        print_address_ref_count(assigned);
      }
    }

    std::cout << "Only original remains.\n";
    std::cout << "  Address: " << original.get()
              << ", value: " << original->get_value()
              << ", copies: " << original.use_count() << endl;

  }

  // Destructor has been called for the double value
  cout << "\nAll shared_ptr out of scope.\n";
  cout << "  Address: " << raw
       << ", value: " << raw->get_value() << endl;

  return 0;
}
```

The output from the above example is

```
Number(41.99) created.
Originally.
  Address: 0x576d4fde2ec0, value: 41.99, copies: 2
Plus one copy.
  Address: 0x576d4fde2ec0, value: 41.99, copies: 3
Plus another copy.
  Address: 0x576d4fde2ec0, value: 41.99, copies: 4
Only original remains.
```

```
  Address: 0x576d4fde2ec0, value: 41.99, copies: 1
Number(41.99) destroyed.

All shared_ptr out of scope.
  Address: 0x576d4fde2ec0, value: 41.99
```

In this example, we have defined a class `Number` to wrap a double value specifically to demonstrate the reference counting behavior of `shared_ptr<T>`. The `Number` class has a constructor and destructor that print out messages when they are called, along with a method `get_value()` that returns the stored value. The output illustrates that the reference count increases when we create copies of `original`. When `original` goes out of scope, its destructor is called. However, notice that `raw` still returns the correct value at the end of the example. This is because the memory has not been reassigned yet, and the original value is still accessible.

It is important to note that further use of `raw` is considered undefined behavior and further motivates the use of smart pointers. It is undefined because eventually the memory will be reassigned, and `raw` will return a possibly incorrect or garbage value. To observe this in action, consider exploring what happens if `Number` is modified to initialize a pointer to a double using `new` in the constructor and then delete the pointer in the destructor. The output will show that the value of `raw` is garbage after the `original shared_ptr<T>` goes out of scope.

7.2.3 Working with smart pointers

Now that we have multiple *types* of pointers, we can more clearly define what a *smart pointer* is. We consider any pointer that has additional functionality beyond the standard asterisk-style pointer to be a smart pointer. In the context of C++, we consider these to be `unique_ptr<T>` and `shared_ptr<T>`.[1]

Next, how do we know which smart pointer to use? The short answer is: **Use `unique_ptr<T>` whenever possible.** There are three main reasons for this:

1. The cleanup and scope of resources are more transparent to other developers.
2. It forces you to think through the *data flow* in your programs.
3. No (global) reference count needs to be maintained behind the scenes.

The first point is essential for maintainability of software—the more transparent the intent of the actual lines of code, the more maintainable it is. The remaining two points come down to performance considerations. Optimizing data flow can yield significant improvements while avoiding even small overhead from reference counting. Nonetheless, when a `unique_ptr<T>` cannot be used, a `shared_ptr<T>` is the next best choice.

With different kinds of pointers intended for different use cases, we can convey information about how a pointer is to be used entirely through the signature of

---

[1] The STL does provide a `weak_ptr<T>`, which we do not discuss here, and libraries sometimes implement and use their own smart pointers as well. Be aware!

a function. For functions that require pointers as input parameters, it is generally advised to use raw pointers or references for the parameters. A reasonable guideline to think of the exceptions to this is when a function modifies the lifetime of its input, a smart pointer should be used. An example of this is now presented.

If a function is to be used as a *sink* for the object, i.e., the function consumes an object to do its work, it should require a unique_ptr<T> as input. A function declaration of the form

```
void sink(unique_ptr<T> p);
```

expresses unambiguously that the function sink is taking ownership of the pointer being passed in. This is in contrast with using a raw pointer for a similar desired result

```
void sink_raw(T* p); // NOTE: deletes object p
```

The use of unique_ptr<T> in the signature requires that either the calling code has to explicitly convert to a unique pointer, or a unique pointer must be moved into the function, for example,

```
double *p = ...;
unique_ptr<double> u = ...;
sink(unique_ptr<double>{p}); // explicit conversion
sink(move(u));               // explicit move
```

Further examples of use cases for different types of smart pointers as function parameters can be found at https://herbsutter.com/2013/06/05/gotw-91-solution-smart-pointer-parameters/

7.3 Functions, functions, functions

There are different ways to look at functions in C++. Sometimes, it is a good idea to design algorithms that take a function as input. For example, it is desirable for an optimization algorithm to accept the function to be optimized as its input. To do this, we need the ability to treat functions as data that can be passed around. In this section, we discuss function pointers, functions objects, and lambda functions. The ability to work with functions as variables is necessary for working with algorithms in the standard template library.

7.3.1 Function pointers

Function pointers are the old-school way of dealing with *functions as variables* with origins from the C language. We show it here for completeness, but we will soon find that the more modern approaches are more favorable from a readability standpoint.

Function pointers are essentially pointers with a special type that encodes the return type and arguments of a specific class of functions. For example,

```
int (*fptr)(double);
```

defines a function pointer that can be used for functions that have a `double` input parameter and an `int` return type. This function pointer cannot be used to point to functions that have any other signature.

Pointing a function pointer to a specific function can either be done at initialization or assigned by reference. For example, a function with signature `int foo(double);` can be assigned to a function pointer in the following ways:

```
int (*fptr)(double){ foo }; // at initialization
fptr = &foo;                // assigned by reference
```

7.3.2 Function objects

The basic idea of function objects (sometimes called *functors*) is simple: define a class that overloads `operator()` so that objects of the class can call something like `object(argument)` to evaluate for some input argument(s). The objects themselves can then be passed around easily. For example, a class defined like

```
class Functor {
  int n;
  public:
    Functor(int N) : n(N) {}
    double operator()(double x, double y){ /* ... */ }
}
```

can be used as

```
Functor F(5);
auto result = F(42.0,33.0);
```

to obtain the double-precision result from the evaluation of `operator()`.

A really nice feature of using function objects is that they can be set up prior to being used and hold some state that can then be used in the `operator()` call. This is demonstrated simply through the constructor in this case, but more advanced functionality can be added to a class of this type.

The type `Functor` can now be used in other places— function arguments, return types, or in other classes—where the expected usage is to be treated as a function.

7.3.3 Lambda functions

Sometimes it might seem like a lot of effort to define a function object just to evaluate a seemingly *simple* function. This situation is where *lambda functions* come into play, allowing us to define functions in line. The syntax is admittedly clunky compared to lambda functions in Python or anonymous functions in Matlab, but the additional complexity allows finer control of their use in C++ compared to the others.

The general format for lambda functions is

```
[ captures ] (parameters) -> returntype { /* body */; };
```

The optional keyword `captures` represents the variables at the current scope and how they should be accessible to the lambda function. There are four ways in which captures can be specified:

1. capture no variables `[]`; use no external variables in the lambda function,
2. capture specific variables, e.g., `[i,&j]` captures `i` by value and `j` by reference,
3. capture all variables by value `[=]`,
4. capture all variables by reference `[&]`.

The `parameters` is the list of input arguments needed when calling the lambda function. This convention is the same as normal function definitions.

The `returntype` is an explicit note of the type of variable that gets returned by the lambda function. The use of this keyword is optional in a lambda definition because the return type can typically be automatically deduced at compile time.

The `body` is the function body that is run when the function is called.

The most basic use of a lambda function looks like

```
auto minimum = [](int a, int b){
  return a < b ? a : b
};
auto m = minimum( 3, 14 );
```

and allows us to have simple function definitions specified wherever we need them, e.g., close to the calling code.

The ability to capture variables from the local scope to be used inside a lambda function gives lambda functions much more power and flexibility (for their relatively simple syntax) over standard functions. Under the hood, the implementation of lambda functions essentially just abstracts the pattern of a `Functor` such that the compiler can automatically build the desired class.

7.4 Header and implementation files

As codebases increase in size, it becomes necessary to think in terms of *interface* code vs. *implementation* code. Interface code should be organized into header files. In C++, these are the files that get `#included` in other code and usually have the extension `.hpp` or `.h`. These header files should define all the appropriate signatures for what can be used from the included code. Alongside a header file, there should be an implementation file, usually with extensions `.cpp`, `.cc`, or `.cxx`, that specifies the implementation details. The implementation file needs to be compiled, and libraries generally take their set of implementation files and compile and link them altogether into a single static (`.a`) or shared (`.so` on Linux or `.dylib` on macOS) library.

This separation has a couple of obvious benefits:

1. The ways in which a library is intended to be used are accessible by looking at the header file.
2. The (nitty-gritty) details are constrained to the implementation file, which can be compiled once on a given platform and linked against from different programs.

When header files are `#included` into a calling code, they are literally copy-pasted to that point in the code. Thus, the compilation of code that has `#includes` in it compiles any of the code in the included header. This is why as little information as is needed to use a library should be included in the header file. The bulk of the details for *how* a code does what it does should be left to the implementation files.

For code that uses templates, this strategy can be problematic. For example, if you set up your class to use arbitrary types through template parameters, when you try to compile the implementation code, you are faced with the question: *which* types should this code get compiled with? There are two possible solutions to this problem:

1. The implementation files can *instantiate* specific versions of the template.
2. All code can be moved to header files.

The first solution is useful when the design is such that there are standard or common types that are used for template parameters. Example code that shows how this is done is found in Listings 7.3 to 7.5. The second solution is useful when the design is such that user-defined types get used as template parameters. This second *header-only* solution can allow for more flexibility in usage, but it has the downside of potentially adding to compilation times significantly when using the library.

Listing 7.3 Header file for Foo

```
template<typename T>
class Foo
{
  T bar;
public:
  void doSomething(T param);
};
```

Listing 7.4 Implementation file for Foo

```
// implementations need to include their own headers
#include "foo.hpp"

// These statements are not needed in the interface
#include<typeinfo>
#include<iostream>
using namespace std;

template<typename T>
void Foo<T>::doSomething(T param){
  cout << "Type: " << typeid(param).name() << endl;
  cout << " bar: " << bar << ", param: " << param << endl;
}

// Instantiate the intended types to be used with Foo<T>
template class Foo<int>;
template class Foo<double>;
```

Listing 7.5 Program using Foo

```
#include "foo.hpp"

int main(int argc, char* argv[]) {

  Foo<int> F1;
  F1.doSomething(42);
  Foo<double> F2;
  F2.doSomething(42.0);

  /*
      This will produce a link time error because Foo<long>
      has not been instantiated in the implementation
   */
  // Foo<long> F3;
  // F3.doSomething(42);

  return 0;
}
```

Chapter 8
The standard template library

In C++, the Standard Template Library (STL) offers various *containers* for holding groups of data, *iterators* for accessing data within containers, and *algorithms* for performing common tasks on groups of data. All these components are implemented as templates in the STL, allowing generic usage with your own types. Everything discussed in this chapter is in the namespace `std`. We also introduce the `std::string` class for its general utility in dealing with strings of characters.

In reality, there are entire courses devoted to data structures and algorithms. In the limited space devoted to this topic that comprises this chapter, we show some examples of common usage of STL components and point to the wealth of resources available for digging down into more detailed information as desired.

The key takeaway from this chapter is that the performance of a program (how long it takes to run and how much memory it uses) can be heavily influenced by the data structures used within a given algorithm. It is thus important to design software that is at least partially agnostic to the particular data structures (or in this chapter, *STL containers*) used within it. Such design can allow the data structures to easily be swapped to compare (and optimize) performance.

8.1 STL containers

There are two main types of containers in the STL: *sequential containers*, which are typically used for sequential access to their elements, and *associative containers*, which can be accessed either sequentially or via a *key*.

8.1.1 Sequential containers

Sequential containers (also referred to as *sequence containers*) are sequential in the sense that their elements are intended to be accessed in a specific order, one after the

© The Author(s), under exclusive license to Springer Nature Switzerland AG 2025

R. J. Spiteri and K. Klenk, *Extreme-Scale Computing*, CMS/CAIMS Books in Mathematics 15, https://doi.org/10.1007/978-3-031-89033-8_8

other. Iterating through elements forwards or backwards is (in principle) fast. The STL sequential containers are:

- `array` - contiguous storage of elements, size known at compile time, fast random access.
- `vector` - contiguous storage, dynamic size, fast add/remove of elements at end, fast random access.
- `list` - non-contiguous storage, doubly linked list, fast add/remove anywhere, slow random access.
- `deque` - non-contiguous storage, double-ended queue, fast add/remove of objects at beginning and end, slow random access.

The `std::array` container serves as a replacement for compile-time static arrays (e.g., `double var[10]`) and aligns more consistently with other STL components. The `std::vector` container is often the most frequently used of the STL containers. **We highly recommend using `vector` as your default choice of container** and only changing to other containers with sufficient justification obtained through profiling your code.[1] This usage is recommended because a `vector` guarantees contiguous memory storage of its elements, resulting in better performance when iterating over the container. We have already observed potential performance implications of iterating over a two-dimensional array in the wrong order in Listing 6.1. A large contribution to the decrease in performance was due to having to constantly fetch data that were far away in memory. A similar sort of thing can happen with one-dimensional containers like `list` because its elements are not guaranteed to be nearby in memory. An example of working with a vector (dynamic size) of arrays (static size) is given in Listing 8.1.

[1] One good justification for using `array` over `vector` is when you know the size of the array will remain fixed after compile time. The use of `array` makes it obvious to others the desired intention of the data structure.

Listing 8.1 A composite container (a vector of arrays), iterating over a container, and applying an algorithm to sort the container in non-descending order. Non-descending order is similar to ascending order but may be more appropriate when the order of the elements is not strictly increasing (i.e., when the container contains equal elements).

```
#include<array>
#include<vector>
#include<iostream>

#include<algorithm>

using namespace std;

int main(int argc, char* argv[]) {

  // NOTE: no allocation of vector size here
  vector<array<double,3>> data;
  for (int i = 0; i < 5; ++i) {
    array<double,3> tmp{static_cast<double>(i),
                        -static_cast<double>(i),
                        10*static_cast<double>(i)};
    // push_back adds to the end of a vector
    data.push_back(tmp);
  }

  // Sort by the 2nd value in the array
  sort(data.begin(), data.end(), [](array<double,3> a, array<double,3> b) {
          return a[1] < b[1];
  });
  cout << "Sorted by second value, non-descending:\n";
  for(const auto &it1 : data) {
    for (const auto &it2 : it1) {
      cout << it2 << " ";
    }
    cout << endl;
  }

  // Sort by the 3rd value in the array
  sort(data.begin(), data.end(), [](array<double,3> a, array<double,3> b) {
          return a[2] < b[2];
  });
  cout << "Sorted by third value, non-descending:\n";
  for(const auto &it1 : data) {
    for (const auto &it2 : it1) {
      cout << it2 << " ";
    }
    cout << endl;
  }

  return 0;
}
```

Listing 8.1 also makes use of iterators and an algorithm (`sort`), upon which we elaborate in Sections 8.2 and 8.3, respectively.

8.1.2 Associative containers

Associative containers are intended to be accessed by a *key*, but they can also be accessed sequentially and *iterated* over. However, their underlying implementations (usually some sort of tree structure) may make such iteration slower than their sequential counterparts. The main associative containers that are in the STL are:

- `set` - unique elements with a specific order; the key and the value are the same.
- `map` - unique elements specified as key-value pairs.
- `multiset` - like `set` but can have non-unique values (and thus, keys).
- `multimap` - like `map` but can have non-unique keys.

All the above associative containers are guaranteed to be sorted by their keys, but they also have unordered variants when prefixed by `unordered_`. The unordered variants have the same functionality, but their keys are not sorted.

As an example, we look at `std::map` in Listing 8.2.

Listing 8.2 Key-value pairs in a `map`. Upon addition of new entries, the `map` is re-sorted by its keys.

```
#include<map>
#include<iostream>
using namespace std;

int main(int argc, char* argv[]) {

  map<int,double> m;
  m[42] = 41.99;
  m[-3] = 36.3;

  // Elements in the map
  cout << "Iterating over a map:\nkey value\n";
  for(auto it : m) {
    cout << it.first << " " << it.second << endl;
  }
  return 0;
}
```

The output of Listing 8.2 is:

```
Iterating over a map:
key value
-3 36.3
42 41.99
```

We note that the key `-3` occurs before the key `42` in the output despite being added to the map later.

8.1.3 Adapters

Adapters can be thought of as *meta*-containers; i.e., containers that use the other containers for their underlying implementation. The adapters are:

- stack - last-in, first-out container
- queue - first-in, first-out container
- priority queue - sorted queue, only highest-priority entry can be accessed directly

We do not discuss further what these adapters do but rather point to `http://www.cplusplus.com/reference/stl/` for further details.

8.2 STL iterators

Iterators are objects that traverse the data in a container class without the caller having to know about the underlying implementation.

Member functions for iterators are:

- `begin()` - produces an iterator pointing at the first element in the container.

- `end()` - produces an iterator pointing to the last element in the container.
- `cbegin()` - produces a `const` iterator pointing at the first element in the container.
- `cend()` - produces a `const` iterator pointing at the last element in the container.

The difference between a `const` iterator and a normal iterator is, as the name is meant to imply, that a `const` iterator cannot be used to change the values to which it points; i.e., `const` iterators are used only for observing data in a container.

With an iterator handy, the common actions one wants to do are:

- dereference: `Operator*` - get the element to which the iterator points.
- increment: `Operator++` - move to the next element in the container.[2]
- comparison: `Operator==` and `Operator!=` - test for equality or inequality of the iterator.
- assignment: `Operator=` - point the iterator at a specific element in the container.

Assignment is generally used to start an iterator by pointing it *somewhere* in a container. Equality comparison is used either to test whether the iterator has reached the desired end point for some loop or to check the return location of an algorithm (see Section 8.3).

There are a few ways to loop through containers using iterators. If the entirety of a container is to be traversed, C++ has a nice succinct notation, referred to as a *range-based* for loop

```
for (auto it : container) { /* ... */ }
```

This loop starts the iterator at the beginning of `container` and increments it after every loop body evaluation. Similarly, one can iterate through `const` references

```
for (const auto &it : container) { /* ... */ }
```

if the members of `container` are not to be modified in the body of the loop.

We recommend that you only use the `auto` keyword when dealing with iterators. The previous two code snippets work for `container` being any STL container type, but the actual type of iterator used is dependent on what type of container it is. For example, if we explicitly use

```
vector<int> container(20);
for( vector<int>::iterator it : container ) { /* ... */ }
```

then changing the `container` type would also require changing the iterator type.

Another way to loop over containers is from `beginning` to end, using the standard `for` loop syntax

```
for( auto it=container.begin();
     it != container.end();
     ++it                        ) { /* ... */ }
```

[2] The pre-increment `++iterator` should be used instead of the post-increment `iterator++`.

This gives more fine-grained control if, for example, you desire to iterate over only part (by replacing the starting point or termination condition).

Sometimes, you may want to iterate over multiple related containers simultaneously. This can be done a number of ways, one being

```
assert(c1.size() == c2.size());
auto it1 = c1.begin();
auto it2 = c2.begin();
for ( ; it1 != c1.end(); ++it1, ++it2) { /* ... */ }
```

which assigns two iterators before the loop (note the empty precondition in the for loop) and then increments both at the end of each loop body evaluation.

8.3 STL Algorithms

Why algorithms? The main goal of the STL, with its containers, iterators, and algorithms, is to make code easier to understand at a higher level. For example, rather than a developer having to go through complex looping code and assess what is going on (the larger the loop body, the more difficult this becomes!), it is easier to reason about code that has collapsed loops into *algorithms* applied over ranges, provided one puts in the time to understand what the standard algorithms do.

STL algorithms are standard operations to be used on ranges of data in containers. Most algorithms are set up to work on items in the range `[first,last)`[3] with `first` and `last` being iterators.

A full list of algorithms in the STL can be found at `https://en.cppreference.com/w/cpp/algorithm`. Here, we only go through a few to see how they are used.

There are four main classes of algorithm in the STL:

- Non-modifying sequence
- Modifying sequence
- Partitioning
- Sorting

We now cover each of these algorithms and provide a candidate example for each.

For each example, we give a basic version of how the algorithm could be programmed. We note that this is generally not how the algorithms are actually coded in a given implementation of the C++ standard because the practical usage often requires much more involved code that can exploit certain hardware instructions (when available).

[3] We explicitly use mathematical notation here, with `first` included in the range but `last` excluded.

8.3.1 Non-modifying sequence

Algorithms that do not modify the ordering of elements in the container—but can potentially change them—are referred to as *non-modifying sequence* algorithms. A particular example of this type of algorithm is

for_each: Apply a single function on every element in the specified range.

The for_each algorithm is equivalent to

```
template<class InputIterator, class Function>
Function for_each(InputIterator first, InputIterator last,
                  Function fn) {
  while (first!=last) {
    fn (*first);
    ++first;
  }
  return fn;       // or as of C++11: return move(fn);
}
```

8.3.2 Modifying sequence

Algorithms that modify (add to, remove from, or move) the elements in a container are referred to as *modifying sequence* algorithms. A particular example of this type of algorithm is

copy: Copy elements in the specified range to an output container.

The copy algorithm is equivalent to

```
template<class InputIterator, class OutputIterator>
OutputIterator copy (InputIterator first, InputIterator last,
                     OutputIterator result) {
  while (first != last) {
    *result = *first;
    ++result; ++first;
  }
  return result;
}
```

We note that the algorithm takes iterators as input, specifying the start and end of the range on which to work, and an iterator that indicates where the range should be copied.

8.3.3 Partitioning

Algorithms that separate the elements in containers into groups based on how they satisfy certain conditions are referred to as *partitioning* algorithms. A particular example of this type of algorithm is

`partition`: Re-arrange the elements in the range such that the first group contains all the elements for which a *unary predicate* returns `true` and the second group for which the unary predicate returns false.

The `partition` algorithm is equivalent to

```
template <class BidirectionalIterator, class UnaryPredicate>
BidirectionalIterator partition (BidirectionalIterator first,
                                 BidirectionalIterator last,
                                 UnaryPredicate pred) {
  while (first != last) {
    while (pred(*first)) {
      ++first;
      if (first == last) return first;
    }
    do {
      --last;
      if (first == last) return first;
    } while (!pred(*last));
    swap (*first,*last);
    ++first;
  }
  return first;
}
```

We note that the algorithm returns an iterator that points to the first element of the second group.

8.3.4 Sorting

Algorithms that order the elements in containers based on a comparison of the elements are referred to as *sorting* algorithms. Similar to partitioning, sorting rearranges the elements in a container, but now instead of a *unary predicate*, a *binary predicate* that compares elements is used as input. A particular example of this type of algorithm is

`sort`: Orders the elements in the range in non-descending order.

Two possible signatures of the `sort` algorithm are

```
template <class RandomAccessIterator>
void sort (RandomAccessIterator first,
           RandomAccessIterator last);

template <class RandomAccessIterator, class Compare>
void sort (RandomAccessIterator first,
           RandomAccessIterator last,
           Compare comp);
```

We note that the first signature uses `operator<` to compare elements and thus works for any type that has a defined `operator<`. The second signature requires a function object with signature

```
bool comp(const Type1 &a, const Type2 &b);
```

The example in Listing 8.1 shows how the elements of a `vector` can be sorted using a lambda function for a custom comparison operation `sort`.

8.4 Other STL components

There are many other useful components in the C++ standard library, including complex numbers `<complex>`, numeric data type limits `<limits>`, random number generation `<random>`, and more! A useful reference to explore them is `https://www.cplusplus.com/reference/std/`.

Of particular note is the `string` class, which can be thought of as a container built specifically for holding 1-byte `char`s. Strings can be manipulated through iterators as well as a number of specific member functions. There is much that can be discussed around strings, so we defer to the reference `https://www.cplusplus.com/reference/string/string/`.

In addition to the member functionality referenced above, `string`s can be manipulated via streaming operations using `<sstream>` from the standard library. Simple usage of how one can produce formatted strings from `sstream` is found in Listing 8.3.

Listing 8.3 Manipulating strings with `sstream` and `iomanip`

```
#include<string>
#include<iostream>
#include<sstream>
#include<iomanip> // for setprecision
using namespace std;

int main(int argc, char* argv[]) {

  stringstream ss;
  double d{41.999999999990};
  int ndigits{14};

  // Create a formatted string through streaming
  ss << "The true meaning of life is "
     << setprecision(ndigits) << d
     << " (to "
     << setprecision(1) << ndigits
     << " digits)." << endl;

  cout << ss.str();

  return 0;
}
```

Chapter 9
Parallel programming patterns

The previous chapter gave an overview of how programming patterns for manipulating data can be provided in a generic form through the STL. This chapter continues the discussion of programming patterns but focuses on programming patterns that can be executed in parallel, starting with an introduction to loop dependence.

The discussion on data dependence in loops is intended to stimulate thought about *data flow* in a given algorithm. What to do when you understand the flow, i.e., how to reformulate to a flow that is better for parallelization, is greatly influenced by whether you are working with a distributed or a shared memory hardware architecture. We defer more detailed discussion to these issues in parts IV and V of this book.

Because the idea of parallel execution for algorithms in the STL has only recently been standardized and implemented in C++ compilers, we do not go into great detail in this chapter. Historically, due to the complexity of how vectorization and multithreading are handled at the hardware level, parallel execution has been handled by external libraries for C/C++.

We note that parallel execution in the C++ standard is for the shared-memory paradigm. This chapter provides a look into how the C++ standard simplifies the usage of parallel algorithms in shared memory. A more detailed presentation of multithreading (and vectorization) is provided in Part V.

9.1 Loop dependence

The key to parallelizing algorithms is to consider how the data change across iterations. For example, are the data in a particular iteration modified in a way that is independent of data in other iterations? Or are the data dependent on the data of (and thus, the modified data of) other iterations?

© The Author(s), under exclusive license to Springer Nature Switzerland AG 2025

R. J. Spiteri and K. Klenk, *Extreme-Scale Computing*, CMS/CAIMS Books in Mathematics 15, https://doi.org/10.1007/978-3-031-89033-8_9

Thinking of algorithms as loops over the data provides a clean way for reasoning about how a given algorithm can be parallelized. For example, the following loops[1]

```
for (int i = 0; i < N; ++i) {
  b[i] = a[i] + 4;
}
for (int i = 1; i < N; ++i) {
  c[i] = b[i-1] - 2;
}
```

do not have any dependence between loops. Every assignment to the `b[i]` (or `c[i]`) variables does not depend on any data that have been modified at a different iteration. The iterations of a loop that carry no data dependence between them are called *loop-independent*. Algorithms with loops that contain only loop-independent iterations are sometimes called *embarrassingly parallel* because the entirety of the loop can be parallelized (run on different compute resources) without any thought behind synchronizing data between iterations.

In contrast, the loops

```
for (int i = 1; i < N; ++i) {
  b[i] = a[i] + b[i-1];
}
for (int i = 2; i < N; ++i) {
  c[i] = b[i-1] - 2*c[i-2];
}
```

do carry dependence between iterations. The first loop has a dependence distance of 1, and the second loop a dependence distance of 2. This distance tells us how far away the data dependence stretches in terms of iterations.

These simple-looking dependencies can simply not be fixed in some cases! Depending on the actual contents of the loop, the calculations may be so dependent on previous iterations that the dependence cannot be removed. In other cases, the dependence may be coming from how the calculations are expressed, and reordering the calculations into an equivalent mathematical formulation may remove dependencies.

Such reformulations often result in additional arithmetic to be performed in an amount that may or may not depend on the number of compute elements processing the loop. This is probably best shown by example (which we do in Section 9.1.1), but we encourage you to think about how you might approach the problem of splitting computations into pieces that *mostly* do not depend on each other.

Sometimes the dependency between iterations may at first look to be unimportant, such as in

```
for (int i = 0; i < N-2; ++i) {
  a[i] = a[i+2] - 50;
}
```

[1] These loops can be written in terms of either `std::transform` or `std::for_each` as shown in Section 9.2.

This is sometimes called an *anti-dependency* because a quick analysis of the loop structure shows it requiring iteration `i` to be working with some pre-existing state of the array a. With that thought in mind, it becomes clear that a requirement for the correct functioning of this loop is that `a[i]` uses the *old* state of `a[i+2]`, and thus `a[i+2]` cannot be updated before `a[i]`. This type of dependency can be removed by having a copy of the a array before entry to the loop. However, if there are additional operations being performed within the loop, there may be other crafty solutions for any given loop.

9.1.1 Resolving loop dependencies

We consider the case where the loop

```
for (int i = 1; i < N; ++i) {
  b[i] = a[i] + b[i-1];
}
```

is to be run on two cores of a CPU. The data dependence complicates this process because written as is, it seems that iteration `i` cannot be started until iteration `i-1` has completed. However, if we think of this as being written as two loops

```
for (int i = 1; i < N/2; ++i) {
  b[i] = a[i] + b[i-1];
}
for (int i = N/2; i < N; ++i) {
  b[i] = a[i] + b[i-1];
}
```

or rather as a nested loop

```
start[0] = 1;   end[0] = N/2;
start[1] = N/2; end[1] = N;
for (int j = 0; j <= 1; ++j) {
  for (int i = start[j]; i < end[j]; ++i) {
    b[i] = a[i] + b[i-1];
  }
}
```

now we have a case where the data flow *between* the two inner loops is almost non-existent; i.e., the `j=1` inner loop could be independently run from the `j=0` inner loop if only it had access to `b[N/2-1]`.

This observation points us in a good direction for proceeding: We can try to be clever around how the `b[N/2-1]` entry is handled. Both of these loops only perform addition, so we can try to work out an alternative (but equivalent) way in which we can obtain the same result in the b array after the computations.

Ensuring that the value of `b[N/2-1]` is zero before the loops start, in this case, we can account for the influence of `b[N/2-1]` on the second half of the array after a first pass over the entries. The following snippet is equivalent to the original loop of the example

```
start[0] = 1;   end[0] = N/2;
start[1] = N/2; end[1] = N;

b[end[0]-1] = 0.0;

// j=0 and j=1 can be executed independently
for (int j = 0; j <= 1; ++j) {
  for (int i = start[j]; i < end[j]; ++i) {
    b[i] = a[i] + b[i-1];
  }
}

// embarassingly parallel loop!
for (int i = start[1]; i < end[1]; ++i) {
  b[i] = b[i] + b[end[0]-1];
}
```

Unfortunately, we have added approximately `N/2` extra additions over the original loop, but these were necessary to handle the dependence between iterates. What we have gained now is the ability to execute the code with two CPU cores. An extension of this process to allow execution on p CPU cores should follow naturally.

The reality of whether or not loop restructuring trickles down to performance improvements depends on many things, in particular whether the program is to run on a *distributed* or *shared* memory architecture. The question of "Does a certain CPU have access to a certain location in memory?" turns out to suggest different ways to structure the data flow of programs. We consider the details to this answer for distributed and shared architectures later in Parts IV and V, respectively.

Rearrangement and addition of operations, while perfectly fine from a purely mathematical point of view, can modify the behavior of floating-point arithmetic in practice. The "equivalent snippet" presented above may contain slight differences in the results due to differences in rounding of intermediate results. This behavior is a near-unavoidable reality of life in the world of computing— and especially in the world of parallel computations where the order of operations is generally not fixed —and we find that it is critical to use algorithms that are *stable* with respect to perturbations from rounding and order of operations generally.

9.2 STL parallel algorithms

We have already seen how much more expressive our code can be when we identify computational patterns that are handled by STL algorithms. When those patterns can be identified and reused, it becomes easier to reason about the code. In addition to that, the STL now provides *parallel execution* for most of its algorithms.

A key point of the previous section was to demonstrate that even simple-looking algorithms (loops) can become complicated to analyze when considering parallel execution. Provided we can rewrite a program to use STL algorithms, we can automatically reap the benefits of parallel execution for certain algorithms.

The following algorithms are examples of those that show excellent performance for numerics when executed in parallel. If your computations can be expressed in terms of the algorithms, they are likely to *scale* well to larger and larger problems run on larger and larger computers. In the `std` namespace,

- `for_each` - applies a function to a given range.
- `transform` - applies a function to a given range and assigns the output to an output range. This type of operation is often called a *map* in the high-performance computing and functional programming worlds, not to be confused with the associative container `std::map`.
- `reduce` - computes a *generalized sum* of an initial value and a binary operation applied over an input range. This type of operation is often called a *reduction* in the high-performance computing world.
- `exclusive_scan` - computes *generalized partial sums* (exclusive) of an initial value and a binary operation applied over an input range.
- `inclusive_scan` - computes *generalized partial sums* (inclusive) of an initial value and a binary operation applied over an input range.

We encourage you to think about how multiple CPUs can work together in carrying out the above operations. We look more closely at implementations of *how* these types of operations are performed in parallel in the later parts of this book.

Parallel execution of STL algorithms comes down to specifying an *execution policy* as the first argument to the (overloaded) algorithm.[2] For example, details of the `transform` operation show that it can be invoked as in Listing 9.1.

Execution policies inhabit the `std::execution` namespace and take one of the following forms (in C++ 17):

- `seq` - (`sequenced_policy`) - algorithm must be run sequentially.
- `par` - (`parallel_policy`) - algorithm can be run in parallel (multithreaded).
- `par_unseq` - (`parallel_unsequenced_policy`) - algorithm can be run in parallel (multithreaded and vectorized).

An example that demonstrates the potential for speed up in `transform` operations (including random number generation) is given in Listing 9.1.

[2] `https://en.cppreference.com/w/cpp/algorithm` specifies all the different overloads for all the algorithms, in all the C++ standards. We realize it is a behemoth, but it allows you to explore any of the algorithms that include an `ExecutionPolicy`.

Listing 9.1 Parallel execution of `std::transform` additionally includes random number generation (to initialize an input vector), `std::function` to simplify passing functions as function arguments, and timing using features from <chrono>. The main program sets up an input vector, and it times the application of the transform.

```
#include <iostream>
#include <vector>
#include <random>     // random number generation
#include <algorithm>
#include <chrono>     // clocks and timing
#include <execution> // parallel execution

void printDuration(std::chrono::high_resolution_clock::time_point start,
                   std::chrono::high_resolution_clock::time_point end) {
  auto diff = end - start;
  std::cout << "Elapsed time "
            << std::chrono::duration <double, std::milli> (diff).count()
            << " ms\n";
}

template<typename T>
void time_transform(const T &policy,
                    const std::vector<double> &data,
                    const int repeat,
                    const std::function<double(double)> fun) {
  for (int i = 0; i < repeat; ++i) {
    std::vector<double> data_out(data.size());

    // Time the transform operation
    const auto start = std::chrono::high_resolution_clock::now();
    std::transform(policy, data.cbegin(), data.cend(), data_out.begin(), fun);
    const auto end = std::chrono::high_resolution_clock::now();
    printDuration(start, end);
  }
  std::cout << '\n';
}

int main() {
  // Test samples and repeat factor
  constexpr size_t samples{5000000};
  constexpr int repeat{10};

  // Fill a vector with random numbers
  std::random_device rd;
  std::mt19937_64 mre(rd());
  std::uniform_real_distribution<double> urd(0.0, 1.0);

  std::vector<double> data(samples);
  for (auto &e : data) {
    e = urd(mre);
  }

  auto f = [](double x) -> double {
           return exp(x)*sin(x)*cos(x)*(1-x*x);};


  // Sort data using different execution policies
  std::cout << "std::execution::seq\n";
  time_transform(std::execution::seq, data, repeat, f);

  std::cout << "std::execution::par\n";
  time_transform(std::execution::par, data, repeat, f);
}
```

Example output timings from Listing 9.1 are:

```
std::execution::seq
Elapsed time 730.493 ms
Elapsed time 724.942 ms
Elapsed time 672.111 ms
Elapsed time 669.092 ms
Elapsed time 674.308 ms
Elapsed time 670.537 ms
Elapsed time 667.676 ms
Elapsed time 671.33 ms
Elapsed time 687.791 ms
```

```
Elapsed time 602.344 ms

std::execution::par
Elapsed time 32.0979 ms
Elapsed time 30.158 ms
Elapsed time 30.2344 ms
Elapsed time 30.2315 ms
Elapsed time 30.2466 ms
Elapsed time 30.2647 ms
Elapsed time 30.2427 ms
Elapsed time 30.2286 ms
Elapsed time 30.2187 ms
Elapsed time 30.2738 ms
```

GCC 9+ provides implementations of STL parallel algorithms using the Intel Threaded Building Blocks (TBB) library.[3]

9.3 Parallel programming patterns primer

The solution to a computational problem can be viewed as a sequence of instructions being applied to a collection of data. From this view, it is intuitive that parallelization of the solution to a computational problem can either come from parallelization of the instructions (task decomposition), parallelization of the data (data decomposition), or both! There is no best approach to parallelizing programs in general; the weighting of task vs. data decomposition depends on the particular problem being solved.

Large and complicated computational problems often need to mix both task and data decomposition methods. Data decomposition is, however, a necessity when problems start to get large. Experience shows that when designing parallel computation software that is intended to scale up, starting from the data decomposition is a good idea if possible. This is because

- applying task decomposition on top of software that uses distributed data structures is easy.
- reworking your data structures to be distributed inside of software that uses task decomposition is hard.

As the saying goes: Parallel computing is easy if you don't care about performance. Task-based problem decompositions that make no consideration of how much data may need to be communicated between tasks (or what kind of network the data are communicated across) may indeed work, but they may not work in a particularly efficient manner.

Parallel programming strategies arise from the physical architectures of the computers on which they run. A useful classification scheme for architectures is that

[3] Open source: `https://github.com/oneapi-src/oneTBB`

of *Flynn's taxonomy*, which classifies parallelization based on two criteria: i) the number of *instruction streams* and ii) the number of *data streams* available within the architecture. Thus, the four main classifications of parallelization are

- SISD - **S**ingle **I**nstruction stream, **S**ingle **D**ata stream.
- SIMD - **S**ingle **I**nstruction stream, **M**ultiple **D**ata streams.
- MISD - **M**ultiple **I**nstruction streams, **S**ingle **D**ata stream.
- MIMD - **M**ultiple **I**nstruction streams, **M**ultiple **D**ata streams.

Modern day large-scale computing almost *always* falls under the MIMD category: many computing components (nodes of a supercomputer, for example), each with their own distinct data. Beyond this, the computing components can themselves be thought of as SISD, SIMD, MISD, or MIMD, depending on their sub-architectures. For example, even modern multicore CPUs are considered to be MIMD devices, with each core of the CPU capable of executing its own instructions and having access to its own local *cache* of data.

Part III now moves on to discussing more details of common architectures encountered in extreme-scale computing.

Additional notes

C-style Allocation of memory

Although it is recommended (and preferred) to use the keywords `new` and `delete` when allocating memory in C++, one may still encounter the usage of `calloc` and `malloc`. Accordingly, it is worth explaining of how these C functions work.

```
void* calloc(size_t num, size_t size);
```

The `calloc` function allocates a block of memory for a specified number of elements and initializes the elements to 0. The parameter `num` is the number of elements one wishes to initialize, and the parameter `size` is the size in bytes to which each element is initialized. A pointer to the allocated memory block is returned on success, and a null pointer is returned on a failure.

```
void* malloc(size_t size);
```

The `malloc` function is similar, with the difference being it does not initialize any memory— it only allocates it. The parameter `size` is the size in bytes of the entire block of memory to be allocated. A pointer is returned to the allocated block of memory on success and a null pointer is returned on a failure.

In order to deallocate memory that has been allocated by either `calloc` or `malloc`, one must use the `free` function:

```
void free(void* ptr);
```

To free allocated memory, one passes the pointer that was returned by either `calloc` or `malloc` as the `ptr` argument.

Exceptions

Exceptions are responses to events that are not meant to occur (e.g., runtime errors). Using exceptions in C++ allows one to detect and handle specific events by using the keywords try, catch, and throw. By using these keywords, one can protect a block of code that may throw an exception. This block of code goes into what is called the try block. Within the try block, one can throw exceptions to handle certain events. Exception handlers or catch blocks are then added to catch potential exceptions that may be thrown within the try block. An example of how exceptions are used is shown in Listing 9.2.

Listing 9.2 A simple example of how to use exceptions in C++. The example below shows a block of code protected by a try block. The try block is meant to ensure that the integer value of the counter variable is greater than or equal to 0. If the value is less than 0, as in this example, the code throws an exception that is then caught and a meaningful error message displayed. Otherwise, the code displays the value of a valid counter variable.

```
#include <iostream>
#include <exception>


int main(int argc, char* argv[]) {
  int counter = -4;
  try {
    if (counter <= 0) {
      throw (counter);
    } else {
      std::cout << "Valid number received" << "\n";
    }
  } catch (int enteredNum) {
    std::cout << "Number was below 0" << "\n";
  }
}
```

Part III
Hardware considerations

Consisting of two chapters (or lecture hours), this part provides an overview of the computing hardware that is frequently encountered in extreme-scale computing. The overview is at such a level to facilitate the understanding of parallel programming strategies (and models) coming up in Part IV and beyond.

We begin by examining common computer hardware components relevant to understanding computational software performance, and in particular, the multi-core CPU architecture that is found in the majority of laptops, desktops, and workstations at the time of writing. Next, we consider *accelerators*, which are specialized compute devices that are not CPUs. In this context, we only consider graphics processing units (GPUs) because they are the most common and widely used accelerators at the time of writing. Finally, we address how both CPUs and GPUs are used in creating the larger cluster computers needed for extreme-scale computation and how this current generation of cluster computers may evolve in the near future.

Chapter 10
Laptops, desktops, and workstations

This chapter explores the main hardware components of present-day consumer computers, including laptops, desktops, and workstations. In Section 10.1, we focus in particular on the sub-structure of the CPU. As an example, we work our way through history, elaborating on Flynn's taxonomy, which we introduced in Section 9.3, all the way to modern day Intel x86 designs. We start with the SISD model of von Neumann architectures and end up at the MIMD model of modern multi-core CPUs. In addition, we highlight another practical aspect that can significantly impact program performance: memory, buses (the connections between different components), and the speeds associated with them. In Section 10.2, we introduce the common hardware components seen in modern consumer computers.

10.1 A conceptual history of the CPU

Today's multi-core CPUs are the result of decades of research and development that started from purely serial architectures. In this section, we explore the conceptual ideas that drove the evolution of CPUs, from the classical von Neumann machine to the modern multi-core CPUs we see in consumer machines today.

10.1.1 The classical von Neumann machine

The classical von Neumann machine is a model of how early serial computers worked; it is an example of the SISD paradigm of Flynn's taxonomy (Section 9.3). The model is divided into two main components, the *central processing unit* (CPU) and *main memory*, which are connected by an interconnect that allows them to exchange instructions and data. The CPU is further divided into a *control unit* and an *arithmetic-logic unit* (ALU). A schematic of the classical von Neumann architecture is given in Figure 10.1.

© The Author(s), under exclusive license to Springer Nature Switzerland AG 2025
R. J. Spiteri and K. Klenk, *Extreme-Scale Computing*, CMS/CAIMS Books in Mathematics 15, https://doi.org/10.1007/978-3-031-89033-8_10

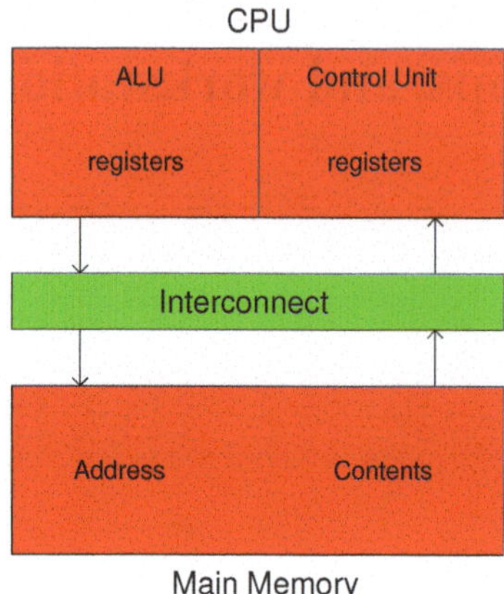

Fig. 10.1 The classical von Neumann machine.

The control unit directs the execution of programs, and the ALU is responsible for carrying out calculations. The purpose of main memory is to store both instructions and data, which are eventually moved into *CPU registers* for their execution. CPU registers are essentially fast memory components located on the CPU itself, making them relatively expensive to manufacture compared to main memory. Consequently, the size of memory in the registers is typically many orders of magnitude smaller than the size of main memory.

Data and instructions move between memory and registers via a *bus* – a generic term in computing that is used to describe anything responsible for transferring data between two different components. It is important to note that no matter how fast the CPU is, the speed at which a von Neumann machine can execute a program is ultimately limited by the rate at which the instructions and data can be transferred through this bus – this limit is famously known as the *von Neumann bottleneck*.

Much effort has gone into modifying the von Neumann machine to alleviate the bottleneck. The key innovations to the von Neumann model include *caching* memory, breaking down complex instructions to smaller functional units, and applying instructions to multiple pieces of data simultaneously.

10.1.2 Extending von Neumann: caching

In a CPU, the cache is another level of memory that exists between main memory and the registers of the von Neumann machine. The cache is intermediate in both respects of size and speed. Having a cache is useful when programs apply a small set of instructions repeatedly or apply instructions to a small set of data. In these two cases, either the instructions or the data can be stored in cache for faster access (relative to main memory).

In a system with caching, when a program needs to apply a certain instruction to a certain datum (piece of data), the CPU first looks in its cache to see if the instruction or datum is present. If the instruction or datum is in cache — good news! That is called a *cache hit*, and the instruction or datum can be used. If it is not in the cache — bad news. That is called a *cache miss* and requires a *fetch* of the instruction or datum from the main memory. This sort of architecture requires more bookkeeping by the CPU to keep track of what is in the different levels of memory and how they relate to each other across the levels. Luckily for us as programmers, this bookkeeping is usually handled at the level of the hardware, at least when using a modern CPU.

When fetching data from main memory to cache, modern CPUs only fetch data in units of *cache lines*. In other words, a cache line is the minimal amount of data that can be fetched from the main memory to cache. For example, a system that has 64-byte cache lines[1] means that a single fetch from main memory would bring 8 double-precision values (each having 8 bytes) into cache. A program that is structured to work with data sequentially would thus require 1/8[2] the number of fetches from RAM when running on this particular cached architecture versus a non-cached architecture. Modern CPUs tend to have separate caches for instructions and data as well as multiple levels of cache.

10.1.3 Extending von Neumann: MISD

Now we look to break away from the SISD paradigm. There are two main approaches to breaking away from SISD. First, we consider the idea of *pipelining*, in which instructions are broken down into finer-grained *functional units* that can be arranged in stages. The idea of pipelining is that the different stages of a complex instruction can be performed simultaneously and in sequence.

For example, suppose that a single addition consists of the following sequence of seven operations (each of which can be executed in a single cycle of the CPU):

1. Get the operands from main memory.
2. Compare exponents.
3. Shift one operand.
4. Add.
5. Normalize the result.
6. Round the result.
7. Store the result in main memory.

Now suppose that there are functional units in the CPU that can perform each of these operations independently. A single piece of data can flow from the output of one functional unit to the input of the next. Because each of the stages can be executed independently of the others, it is not necessary to wait for the entire pipeline

[1] Cache-line size is system dependent, but 64-bytes is a common size at the time of writing.

[2] Cache sizes are generally orders of magnitude larger than cache-line sizes.

to complete before another piece of data is fed into the beginning. Assuming you would like to add N pairs of numbers, the following would be observed:

- Nothing stored in main memory for the first 6 cycles.
- 1 result stored in memory for each of the next N cycles.

That is, it would take $N + 6$ cycles to add N pairs of numbers through this pipeline — much better than the $7N$ cycles (when $N > 1$) that would be required if the architecture had no pipeline.

Combining this idea of pipelining with caching memory closer to the registers, modern CPUs also include a *pre-fetcher* in their pipelines. Recall that it is the fetching from main memory that causes the von Neumann bottleneck. The pre-fetcher tries to anticipate what memory is to be accessed next and fetch the necessary cache line from main memory *before* the data are actually needed. The ability of a pre-fetcher to accurately fetch the correct data from main memory is generally limited to sequential (and sometimes *strided*) data access. Again, we emphasize the importance of *sequential* data access in this discussion.

10.1.4 Extending von Neumann: SIMD

The next extension we consider is to allow for more data to flow into the registers and be operated on simultaneously. Instructions that perform the same operation on multiple pieces of data are called *vector* instructions. Modern CPUs now include vector registers that load multiple pieces of data into them and perform a single vector instruction across them. For example, vector registers that are 128-bits wide can hold and operate on two double-precision (or four single-precision) values simultaneously.

A downside to vector instructions (and pipelining as well!) is that they exhibit limited scaling. This is because these features reduce—but do not eliminate—the von Neumann bottleneck.

10.1.5 Putting it all together

The discussion above paints the picture of how caching, pipelining, and vector MIMD instructions come together in a modern CPU. It does not stop there, however. The modern CPU also includes multiple *cores* that can act independently. Section 10.2.2 discusses how these different elements of von Neumann extensions all come together in more detail.

10.2 Components of a modern computer

Typical laptops, desktops, and workstations are usually centered around a *motherboard*, whose purpose is to orchestrate the connections between the different components of the computer. From an extreme-scale computing point of view, the main components with which we concern ourselves are:

- Central processing units (CPUs) – the primary "brain" of the computer.
- Random access memory (RAM) – the short-term ("active" or "main") memory of the computer.
- Persistent storage (hard disk drives or solid-state drives) – the long-term memory of the computer.
- Peripheral component interconnect express (PCIe) connections – a general connection *bus* to attach miscellaneous devices to the motherboard.
- Graphics processing units (GPUs) – the primary component for processing, displaying, and outputting graphics.

There are different speeds associated with moving data between the different components. The time it takes for a specific operation to move its required data between components is called *latency*. This section introduces the significance of latency on performance and emphasizes the importance of optimizing the positioning of data during program execution. In the remaining sections of this chapter, we restrict discussion to the CPU and memory, deferring more detailed discussion of PCIe and GPUs to the next chapter.

10.2.1 Motherboards

As mentioned, the motherboard is the main circuit board of the computer, and its role is to organize the connections among the different components.

Motherboards house a CPU in a *socket*. Most consumer computers have a single socket, allowing for a single CPU, whereas higher-end workstations can have multi-socket motherboards, allowing for multiple CPUs. Most desktop and workstation motherboard sockets permit replaceable CPUs. Some laptops do too, but the trend is moving towards *soldering* CPUs directly to the laptop's motherboard.

RAM modules are connected to the motherboard through pin connectors. They are usually removable and replaceable, and a given motherboard generally requires specific types of RAM modules for its RAM slots. Similar to CPUs, the trend, especially in laptops, is moving towards soldering RAM directly to the motherboard.

When considering a multi-socket motherboard, the RAM modules are routinely wired such that only certain RAM modules are *directly* connected to a CPU; e.g., in a dual-socket system, half of the RAM modules are connected to one CPU and the other half to the other. This results in what is called *non-uniform memory access* (NUMA) and has measurable implications for performance (see Part V) when a CPU must interact with data in RAM that is not directly attached to it.

Storage devices are generally interchangeable and replaceable, with the trend again moving towards soldered devices in laptops. Depending on the storage device, it may connect to the motherboard through a SATA connection (a common interface for hard-disk drives and older solid-state drives) or an M.2 connection (a newer interface for NVMe solid-state drives).

Desktop and workstation motherboards generally have multiple PCIe slots for connecting additional peripherals. Each PCIe slot will have a number of PCIe lanes(1, 4, 8, 16, or even 32) associated with it for data transfer between the peripheral and the CPU. Normally, GPUs occupy these slots; however, you can also find a wide array of hardware for these slots, including field-programmable gate arrays (FPGAs), additional storage devices, and specialized hardware for certain tasks (TV tuners, fast network cards, RAID cards, etc.).

GPUs of some form are needed for graphical output to, for example, a monitor. Modern consumer CPUs generally have *integrated graphics* cards, which represent a small part of the CPU that can take over some of the system RAM for graphical purposes. You can also find more powerful *discrete graphics* cards that plug into the system via PCIe. Discrete graphics cards have their own internal memory, which is typically faster than system RAM.

10.2.2 CPUs

Current CPUs combine the best parts of all aspects discussed in Section 10.1. Through many years of experimental architectures, the common form of the Intel CPU now includes the following:

- Multiple cores in a single CPU.
- Multiple levels of cache. Common among Intel processors are three levels:
  - L1: split into separate L1i and L1d (instruction and data caches) that are closest to the registers and private to each core.
  - L2: larger and slightly slower than L1 and private to each core.
  - L3: larger and slightly slower than L2 and shared among all cores.
- Multiple *hyperthreads* per core.
- Hardware pre-fetchers that can bypass higher-level caches.
- Vector registers up to 512 bits wide (AVX-512) in each core.

A figure of a modern CPU in comparison to a GPU is presented in the next chapter (see Figure 11.1). The last two points imply that each core of a modern CPU is a vector processor with pipelining. That is, pre-fetching is performed on a per-core basis, and it can be done independently of performing arithmetic operations such as vector additions on pairs of data.

Although this description specifically refers to Intel processors, most processors follow this general pattern: multiple cores, multiple levels of cache (some private, some possibly shared), and some degree of vector instructions within each core.

Some useful commands for inspecting the CPUs on a Linux system are:

- `lscpu` – Show the number of sockets and details of the CPUs present in each.
- `likwid-topology -g` – Show a more detailed *topology* of the motherboard and CPUs: which hyperthreads are mapped to which cores, which cores are mapped to which CPUs, which CPUs are mapped to which sockets, and how the cache levels are laid out. The `-g` option provides a simple ASCII graphic of the layout.

Examples of the above two commands are found in Figures 10.2 and 10.3. The `likwid` toolkit can either be downloaded from your Linux distribution's package manager or from `https://github.com/RRZE-HPC/likwid`.

```
Architecture:                    x86_64
CPU op-mode(s):                  32-bit, 64-bit
Byte Order:                      Little Endian
Address sizes:                   39 bits physical, 48 bits virtual
CPU(s):                          16
On-line CPU(s) list:             0-15
Thread(s) per core:              2
Core(s) per socket:              8
Socket(s):                       1
NUMA node(s):                    1
Vendor ID:                       GenuineIntel
CPU family:                      6
Model:                           165
Model name:                      Intel(R) Core(TM) i7-10700 CPU @ 2.90GHz
Stepping:                        5
CPU MHz:                         2900.000
CPU max MHz:                     4800.0000
CPU min MHz:                     800.0000
BogoMIPS:                        5799.77
Virtualization:                  VT-x
L1d cache:                       256 KiB
L1i cache:                       256 KiB
L2 cache:                        2 MiB
L3 cache:                        16 MiB
NUMA node0 CPU(s):               0-15
********************************************************************************
Graphical Topology
********************************************************************************
Socket 0:
+---------------------------------------------------------------------------------------+
| +--------+ +--------+ +--------+ +--------+ +--------+ +--------+ +--------+ +--------+ |
| |  0 8   | |  1 9   | |  2 10  | |  3 11  | |  4 12  | |  5 13  | |  6 14  | |  7 15  | |
| +--------+ +--------+ +--------+ +--------+ +--------+ +--------+ +--------+ +--------+ |
| +--------+ +--------+ +--------+ +--------+ +--------+ +--------+ +--------+ +--------+ |
| |  32 kB | |  32 kB | |  32 kB | |  32 kB | |  32 kB | |  32 kB | |  32 kB | |  32 kB | |
| +--------+ +--------+ +--------+ +--------+ +--------+ +--------+ +--------+ +--------+ |
| +--------+ +--------+ +--------+ +--------+ +--------+ +--------+ +--------+ +--------+ |
| | 256 kB | | 256 kB | | 256 kB | | 256 kB | | 256 kB | | 256 kB | | 256 kB | | 256 kB | |
| +--------+ +--------+ +--------+ +--------+ +--------+ +--------+ +--------+ +--------+ |
| +-----------------------------------------------------------------------------------+ |
| |                                       16 MB                                       | |
| +-----------------------------------------------------------------------------------+ |
+---------------------------------------------------------------------------------------+
```

Fig. 10.2 Topology of a single socket desktop with an 8-core CPU. First output is from the Linux command `lscpu` giving info about the CPU. Second output is from `likwid-topology -g`, showing how the threads, cores, and caches are set into hierarchies. Each core has its own L1 and L2 cache (256 KiB and 2 MiB, respectively), and the 16 MiB L3 cache is shared across all cores. This is standard design for modern Intel processors. See also Figure 10.3.

```
********************************************************************************
Graphical Topology
********************************************************************************
Socket 0:
+---------------------------------------------------------------------------------------------------------------------------------------------------------------------------+
| +--------+ +--------+ +--------+ +--------+ +--------+ +--------+ +--------+ +--------+ +--------+ +--------+ +--------+ +--------+ +--------+ +--------+ +--------+ +--------+ |
| |  0 32  | |  1 33  | |  2 34  | |  3 35  | |  4 36  | |  5 37  | |  6 38  | |  7 39  | |  8 40  | |  9 41  | | 10 42  | | 11 43  | | 12 44  | | 13 45  | | 14 46  | | 15 47  | |
| +--------+ +--------+ +--------+ +--------+ +--------+ +--------+ +--------+ +--------+ +--------+ +--------+ +--------+ +--------+ +--------+ +--------+ +--------+ +--------+ |
| +--------+ +--------+ +--------+ +--------+ +--------+ +--------+ +--------+ +--------+ +--------+ +--------+ +--------+ +--------+ +--------+ +--------+ +--------+ +--------+ |
| |  32 kB | |  32 kB | |  32 kB | |  32 kB | |  32 kB | |  32 kB | |  32 kB | |  32 kB | |  32 kB | |  32 kB | |  32 kB | |  32 kB | |  32 kB | |  32 kB | |  32 kB | |  32 kB | |
| +--------+ +--------+ +--------+ +--------+ +--------+ +--------+ +--------+ +--------+ +--------+ +--------+ +--------+ +--------+ +--------+ +--------+ +--------+ +--------+ |
| +--------+ +--------+ +--------+ +--------+ +--------+ +--------+ +--------+ +--------+ +--------+ +--------+ +--------+ +--------+ +--------+ +--------+ +--------+ +--------+ |
| |  1 MB  | |  1 MB  | |  1 MB  | |  1 MB  | |  1 MB  | |  1 MB  | |  1 MB  | |  1 MB  | |  1 MB  | |  1 MB  | |  1 MB  | |  1 MB  | |  1 MB  | |  1 MB  | |  1 MB  | |  1 MB  | |
| +--------+ +--------+ +--------+ +--------+ +--------+ +--------+ +--------+ +--------+ +--------+ +--------+ +--------+ +--------+ +--------+ +--------+ +--------+ +--------+ |
| +-------------------------------------------------------------------------------------------------------------------------------------------------------------------------+ |
| |                                                                        22 MB                                                                                            | |
| +-------------------------------------------------------------------------------------------------------------------------------------------------------------------------+ |
+---------------------------------------------------------------------------------------------------------------------------------------------------------------------------+
Socket 1:
+---------------------------------------------------------------------------------------------------------------------------------------------------------------------------+
| +--------+ +--------+ +--------+ +--------+ +--------+ +--------+ +--------+ +--------+ +--------+ +--------+ +--------+ +--------+ +--------+ +--------+ +--------+ +--------+ |
| | 16 48  | | 17 49  | | 18 50  | | 19 51  | | 20 52  | | 21 53  | | 22 54  | | 23 55  | | 24 56  | | 25 57  | | 26 58  | | 27 59  | | 28 60  | | 29 61  | | 30 62  | | 31 63  | |
| +--------+ +--------+ +--------+ +--------+ +--------+ +--------+ +--------+ +--------+ +--------+ +--------+ +--------+ +--------+ +--------+ +--------+ +--------+ +--------+ |
| +--------+ +--------+ +--------+ +--------+ +--------+ +--------+ +--------+ +--------+ +--------+ +--------+ +--------+ +--------+ +--------+ +--------+ +--------+ +--------+ |
| |  32 kB | |  32 kB | |  32 kB | |  32 kB | |  32 kB | |  32 kB | |  32 kB | |  32 kB | |  32 kB | |  32 kB | |  32 kB | |  32 kB | |  32 kB | |  32 kB | |  32 kB | |  32 kB | |
| +--------+ +--------+ +--------+ +--------+ +--------+ +--------+ +--------+ +--------+ +--------+ +--------+ +--------+ +--------+ +--------+ +--------+ +--------+ +--------+ |
| +--------+ +--------+ +--------+ +--------+ +--------+ +--------+ +--------+ +--------+ +--------+ +--------+ +--------+ +--------+ +--------+ +--------+ +--------+ +--------+ |
| |  1 MB  | |  1 MB  | |  1 MB  | |  1 MB  | |  1 MB  | |  1 MB  | |  1 MB  | |  1 MB  | |  1 MB  | |  1 MB  | |  1 MB  | |  1 MB  | |  1 MB  | |  1 MB  | |  1 MB  | |  1 MB  | |
| +--------+ +--------+ +--------+ +--------+ +--------+ +--------+ +--------+ +--------+ +--------+ +--------+ +--------+ +--------+ +--------+ +--------+ +--------+ +--------+ |
| +-------------------------------------------------------------------------------------------------------------------------------------------------------------------------+ |
| |                                                                        22 MB                                                                                            | |
| +-------------------------------------------------------------------------------------------------------------------------------------------------------------------------+ |
+---------------------------------------------------------------------------------------------------------------------------------------------------------------------------+
```

Fig. 10.3 Topology (from `likwid-topology -g`) of a dual-socket workstation with two Intel(R) Xeon(R) Gold 6130 CPUs. Notice again the organization of two threads for each core, private L1 and L2 caches for each core (32 kB and 1MB, respectively), and a 22 MB L3 cache that is shared across all cores on a given CPU.

10.2.3 Random access memory

The RAM of a computer is the *main memory* of the von Neumann machine. This is where the active bits of a running program are located. Most commonly, double data rate (DDR) RAM is used in modern computers.

The swappable modules mentioned earlier generally come in the DIMM (dual in-line memory module) form factor for desktops and workstations or SO-DIMM (small outline DIMM) for laptops. The *dual* in the names refers to their 64-bit data paths (compared to the *single* of the increasingly forgotten age of 32-bit computers).

The speed of RAM is often reported in megatransfers per second (MT/s)—the number of transfers the module is capable of performing every second—or in megahertz (MHz)—the actual frequency of the module's cycles. For DDR RAM, the measure in MT/s is twice that of its MHz because the "double" data rate means it is performing two transfers per cycle.

For example, DDR5 6000 refers to DDR RAM capable of 6000 MT/s; i.e., it is running at 3000 MHz. Considering that the RAM is transferring 64-bit data, this means that this RAM is capable of transferring a theoretical maximum of `6000*64/8 = 48000 MB/s` to a CPU. The actual transfer speed varies; it is dictated by a given combination of motherboard, RAM, and CPU.

In addition to normal RAM, there is *error-correcting code* RAM, known as ECC RAM. This type of RAM is generally only found in workstations (and servers/clusters). The main idea behind ECC RAM is that it contains additional protections to ensure that information written to RAM is the same information that gets read back from RAM. ECC RAM can detect and correct single-bit errors, but it can only detect (and not correct) double-bit errors.

Certain CPUs allow multiple *channels* of RAM to be connected to the CPU, allowing the CPU to obtain more data from RAM in parallel. The theoretical maximum transfer rate is multiplicative with the number of channels. That is, using our DDR5 6000 example above, combined with a CPU capable of handling six channels of RAM, we can obtain a theoretical maximum data transfer rate of 288 GB/s between RAM and the CPU.

10.2.4 Persistent storage devices

Persistent storage devices are used for long-term storage within a computer. They come in two main forms: hard disk drives (HDDs) and solid-state drives (SSDs). HDDs store data on magnetic plates and use an actuator arm to read the data off of the disk as they spin. SSDs store data by setting the state of transistors, similar to how RAM or USB flash drives work. There are many form factors for SSDs, and the most common are 2.5-inch SATA connected drives and NVMe M.2 drives.

Because of the NAND gate transistors[3] used in SSDs, they are much faster than HDDs. Modern SSDs can reach read/write speeds of a few gigabytes per second, with newer NVMe drives reaching speeds of 5–7 GB/s. Recent technological advancements have also made SSDs much more reliable than ever, making them the default choice for modern computers.

In relation to persistent storage, you may come across the term RAID ("redundant array of independent disks"). RAID comes in many formats (levels) with different purposes. For example, RAID 0 is used to get multiple disks to work together to increase the overall read/write performance, but it offers no data redundancy. RAID 1 duplicates data across multiple disks for redundancy while still providing a performance boost, although not as much as RAID 0. The other RAID levels incorporate additional features to assist in creating redundant layouts and perform error correction.

Within the Linux and Unix world, persistent storage devices are also used for *swap* space. Swap space can be thought of as virtual memory that the computer can use if it runs out of RAM space while executing applications. In particular, *swapping* refers to the process of moving entire processes between RAM and swap space, whereas *paging* refers to the process of moving individual pages of memory between RAM and swap space. Swap space should typically be reserved for emergency situations because frequent swapping/paging can significantly impact performance. To understand why heavy swapping/paging is not ideal, compare the speeds of RAM and persistent storage devices outlined in this chapter. In fact, shared computing resources generally do not even allow compute jobs to use any swap space – such jobs will simply be killed if they go beyond their requested (and allotted) RAM.

[3] NAND stands for "not AND" and is named for the type of logic gate present in solid-state memory.

Chapter 11
Accelerators and cluster architectures

This chapter continues the discussion on hardware components, initially focusing on GPUs and their connection through PCIe. We then shift our attention away from individual computer components and provide an overview of how cluster architectures are designed.

11.1 Peripheral connections and accelerators

As mentioned in the previous chapter, PCIe provides a standardized means for connecting additional devices to the motherboard. Examples relevant to supercomputers include GPUs, high-speed networking cards, and additional storage devices.

11.1.1 PCIe

PCIe allows *bi-directional*, *serial* communication of data across its bus, meaning that data can be transferred to/from PCIe-connected devices, and the data are sent in a sequential stream, bit by bit. Although a full dive into the technical nature of PCIe is beyond the scope of this discussion, an understanding of its fundamental properties around its connection interface is useful for understanding performance and potential performance bottlenecks.

Although the communication is performed serially, PCIe can communicate across up to 32 *lanes*, increasing the capacity and speed of data transmission through the bus. This behavior can be considered as a form of parallel communication, but it should not be confused with parallel communication busses. PCIe is better thought of as a *multi-channel serial bus*, where each of the channels (the lanes) acts independently of the others. In that sense, it is more like *concurrent computation* than parallel computation; see Chapter 36.

© The Author(s), under exclusive license to Springer Nature Switzerland AG 2025

R. J. Spiteri and K. Klenk, *Extreme-Scale Computing*, CMS/CAIMS Books in Mathematics 15, https://doi.org/10.1007/978-3-031-89033-8_11

PCIe devices generally note how many PCIe lanes they use: "PCIe x1," "PCIe x2," "PCIe x4," "PCIe x8," "PCIe x16," or "PCIe x32." In consumer motherboards, PCIe x16 devices are the most common at present. This is due in part to the logistics of working with the larger connections of x32 in the smaller form factor of a desktop or workstation. PCIe x32 devices are generally reserved for high-speed networking in server and cluster architectures.

The more lanes in a PCIe connection, the higher the capacity for data transmission. The capacity of each lane is dependent on the version of PCIe being used. The throughputs of the modern versions of PCIe are summarized in Table 11.1. The

Version	Standardized	Transfer rate (GT/s)	Throughput (GB/s/lane)
3.0	2010	8.0	0.985
4.0	2017	16.0	1.969
5.0	2019	32.0	3.938
6.0	2022	64.0	7.563
7.0 (planned)	2025	128.0	15.125

Table 11.1 PCIe versions and their associated transfer capacities. The transfer rate is the serial rate of the encoded data (encoded with 128b/130b). The throughput is the pre-coded bit rate per lane.

throughput of a PCIe lane is determined by its transfer rate multiplied by its *line encoding*. PCIe versions 3.0–5.0 use a line encoding of 128b/130b, which means that 128 *useful* bits can be transferred while 2 bits are used to ensure reliable transmission. PCIe 6.0 and beyond are using a more nuanced line encoding scheme, based on 3-way forward error correction (FEC) and a 256 byte Flow Control Unit (FLIT) to transfer 242 bytes of data, with the remaining 14 bytes used for error correction.

The PCIe version space has recently been evolving rapidly, where we can speculate that the successor to PCIe 7.0 (most likely PCIe 8.0) will be released and inside consumer hardware near the close of this decade (2030). At the time of writing, PCIe 3.0 can still be considered fairly common in older hardware that is still actively used. However, hardware introduced in 2020 or later is likely to be at least PCIe 4.0, with PCIe 5.0 devices quickly becoming available to the general consumer since 2023. PCIe 6.0 devices are expected to make their splash in 2024, most likely starting in high-performance hardware before making their way to the general consumer shortly thereafter. Finally, PCIe 7.0 is still under the standardization process.

With each PCIe release, it can be seen that the transfer rate is doubling. At PCIe x16 with version 3.0, the theoretical peak throughput is 15.75 GB/s. With version 5.0, PCIe x16 provides a theoretical peak throughput of 63 GB/s. This is a significant increase in throughput and is *much* faster than the read/write speeds of traditional disk drives and even solid state drives (SSDs). Furthermore, PCIe speeds are encroaching on the transfer speeds between the CPU and RAM. Although, RAM and PCIe are not directly comparable because they serve different purposes, the PCIe bottleneck is becoming less of a concern for extreme-scale computing.

11.1.2 GPUs

Displaying pixels on a monitor essentially comes down to a massive embarrassingly parallel task. GPUs are designed specifically to perform this task, but when framed as "hardware capable of performing large-scale embarrassingly parallel tasks," the use of GPUs for general-purpose computing (called *GPGPU computing*) emerges naturally. GPU cores typically have slower clock speeds than CPU cores, but the GPUs more than make up for that through a much larger number of cores.

Since the late 2000s, GPUs have been steadily gaining ground in the ESC space. GPU acceleration is how many of the Top500 supercomputers currently achieve their performance. In fact, it enabled the achievement of exascale computing in 2021 and at present is the de facto roadmap for extreme-scale computing moving forward. The reason for this is because GPUs tend to offer higher parallel performance with lower power consumption,[1] a classic example of a win-win situation.

GPUs excel at SIMD operations, similar to what was discussed regarding vector instructions previously but conceptually with even wider vector capabilities. On top of this, modern GPU architectures make use of *streaming multiprocessors*. Streaming multiprocessors are intended to help organize the flow of data and schedule the threads on the lower-level cores.

A visual summary of the architectural differences of GPUs vs. CPUs is given in Figure 11.1. The main difference is that GPUs have many cores for a given control unit—many cores are assumed to be doing the same operations on different data—whereas general-purpose CPUs tend to have a one-to-one correspondence between cores and control units.

Consumer GPU hardware generally supports single-precision floating-point arithmetic because that is usually all that is necessary for graphics. For scientific and computational workloads, however, hardware support for double-precision floating-point arithmetic is often necessary. Certain workstation GPUs and GPUs designed specifically for GPGPU include hardware for double-precision arithmetic.

With the rise of machine learning and neural networks, specialized cores for fast matrix multiplication have been evolving—the so-called *tensor core*. Tensor cores generally allow for mixed-precision arithmetic as well. Some more generic GPUs also support the machine-learning-friendly half-precision and BFloat16 floating-point precisions. The modern GPU thus potentially has many different types of cores optimized for different but related tasks.

GPUs have a memory hierarchy as well, noting the difference of lowest-level cache (compared to CPUs) pointed out in Figure 11.1. GPUs contain their own interior RAM, which is generally faster than the RAM found on the motherboard. For example, high-bandwidth memory 2 (HBM2) has a 252 GB/s per *package* theoretical peak, and a GPU can have multiple packages running in parallel, similar to multi-channel RAM on a motherboard. The NVIDIA Volta GV100, for example, uses HBM2 and has a peak bandwidth of 900 GB/s. This is much higher than the

[1] The notable exception being the current second-fastest system, Fugaku, which uses a specialized ARM CPU but adopts GPU memory technologies for its RAM.

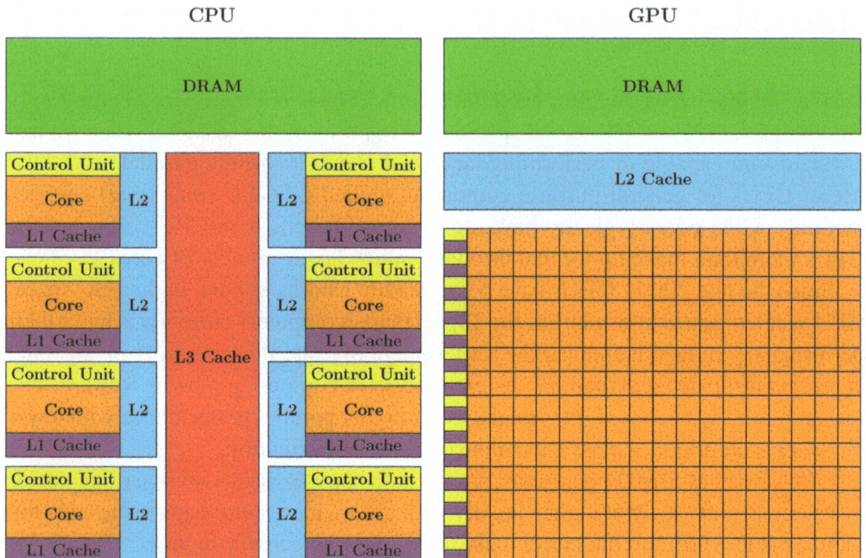

Fig. 11.1 Schematic CPU vs. GPU architectural differences. General-purpose multi-core CPUs tend to have cores with private caches and their own control units. With GPUs, the situation is that lowest-level caches are generally shared across many cores, and control units are responsible for many cores.

memory bandwidth of system RAM, and it also means that moving data *within* a GPU can be done more than an order of magnitude faster than moving data *to/from* a GPU through a PCIe x16 3.0 connection.

11.1.2.1 Multiple GPU machines

It is possible to have multiple GPU devices working together on a single machine. In the context of GPGPU computing, this means GPUs working together on the same problem. In such a case, data transfer between the devices is highly likely to be bottlenecked by the connections between them. Two GPUs connected to a motherboard via PCIe can only work as fast as the PCIe connections allow them to communicate. With luck, they would be working on a problem that does not require too much communication.

The PCIe bottleneck is becoming less of a concern for modern systems because the throughput of PCIe connections is increasing rapidly. However, today's GPUs often feature network interface controllers (NICs) directly connected to them to accelerate communication between GPUs and other devices even more. NVIDIA provides a special connection called NVLink to assist with this connection bottleneck on certain NVIDIA GPUs. NVLink-connected (second-generation) Volta GPUs have a peak bandwidth capacity of 300 GB/s. This is almost 5 times that of PCIe x32 4.0 and about double the memory bandwidth of the motherboard RAM.

11.2 Cluster architectures

With all the relevant components highlighted, we now look into how they are combined to create extreme-scale computing environments and "supercomputers".

11.2.1 Large shared-memory machines

Large shared-memory machines often make use of large multi-socket motherboards, beyond what is found even in workstations. Such machines often have four or more CPU sockets, each populated with a multi-core CPU, and terabytes of RAM.

As mentioned in the previous chapter, with the use of multiple sockets comes the consequence of non-uniform memory access (NUMA). Although the memory space is fully addressable by any given core, the temporal properties of how long it takes to access a given piece of memory depend on where that memory is located. This location dependence can mean inconsistent or poor performance of software that does not take this property into account, and so care should be taken to program such machines.

A couple of downsides of large shared-memory machines are

- Multi-socket motherboards are relatively expensive.
- Ignoring NUMA results in poor performance, particularly as systems scale up.

An approach to dealing with the second challenge is to be more aware of how a program deals with its memory. This awareness generally leads to a style of programming where a given core only looks in a relatively small area of the total available memory, i.e., emphasizing data locality yet again. Applying such a restriction to your programs enables the move to *distributed* programs (where every core only has access to a subset of total memory and must explicitly message other cores for their data) to be quite natural. This approach then assists with the first downside, exchanging large memory machines with a conceptual network of smaller machines.

11.2.2 Distributed clusters

Distributed clusters consist of many compute nodes connected through high-speed network *switches*. This architectural approach reduces costs by utilizing smaller compute nodes instead of a few large shared-memory machines. However, there is a trade-off: the network infrastructure can become expensive because the connections between nodes must be fast enough to handle substantial communication. When designing distributed clusters, this cost factor must be carefully considered. In the following discussion, we explore how the layout of switches can mitigate this cost, but first, we introduce some notes on high-speed networking.

11.2.2.1 High-speed networking

For cluster computing, networking that is *faster* than gigabit Ethernet is generally required. The most commonly found networking solutions in this space at present are via the InfiniBand standard.

InfiniBand is a networking standard designed for high-speed networking, allowing theoretical speeds of up to 1200 Gb/s. To ensure reliable transmission, 64b/66b line encoding is used, which is not as efficient as PCIe, but it is necessary because data corruption becomes more likely over longer distances.

The speeds associated with InfiniBand are referred to by a three-letter initialism. The first release SDR (single data rate) had a theoretical peak of 2.5 Gb/s for one link. Additional links allowed for higher bandwidth and faster transfer speeds. Links are similar to PCIe lanes in the sense that data are transferred in parallel. Each standard of InfiniBand can use between 1 and 12 links.

The current release of InfiniBand is NDR (next data rate), which has a single link theoretical peak of 100 Gb/s. Future releases XDR (extended data rate) and GDR (generation data rate) are planned to effectively double the bandwidth of NDR at each release. This plan means that the full realization of GDR will have a theoretical peak of 4800 Gb/s for a 12-link implementation!

11.2.2.2 Connection topologies

Crossbar connections are ideal for performance. In the crossbar connection topology, there are sufficiently many switches so that every node can directly communicate with every other node. Crossbar connectivity is expensive to scale, however, because it requires $O(p^2)$ switches to implement for p compute nodes. This topology is visualized in Figure 11.2. The crossbar topology is the best-case for performance, but it has prohibitively expensive networking costs as p increases.

Tree-based connection topologies provide scaling that is slightly better than quadratic scaling for connecting the nodes of a cluster. The *fat-tree* connection topology visualized in Figure 11.3 can be interpreted the following way. A message between any two compute nodes may need to be routed through multiple switches. The switches near the root of the tree structure are generally populated with denser, faster networking capabilities than the switches near the nodes (the leaves of the tree structure) because the switches near the root are required to handle more communication traffic; this is where the *fat* name comes from – their thick roots. Tree topologies generally require $O(p \log p)$ network switches. This scaling is better than

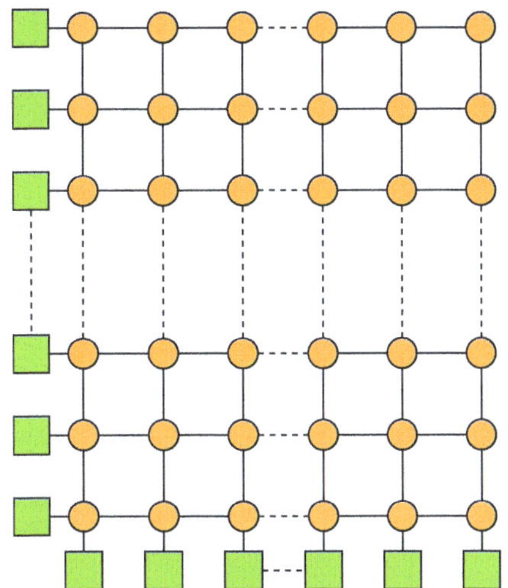

Fig. 11.2 Crossbar connection topology connecting p inputs (green squares on left) to p outputs (green squares on bottom) via p^2 network switches (orange circles).

the quadratic scaling of a full crossbar connection, but it is still superlinear in its scaling.

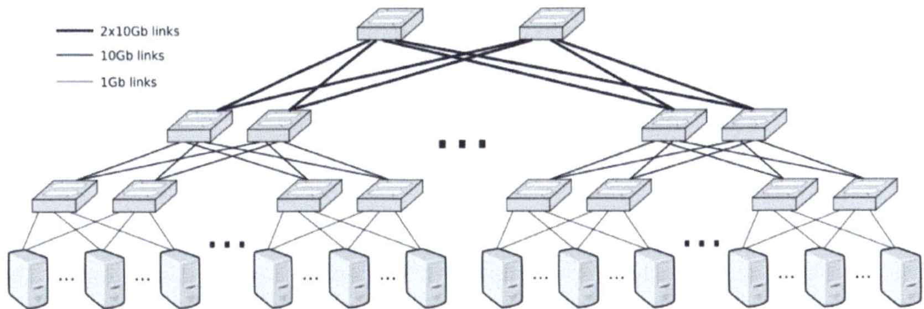

Fig. 11.3 Fat-tree connection topology. Networking near the root of the tree is designed to handle more traffic than the networking near the nodes. Communication between any two nodes may need to be routed through multiple network switches.

Torus connection topologies are built on nearest neighbor connections. In addition to these connections, periodic boundary conditions are applied such that nodes on one end are directly connected to nodes on the opposite end of the layout. Nodes existing in an N-dimensional torus topology can be thought of as living on N *loops* of connections, one loop for each dimension of the connections. Connections for 1D and 3D tori are shown in Figure 11.4. Each node in a torus topology has a network switch within it that is connected to the neighbors. Thus, the scaling of required network infrastructure is $O(p)$—finally, a topology with linear scaling in high-speed networking resources. Even this topology is not without its drawbacks, however. There are a few things to consider within torus topologies:

- The number of neighboring nodes is 2^N for an ND torus.

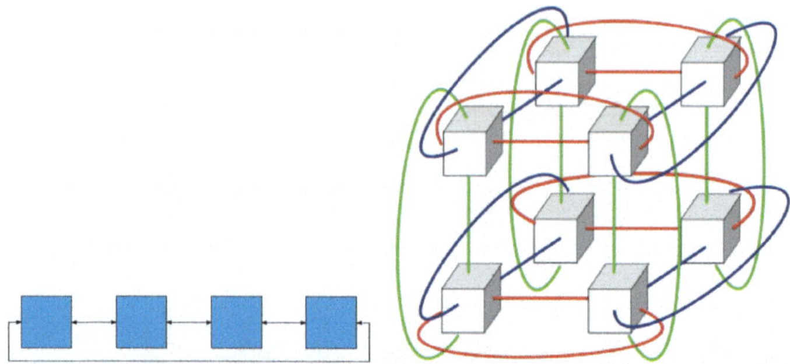

Fig. 11.4 Left: a 1D torus connection topology. Right: A 3D torus connection topology. Nodes are connected to nearest neighbors in each of N dimensions of a ND torus. In addition, nodes on the ends are connected to nodes on the opposite end of that dimension. Each node of the 3D torus, for example, exists on three different colored loops, one in each direction.

- Point-to-point communication between any two arbitrary nodes may require many *hops* through the switches of intermediate nodes.

With these drawbacks in mind, communication on torus networks may suffer from congestion, especially if many nodes require data from further than one hop away. Again, this stresses the importance of data locality!

Torus connection topologies provide a great balance between performance and cost at extreme scales. Many of the largest supercomputers in the world are built using 3D torus configurations. Fugaku, the fourth fastest supercomputer in the world[2] (at the time of this writing) has a 6D torus topology connecting its nodes.

11.2.3 Distributed file systems

When working with clusters, it is important to consider the organization of the underlying file system. The documentation for a given cluster is the best place to look for this information. Most often, the default file system is a distributed (or network) file system, meaning that the storage of files is spread across specific storage devices that get mounted and made available to all nodes. Fortunately, users can log into a cluster and have access to all their files without explicit knowledge of where their files are located.

A common organizational structure clusters employ is to have three primary locations that access the distributed file system: *home*, *scratch*, and *project*. Each user has their own home directory with a fixed storage capacity. This space is where users should store their source code, executables, and configuration files. The scratch space offers increased storage and faster performance, making it ideal for storing files generated during the execution of a job. The project space is where files should

[2] `https://www.fujitsu.com/downloads/TC/sc10/interconnect-of-k-computer.pdf`

be stored for long-term storage. This space is usually slower than the scratch space, so jobs should write data to scratch and users should move files to the project space when the job is complete. The project and scratch spaces are usually shared by all users, with users separated into distinct groups that are given specific allotments of storage in each space.

Furthermore, clusters may have additional storage locations for users requiring increased performance. This can include access to *burst buffer* storage (the local storage of a node). Burst buffer storage provides faster performance than the general distributed file system by utilizing faster storage devices (e.g., SSDs) or by using *ramdisk* (using RAM as a temporary disk drive) storage. Access to local storage depends on the scheduler being used. For example, clusters using the SLURM scheduler may have access to the environment variable *SLURM_TMPDIR*, which accesses the local storage of the compute node currently running a job.

Finally, the two most popular software platforms at the time of writing for enabling distributed file systems are Lustre[3] and GPFS[4]. Lustre is an open-source solution, whereas GPFS is a proprietary solution offered by IBM. Both are highly scalable and offer high performance. Generally, knowing the details of the file system software is unnecessary because the file system will behave like any typical Linux file system. However, it is important to know that options like local node storage and burst buffer storage are available for workloads requiring faster access to files.

11.3 Additional Resources

- `https://www.lustre.org/`
- `https://www.ibm.com/products/spectrum-scale`

[3] `https://www.lustre.org/`

[4] `https://www.ibm.com/products/spectrum-scale`

Part IV
Distributed-memory programming

Consisting of six chapters (or lecture hours), this part kicks off a phase of the journey that is at the heart of extreme scientific computing: distributed memory programming. In practice, extreme-scale computing thrives in a distributed memory paradigm. The oldest (and at present still the most prevalent) method for distributed memory programming is the Message Passing Interface (MPI).

MPI is a standard that allows multiple processes of the operating system to communicate data among each other. This is useful on standard computers, where different processes can be mapped to different CPU cores, but it really starts to shine when those processes are distributed between different compute nodes, allowing access to more CPU cores and more RAM to work together on a given problem.

We present the view of *distributed*-memory programming via MPI before *shared*-memory programming in this book because the style of data organization required for multi-process distributed programs turns out to be highly useful (to the point of being de facto necessary) for creating efficient shared-memory programs.

Chapter 12
Introduction to MPI

This chapter introduces MPI. We give a brief history of the evolution of the MPI standard, common implementations of the standard, and compiling and linking MPI programs. Specific functionality that is present in the different versions of the standard are discussed in later chapters where necessary.

12.1 A brief history

What is MPI? The Message Passing Interface (MPI) is a library that provides tools for coordinating interactions between different operating system processes. It is general enough to allow for these processes to exist either on a single machine using shared memory or on different physical machines in a network that is configured to use MPI.

Where did MPI come from? In the early 1990s, supercomputers composed of many smaller compute *nodes* started to become more commonplace. At this time, managing the computation across different nodes was generally handled with a custom implementation for any given machine. In 1994, MPI was introduced as an attempt to standardize the interface for setting up multiple processes and managing communication between them.

The MPI standard is not a formal standard in the way that ISO C++ is, but the core components of the standard have remained unchanged since that initial version from over 25 years ago. The longevity has to do with the well-designed nature of the base functionality of MPI that was general enough to allow for decades of supercomputer advancement while using the same interface for communications within their systems.

MPI-2 was finalized in 1997 and brought with it official language bindings for C++ and Fortran90. The C++ bindings were considered a failure because they essentially just reproduced the functionality of the C bindings in a 1:1 manner. This limited mapping resulted in more code to maintain over the long-term while providing no functional benefit. The C++ bindings were thus deprecated as of MPI-2.2 in 2009 and

© The Author(s), under exclusive license to Springer Nature Switzerland AG 2025

R. J. Spiteri and K. Klenk, *Extreme-Scale Computing*, CMS/CAIMS Books in Mathematics 15, https://doi.org/10.1007/978-3-031-89033-8_12

were removed entirely from the MPI-3 standard. The most useful feature added in MPI-2 is the interface for performing parallel I/O operations. The ability to perform one-sided communications, where a given process can write to (or read from) a pre-specified portion of another process's memory without that other process having to participate in the communication, was also a useful introduction.

MPI-3 arrived in 2012 and brought additional functionality for working with *collective communication*, as well as refined the interface for working with one-sided communication. MPI-4.1 was finalized in November 2023, refining the elements of MPI-4, and the most recent finalized standard at the time of writing. The MPI-5 standard is in the process of being developed. Detailed—excruciatingly so—information on all the MPI standards can be found in the standard documents themselves at `https://www.mpi-forum.org/docs/`.

MPI being defined as an *interface*, in particular, allows for different implementations to exist, and implementations optimized for certain supercomputing architectures can be used as needed. The two most common open source implementations of the MPI standard at present are OpenMPI `https://www.open-mpi.org/` and MPICH `https://www.mpich.org/`. These are quite portable and can easily be installed on macOS and most Linux distributions via package managers. Alternatively, the source is freely available to download, build, and optimize to your needs.

Many supercomputers at present simply make use of either OpenMPI or MPICH. In addition to these implementations, there are additional open-source implementations such as MVAPICH `https://mvapich.cse.ohio-state.edu/` that perform similarly, all things considered, but just do not quite have the market share of OpenMPI or MPICH. Furthermore, hardware vendors may provide their own implementations of the MPI standard. Intel MPI `https://software.intel.com/content/www/us/en/develop/tools/oneapi/components/mpi-library.html` is an example of this.

12.2 Practical considerations

As mentioned, the suggested method for using MPI in C++ code is via the C API.

The Boost::MPI library `https://www.boost.org/doc/libs/1_75_0/doc/html/mpi.html` provides a C++ interface that implements the most commonly used functionality of the MPI-3.x standard in a modern C++ style. In addition to this, it provides useful components to ease communicating general objects and STL containers. Although we do not cover Boost::MPI in detail in this book, we do consider the concept of *serializing* objects by defining custom MPI datatypes in Chapter 16 to convey some understanding of what Boost::MPI is doing to ease things behind the scenes.

12.2.1 Compiling MPI programs

Compiling MPI programs is done with `mpicc` for C, `mpicxx` for C++ programs, and `mpif90` for Fortran. These commands tend to use standard compilers that are on the system but provide additional arguments to the compiler. For example, you can take a look at the specific command-line call that invoking `mpicxx` provides by using the command `mpicxx --showme`. For example, running this on a system that has MPI installed may result in something that looks like

```
$ mpicxx --showme
icpc -I/path/to/mpi/include -L/path/to/mpi/lib -lmpi_cxx -lmpi
```

where `icpc` is the compiler being used, `-I/path/to/mpi/include` tells the compiler where to look for the MPI header files, `-L/path/to/mpi/lib` tells the linker where to look for the compiled MPI libraries, and `-lmpi_cxx -lmpi` tells the linker to link in both the C and C++ builds of the libraries.

Because the wrapper scripts just add arguments to ensure that the compiler can find and link with an MPI library, *normal* compiler options can also be specified when using them. For example,

```
$ mpicxx -O3
```

for setting optimization level 3,

```
$ mpicxx -c
```

to compile object files,

```
$ mpicxx -o <program name>
```

to specify the output of either a compilation or linking of object files, among many others which can be explored with the documentation of the underlying compiler.

Because options are generally passed through the compiler wrappers to the underlying compiler, running

```
$ mpicxx --verion
```

shows the version info for the compiler with which the wrapper is configured. To get the version of MPI that the compiler wrapper is using, you can run

```
$ mpicxx --showme:version
```

Investigating the man page for `mpicxx` shows that really the only options that get handled by the wrappers are variants of `--showme`. All other options get passed through to the underlying compiler.

12.2.2 Running MPI programs

Once an MPI executable has been created, it can be run as a single process with the usual `./program` or as multiple processes using the `mpirun` command.

Specifying just `mpirun ./program` runs the program using the default number of processes with which the MPI library has been configured, usually the number of physical cores on the system. You can gain more control by using, e.g., `mpirun -np 16 ./program`, which will run the executable as 16 processes. Similar run functionality can be achieved with `mpiexec`. In fact, in OpenMPI, `mpirun` and `mpiexec` are exactly the same thing, both of them being symbolic links to the `orterun` program:

```
$ which mpiexec
/usr/bin/mpiexec
$ readlink -f /usr/bin/mpiexec   # "follow" symbolic links
/usr/bin/orterun
```

The `orterun` program takes its name from the Open Run-Time Environment, which OpenMPI uses to manage processes that are run with MPI. We note that all MPI implementations use a run-time environment to manage the communication between MPI processes, and this run-time environment differs across implementations.[1]

On shared computing resources, computational jobs must be submitted through a job scheduler. The SLURM workload manager[2] is a popular choice used on many supercomputers for managing and scheduling system resources. In practice, you may encounter other resource managers; however, the general idea of submitting jobs to a scheduler remains the same.

A quick note on performance. The number of *processes* specified in the `mpirun -np` command is the number of operating system processes. When installed on a system, MPI can either be configured to allow as many processes as desired or have a strict upper limit. Installing OpenMPI through the Ubuntu repositories, for example, allows you to run an MPI executable with many more processes than you have physical cores (or even *hardware threads*) on your system. However, it is often the case that performance gains are not observed beyond the number of physical processing cores on the system, and in fact performance degradation starts to occur when using more processes than cores. This situation means that each physical core has to work with more than a single process. For a fixed problem size, this means there will *only* be extra overhead (over using the same number of processes as cores) in managing the additional processes. However, this situation can nonetheless be useful for testing that MPI code does not break down before moving on to a production run with many processes.

Alternatively, shared computing resources often strictly limit the number of MPI processes a job can run on a compute node to the number of physical cores on that node. There are further considerations in managing the run-time environment that emerge in Chapter 16 when we discuss measuring performance of MPI programs.

---

[1] For example, MPICH's run-time environment is called *Hydra*, its namesake being the mythological creature. Fortunately for our operating systems, however, its heads (processes) do not multiply when one crashes.

[2] `https://slurm.schedmd.com/`

Chapter 13
MPI Basics

This chapter goes over the very basics of MPI programming. Using a *hello, world* example, we discuss MPI naming conventions, various MPI entities that come with the library, MPI communicators, and MPI datatypes. This material serves as a primer for the greater detail to come in the following chapters.

13.1 Basic Program Structure

The basic structure for an MPI program is given in Listing 13.1.

Listing 13.1 MPI *hello, world* in C++. Demonstrates how to include the MPI library, initialize MPI (potentially) with command-line arguments, query a communicator, and clean up the MPI program.

```
#include <iostream>
#include <mpi.h>  // This is the header file for MPI functions

/*
  int main(int argc, char* argv[])

  Inputs: none

  Outputs: Prints "Hello World #" where # is the rank of the process.
*/

int main(int argc, char* argv[]) {
  // First initialize the MPI interface.
  MPI_Init(&argc, &argv);

  // Determine the rank of the current process
  int rank;
  MPI_Comm_rank(MPI_COMM_WORLD, &rank);

  // Determine the number of processes
  int size;
  MPI_Comm_size(MPI_COMM_WORLD, &size);

  // All processes print hello
  std::cout << "Hello from rank " << rank << " out of " << size << std::endl;

  //  Shut down the MPI system when done.
  MPI_Finalize();

  return 0;
}
```

© The Author(s), under exclusive license to Springer Nature Switzerland AG 2025

R. J. Spiteri and K. Klenk, *Extreme-Scale Computing*, CMS/CAIMS Books in Mathematics 15, https://doi.org/10.1007/978-3-031-89033-8_13

To help us get a basic understanding of how to set up a basic MPI program, we now walk through the above example. First, in order to access MPI entities in a C++ program, an `include` statement must appear in any source file that contains MPI function calls or constructs. Specifically, in C and C++, this looks like `#include <mpi.h>`. All MPI entities (subroutines, constants, types, etc.) begin with `MPI_`, making them easily identifiable within MPI code. In order to use the MPI entities, we must first initialize the MPI environment. Without this, MPI calls are not recognized. To initialize the environment, we call the `MPI_Init(&argc, &argv)` function. As you may have noticed, our `MPI_Init` call takes two arguments that look similar to those of our main function. If you guessed that this allows us to pass command-line arguments to MPI, you would be correct, and this topic is covered more in depth in later chapters.

Compiling Listing 13.1 (assuming it is stored in the file `MPI_hello.cpp`) can be done with the command

```
$ mpicxx -o hello MPI_hello.cpp
```

and run with (say) 8 processes using

```
$ mpirun -np 8 ./hello
```

If you are working on a system with fewer than 8 physical CPU cores, you may receive an error message that includes

```
There are not enough slots available in the system to
satisfy the 8 slots that were requested by the application
...
```

To (potentially) work around this,[1] we add the `--oversubscribe` option

```
$ mpirun -np 8 --oversubscribe ./hello
```

As an aside, to *oversubscribe* cores in an MPI execution means to run more MPI processes than there are physical CPU cores available. Although this does not give better performance (in fact, it will generally be worse due to how the operating system needs to manage the processes), you may find it useful for debugging MPI programs on smaller machines like laptops, desktops, or workstations because it allows you to effectively simulate a larger machine.

Example output of running Listing 13.1 with 8 processes looks like:

```
Hello from rank 1 out of 8
Hello from rank 5 out of 8
Hello from rank 2 out of 8
Hello from rank 4 out of 8
Hello from rank 7 out of 8
Hello from rank 3 out of 8
Hello from rank 6 out of 8
Hello from rank 0 out of 8
```

[1] Consider reading more into `man mpirun`.

The (literally) random ordering of the output from a parallel program often comes as a shock the first time it is encountered. Take a moment to compare the output with the code of Listing 13.1 to try to figure out what is happening before continuing.

To understand what is going on in the body of the code in Listing 13.1, it is helpful to introduce the following key terms:

- *Communicator*: A communicator is a handle representing a group of processes that can communicate with one another. A useful pre-defined communicator is MPI_COMM_WORLD, which handles all of the processes available to a given execution of a program.
- *Rank*: The rank of a process is the unique identifier of a process within a communicator.

Using the above definitions, we can now see that the line that contains the function call `MPI_Comm_rank(MPI_COMM_WORLD, &rank)` fills the rank variable with an integer of the rank of the current process that calls this function. The next function call simply fills the size variable with the current number of processes within the communicator. Lastly, we use `MPI_Finalize()` to indicate we are done with the MPI environment, meaning any MPI function calls other than INIT will not work after this point.

Within an MPI program, you can think of the `MPI_Init()` routine as setting up the MPI_COMM_WORLD communicator with the number of ranks (its size) that have been passed to the program via the command line and giving them each a unique identifier. After `MPI_Init()`, *every* rank in the program executes each statement until after `MPI_Finalize()` is called. `MPI_Finalize()` cleans up all MPI data structures and cancels incomplete operations. If any processes do not call `MPI_Finalize()`, the program hangs.

And there it is, we have successfully walked through a simple MPI program.

13.2 MPI return codes

All MPI routines return an error code that can be checked for the successful execution of the routine. The true function signature of `MPI_Init()` is actually

```
int MPI_Init(int *argc, char ***argv);
```

so it indeed does return an `int` that can be checked to ensure the routine has executed successfully. The error code returned is equal to the pre-defined integer constant MPI_SUCCESS if the routine ran successfully. If an error occurred within an MPI routine, `ierr` returns an implementation-dependent value indicating the specific error. Examples of error code values are

MPI_SUCCESS: No error. MPI routine completed successfully.

MPI_ERR_COMM: Invalid communicator. It is common for this error to occur because a null communicator is used in a communication call.

MPI_ERR_COUNT: Invalid count argument. Count arguments must be non-negative; a count of zero is often valid.

MPI_ERR_TYPE: Invalid datatype argument. This error can also occur if an uncommitted MPI_Datatype (see MPI_Type_commit) is used in a communication call.

MPI_ERR_TAG: Invalid tag argument. Tags must be non-negative; tags in a receive (MPI_Recv, MPI_Irecv, MPI_Sendrecv, etc.) may also be MPI_ANY_TAG. The largest tag value is available through the attribute MPI_TAG_UB.

MPI_ERR_RANK: Invalid source or destination rank. Ranks must be between zero and one less than the communicator size; ranks in a receive (MPI_Recv, MPI_Irecv, MPI_Sendrecv, etc.) may also be MPI_ANY_SOURCE.

We have already touched on the meaning of communicators and ranks, and so now we finish the chapter with data types and message communications to help put the above error return codes into context.

13.3 MPI Data types

MPI provides its own data types (to be used in communication routines) that are essentially standard C types with different names. Examples of this are MPI_INT and MPI_DOUBLE. A complete list of the pre-defined types for MPI in C is found in Table 13.1. MPI allows custom data types to be defined and used as well. Custom

C++ type	MPI type
char	MPI_CHAR
unsigned char	MPI_UNSIGNED_CHAR
char	MPI_SIGNED_CHAR
short	MPI_SHORT
unsigned short	MPI_UNSIGNED_SHORT
int	MPI_INT
unsigned int	MPI_UNSIGNED
long int	MPI_LONG
unsigned long int	MPI_UNSIGNED_LONG
long long int	MPI_LONG_LONG_INT
float	MPI_FLOAT
double	MPI_DOUBLE
long double	MPI_LONG_DOUBLE
unsigned char	MPI_BYTE

Table 13.1 Table of MPI types and their correspondence to underlying C++ types.

data types are known as *derived types* and can be used to construct more elaborate messages to be communicated. Derived types are covered in more depth in Chapter 16.

13.4 Communicating Messages

To reiterate, the key feature of MPI is that it allows multiple processes to explicitly communicate data between them. This is done within code by calling one of MPI's many communication routines.

Processes in an MPI program can explicitly communicate with each other only if they share a communicator. By default, any two ranks can communicate through the `MPI_COMM_WORLD` communicator. A more advanced aspect of MPI involves defining new communicators to help organize data flow in a program, a topic that is discussed in Chapter 16. If a process belongs to more than one communicator, its rank can differ in each communicator.

There are various forms of communication that can be performed within an MPI program. These range from blocking point-to-point communication—where two ranks can directly exchange data—to non-blocking collective communications—involving all ranks in a communicator exchanging data together. So-called *blocking* communications take over a process's execution while it waits for the communication to complete, whereas *non-blocking* communications work more on a set-it-and-forget-it basis, where a given rank can carry on doing other work while the messaging is being carried out. Because the network communication of MPI data can be performed independently from CPU computations, the use of non-blocking communications can be highly effective for *overlapping* computation and communication, effectively *hiding* the costs of communication.

Non-blocking communication requires *tagging* of messages so that the underlying MPI implementation can periodically check whether the message has completed.

The next two chapters discuss communication in much greater detail.

Chapter 14
Point-to-point communication

Now that we have seen a simple example on how to write, compile, and run a basic MPI program, we can continue by demonstrating how to actually send messages with MPI. The true power of MPI is in its message-sending capabilities, which allow processes to explicitly communicate data among one another. In order to achieve this level of communication, MPI has *sending* and *receiving* operations providing both *blocking* and *non-blocking* communications, the details of which are discussed in this chapter.

14.1 Blocking communication

Different ranks in an MPI communicator can explicitly send data between each other using what is called *point-to-point* communication. We first consider the simplified version of this via the *blocking* point-to-point messaging routines `MPI_Send()` and `MPI_Recv()`. In this context, the term *blocking* refers to what happens to the program execution while the communication is happening. An MPI rank using blocking communication suspends its execution until the communication has completed.

The first part of blocking point-to-point communication is the `MPI_Send()` routine, which is used for identifying the data to be sent from the calling rank, choosing the destination rank to send the data, and identifying the message with a tag. The function signature is

```
int MPI_Send(
      void*        data,
      int          count,
      MPI_Datatype datatype,
      int          destination,
      int          tag,
      MPI_Comm     communicator);
```

The first three arguments of `MPI_Send()` determine the contents of the message, and the last three arguments add additional information that allows MPI to match up the communication with a destination. Let us take a look at each argument individually and specify exactly what makes up an `MPI_Send()` call. The first

© The Author(s), under exclusive license to Springer Nature Switzerland AG 2025
R. J. Spiteri and K. Klenk, *Extreme-Scale Computing*, CMS/CAIMS Books in Mathematics 15, https://doi.org/10.1007/978-3-031-89033-8_14

argument is a pointer to the buffer into which the message is packed. We note the void* type here. This is a pointer to the beginning of the data to be sent. The next two arguments determine how MPI interprets the memory after that point: count indicates the number of variables of type datatype that are to be sent in the message. For the pre-defined MPI types found in Table 13.1, MPI_Send() sends contiguous chunks of memory. To pass non-contiguous data, a custom type must be defined. We discuss more on custom data types in Chapter 16.

The final three arguments are interpreted as follows: destination is a rank on communicator, and tag is an integer to identify the message, where communicator must be a communicator that includes both the calling rank of MPI_Send() and the destination rank. The tag is an identifier that MPI uses to pair this MPI_send() up with a corresponding MPI_recv(), as we now discuss.

The call to MPI_Recv() is more or less the same as MPI_Send() with just one more argument to help with diagnostic info about the message. The MPI_Recv() function signature is

```
int MPI_Recv(
        void*         data,
        int           count,
        MPI_Datatype  datatype,
        int           source,
        int           tag,
        MPI_Comm      communicator,
        MPI_Status*   status);
```

The first three arguments of MPI_Recv() specify the characteristics of the memory where the message data are to be stored. The second three arguments specify where the message was coming from: the source rank on communicator that sent the message and the identifying tag of the message. The final argument holds the status of the message in an MPI_Status struct. The MPI_Status struct is discussed in more detail in the next section.

Some key things to keep in mind when using point-to-point blocking communication routines in MPI are:

- Sending processes **must** specify a destination process rank and a non-negative tag.
- The receiving data buffer size must be at least as large as the sending buffer size, and it must be allocated on the receiving process before being used in MPI_Recv().
- A wildcard can be used to receive a message from any sending process; more on this in Section 14.2.

Now that we have covered the basics of MPI_Send() and MPI_Recv() , let us dive into the example code of Listing 14.1.

Listing 14.1 Basic point-to-point communication with `MPI_Send()` and `MPI_Recv()`. We make special note of the counterintuitive nature of receives occurring before the sends in the layout of the source code. The *blocking* nature of `MPI_Send()` makes this sort of source code ordering common in MPI programs.

```
#include <iostream>
#include <mpi.h>

int main(int argc, char* argv[]) {
  MPI_Init(&argc, &argv);

  // Determine the rank of the current process
  int rank;
  MPI_Comm_rank(MPI_COMM_WORLD, &rank);

  // Determine the number of processes
  int size;
  MPI_Comm_size(MPI_COMM_WORLD, &size);

  int data;
  // If not the first process, wait for permission to proceed
  if (rank > 0) {
    // Wait for a message from process rank-1
    MPI_Recv(&data, 1, MPI_INT, rank-1, 0, MPI_COMM_WORLD, MPI_STATUS_IGNORE);
    std::cout << "Rank "<< rank << " has received message with data " << data
              << " from rank " << rank-1 << std::endl;
  }

  // All processes print hello
  std::cout << "Hello from rank " << rank << " out of " << size << std::endl;

  // All processes send the go ahead message except the last process
  if (rank < size-1) {
    // Send the go ahead message to rank+1.  Using rank^2 as the data
    data = rank*rank;
    MPI_Send(&data, 1, MPI_INT, rank+1, 0, MPI_COMM_WORLD);
  }

  MPI_Finalize();
  return 0;
}
```

As discussed in the previous chapter with Listing 13.1, all MPI calls must be between an `MPI_Init()` and an `MPI_Finalize()`. Next, we get some information about the `MPI_COMM_WORLD` communicator: each process gets its rank in the communicator and the total size of the communicator. One thing that may seem counterintuitive at first is the location of the `MPI_Recv()` guarded by an `if` statement. This construct is needed because a process must wait to receive a sent message. We can think of two processes executing through the code line by line at the same time. Once processes reach the `if` statement in the example, the one with rank 0 will skip over the `if` and continue execution. The process with rank 1 will enter the `if` statement and wait at `MPI_Recv()` for an incoming message. When rank 0 reaches the `MPI_Send()` and sends the message to rank 1, rank 1 eventually receives the message and resumes its execution.

This is, in general, a good approach to tracing MPI code. We think of the execution of a single process through the code—say, rank 0. We trace its execution until it gets to a place where it is waiting for something from another process, then switch to the rank that rank 0 is waiting on and trace that rank's execution. Then rinse and repeat to get a feel for how the different ranks are navigating the lines of the code. When considering multiple processes executing the code, this additional layer of rank-specific execution order can give some sense of *non-linear* execution when viewed as a whole. There is no need to fret, however, because each core's execution can be traced out independently, following its rank-specific iteration and branching path, only switching to other ranks as discussed above.

As an exercise, can you determine what the output of Listing 14.1 will be when there are exactly 2, 5, or 10 processes executing the code? For reference, the output when run with 4 ranks should be

```
Hello from rank 0 out of 4
Rank 1 has received message with data 0 from rank 0
Hello from rank 1 out of 4
Rank 2 has received message with data 1 from rank 1
Hello from rank 2 out of 4
Rank 3 has received message with data 4 from rank 2
Hello from rank 3 out of 4
```

We note the use of `MPI_STATUS_IGNORE` as the final argument to `MPI_Recv()`. `MPI_STATUS_IGNORE` is a placeholder `MPI_Status*` that MPI implementations provide to use as a sort of null argument for message statuses. It can be used when one is not concerned with the status of a message, as in this example. However, we should generally be interested in the details of our messages, and the rest of this chapter addresses the what, why, and how involved in taking up such an interest.

14.2 Tags and the `MPI_Status` structure

What's the deal with message statuses?

There are two elements with which one should be familiar when communicating via MPI. These elements are: *tags*, which we have already seen in a basic form in Listing 14.1, and the `MPI_Status` struct. In the context of blocking communications, tags are optional (note how *every* message in Listing 14.1 is tagged with an id of 0) and are represented by non-negative integers. Tags are used for differentiating between different messages. When an `MPI_Recv()` has a specific tag noted, it waits (blocking that rank's execution) until it sees a message with that specific tag. For a sent and received message to be matched, it is necessary that their communicators match and that the source and destination processes are consistent. As briefly mentioned above, a process can receive a message without knowing the sender, but it also does not need to know the tag. The former can be handled by using the placeholder `MPI_ANY_SOURCE` and the latter with placeholder `MPI_ANY_TAG`.

This information can be recovered via the status argument. The `MPI_Status` struct is defined as:

```
typedef struct _MPI_Status {
  int MPI_SOURCE;
  int MPI_TAG;
  int MPI_ERROR;
  // other members (potentially) depending on implementation
} MPI_Status, *PMPI_Status;
```

noting that this form of defining a struct gives us easy access to the type aliases MPI_Status and PMPI_Status,[1] the latter being equivalent to pointers to MPI_Status, meaning that the following two lines would define the same thing

```
MPI_Status *p_status;
PMPI_Status p_status;
```

The MPI standard requires MPI_Status to have at least MPI_SOURCE, MPI_TAG, and MPI_ERROR as its members.[2] The three fields contain exactly what one may expect them to: the source, tag, and an error code of the message that has been received.

14.3 Non-blocking point-to-point communication

Why do we need to know more about the messages?

If a process tries to receive a message for which there is no matching send, it will block forever. This is called *deadlock*, and as you can probably guess, it is not a good thing. Deadlocks are almost always the result of an incorrectly structured program, i.e., the fault of the programmer. It may be the case that a distributed algorithm can be structured such that blocking communications can be used just fine, but that is not always the case. In order to get around the possibility of deadlock, we use *non-blocking* (also called *asynchronous*) communication.

There are a couple of additional benefits of using non-blocking communication. The first is that it allows the overlapping of communication and computation. The second is in the efficiency of the throughput of the physical messages when using shared cluster resources.

The concept of overlapping communication and computation goes like this: rank 0 sets up the minimal data needed for its message and sends that off to rank 1. Without waiting to confirm reception of the message, rank 0 carries on with additional computations until it gets to a point where it *needs* data from rank 1. Only now does rank 0 wait for a message from rank 1. On the other hand, rank 1 is doing a similar thing: it sets up the minimal data needed for its message to rank 0, sends the message, and carries on with its own computations until it gets to a point where it needs to use data coming from rank 0. Hopefully, by the time rank 0 needs data from rank 1 and rank 1 needs data from rank 0, the messages will have completed, and they can both carry on with their computations. This sort of flow in the program can be more difficult to program, but depending on the algorithm being used, it may essentially be able to *hide* all the communication time by overlapping it with computation.

Efficiency and throughput of messages in a cluster are a little more difficult to speak precisely about. As we saw in Chapter 11, compute clusters are often designed with non-uniform network resources. This non-uniform design is further embedded

[1] This way of defining types and pointers to types—with PXXX for the pointer variant—is commonly used in C/C++.

[2] The implementations of the standard may have additional members. These members are used to help optimize messaging within the library and should generally not be used (or in particular *relied upon*) for code using MPI because the resulting code is likely to be non-portable.

in a communication network where there can be other parallel jobs alongside yours. Combined, the upshot is that even if you are able to balance the quantities of data in your messages *perfectly* for your application, the messages themselves can vary wildly in their time to completion, depending on which nodes particular ranks are allocated and the other jobs running on the cluster.

In the end, we can strive to have each rank take an approach of "Send the data that other ranks need as soon as you can; compute as much as you can before you need to work with data from other ranks." This approach of sending as early as possible and computing as much as possible before checking for incoming data generally leads to better performance independent of the specific network architecture being used.

How do we better deal with the management of messages?

The `MPI_Send()` and `MPI_Recv()` routines have non-blocking variants `MPI_Isend()` and `MPI_Irecv()`, respectively. `MPI_ISend()` looks similar to its blocking counterpart but differs slightly in its function signature:

```
int MPI_Isend(
        void*        data,
        int          count,
        MPI_Datatype datatype,
        int          destination,
        int          tag,
        MPI_Comm     communicator,
        MPI_Request* request);
```

The first six arguments of the non-blocking send are the same as the blocking variant. The difference comes with the addition of one more argument: `MPI_Request*`. An `MPI_Request` is called a communication request object and is allocated on a call to `MPI_Isend()`. The request object allows investigatation into (at a later time) whether a message has been completed or to block until it has.

`MPI_Irecv()` is again similar to its blocking counterpart, with the last argument being the only part of its function signature that differs. That is, instead of the status argument, there is an `MPI_Request*` argument. The function signature is as follows:

```
int MPI_Irecv(
        void*        data,
        int          count,
        MPI_Datatype datatype,
        int          source,
        int          tag,
        MPI_Comm     communicator,
        MPI_Request* request);
```

Once the non-blocking send and receive routines have been called, we can introduce a synchronization point by using `MPI_Wait()`:

```
int MPI_Wait(
        MPI_Request* request,
        MPI_Status*  status);
```

The `MPI_Wait()` routine tells the MPI implementation to periodically check the status of the `request`, only allowing a rank to proceed after it has completed. Similar to the blocking `MPI_Recv()` routine, the `status` variable gets filled with information about the message.

In addition to `MPI_Wait()`, which does not proceed until the message completes, the `MPI_Test()` routine can be used to explicitly test the status of a message. This behavior is useful in that it provides finer control over incremental calculations, where the code is looping over some calculation but can possibly do more if a given communication has not completed. The signature for `MPI_Test()` is

```
int MPI_Test(
      MPI_Request* request,
      int*         ready,
      MPI_Status*  status);
```

The extra argument over `MPI_Wait()` is the input/output pointer `ready`, which is populated with either 1 or 0 depending on whether the message has completed or not, respectively.

An example code that uses non-blocking communication routines is found in Listing 14.2. We note how this example requires a little more code than Listing 14.1, but it follows a more linear path that is similar among all executing ranks when traced.

Listing 14.2 Basic non-blocking point-to-point communication with `MPI_Isend()` and `MPI_Irecv()`. After the non-blocking communication routines have been called, a given rank does not proceed until the communication is completed. This behavior is achieved with `MPI_Wait()`, used as a synchronization point. Compared to Listing 14.1, tracing a single rank's execution is more linear and thus may be more amenable to manual tracing to debug higher process counts.

```
#include <iostream>
#include <mpi.h>

int main(int argc, char* argv[]) {
  MPI_Init(&argc, &argv);
  int rank, size;
  MPI_Comm_rank(MPI_COMM_WORLD, &rank);
  MPI_Comm_size(MPI_COMM_WORLD, &size);
  int data;

  // This example will hang if run on only 1 rank
  // - MPI_Irecv never gets called -> request never gets satisfied
  if(size == 1) {
    std::cout << "\n Example must be run as more than 1 process." << std::endl;
    MPI_Abort(MPI_COMM_WORLD, -1);
  }

  // Both of these are required for dealing with non-blocking communications
  MPI_Status status;
  MPI_Request request;

  // All processes send the go ahead message except the last process
  if (rank < size-1) {
    // Send the go ahead message to rank+1, using rank^2 as the data
    data = rank*rank;
    MPI_Isend(&data, 1, MPI_INT, rank+1, 1000*rank, MPI_COMM_WORLD, &request);
  }

  // Corresponding non-blocking receive
  if (rank > 0) {
    MPI_Irecv(&data, 1, MPI_INT, rank-1, 1000*(rank-1), MPI_COMM_WORLD,
              &request);
  }

  // All processes print hello
  std::cout << "Hello from rank " << rank << " out of " << size << std::endl;

  // If not the first process, wait for permission to proceed
  if (rank > 0) {
    // This will block each rank until the request has completed
    MPI_Wait(&request, &status);
    // Note the use of MPI_Status to output the sending rank
    std::cout << "Rank "<< rank << " has received message with data " << data
              << " from rank " << status.MPI_SOURCE << std::endl;
  }

  MPI_Finalize();
  return 0;
}
```

Some key features of this example to contrast with the earlier Listing 14.1 are:

- The sending and receiving code are physically close together. This proximity reduces the cognitive effort of the programmer to match up the communications.

- Tracing the code path of any single rank takes it to the same synchronization point as any other rank.
- Overlapping of communication and computation is apparent. This structure of
  - `MPI_Isend()/MPI_Irecv()` the necessary data,
  - perform computations that do not rely on messages (output "`Hello from...`"),
  - wait for messages to complete, and
  - perform computations that are dependent on the messages

  is a familiar pattern in parallel computing. It also delineates which parts of the code execute before the communication completes and which execute after.

Example output of running Listing 14.2 with 8 processes is

```
Hello from rank 0 out of 8
Hello from rank 2 out of 8
Hello from rank 7 out of 8
Hello from rank 6 out of 8
Rank 7 has received message with data 36 from rank 6
Hello from rank 4 out of 8
Rank 2 has received message with data 1 from rank 1
Hello from rank 1 out of 8
Rank 1 has received message with data 0 from rank 0
Rank 6 has received message with data 25 from rank 5
Hello from rank 5 out of 8
Rank 4 has received message with data 9 from rank 3
Rank 5 has received message with data 16 from rank 4
Hello from rank 3 out of 8
Rank 3 has received message with data 4 from rank 2
```

We notice how the output is no longer as nicely ordered as the example output of Listing 14.1. The blocking communication of the former example led to a strictly enforced order in which the output occurred. This latter example is structured to have each rank output to `stdout` whenever it can.

The computation that is performed here while messages are in transit is trivial by design. Problems that do more meaningful work while messages are in transit typically require more components from the MPI library and thus require the topics of the next chapters as well.

Chapter 15
Collective communication

The previous chapter discussed point-to-point communication: how to send and receive messages between individual processes in an MPI program. This chapter deals with *collective communication*: how to send and receive messages among many different processes at the same time. Collective communications include various forms of one-to-all, all-to-one, and all-to-all communication patterns. We now launch into the basic functionality MPI has for dealing with these different collective communication scenarios.

15.1 The why

With access to point-to-point communication routines, the existence of collective communication routines—say a one-to-all communication—may be met with the question: Why not just implement this with a series of point-to-point communications? Part of the reason for this comes down to expressivity of algorithm. It is common that algorithms require data to be moved among all processes in a communicator, but code to perform such tasks can be complex. Fortunately, MPI provides communication routines to do this by default through code that can be invoked in a clear and (relatively) concise manner. Another compelling bonus of this approach, however, is that it benefits performance as well because the MPI implementation is aware of (and optimized for) the hardware architecture on which it is running (as discussed in Section 11.2), conferring significant benefits to developer productivity because communication between all processes does not have to be managed explicitly. We adopt the viewpoint that there is often little to be gained from the effort of re-inventing a wheel that is less round than the available ones.

This chapter discusses the main use cases for collective communication: *broadcasting* (sending) data from one rank to many, *reducing* (summarizing) data from many ranks to one, *scattering* (splitting up) data from one rank to many, and *gathering* (collecting) data from many ranks to one.

© The Author(s), under exclusive license to Springer Nature Switzerland AG 2025
R. J. Spiteri and K. Klenk, *Extreme-Scale Computing*, CMS/CAIMS Books in Mathematics 15, https://doi.org/10.1007/978-3-031-89033-8_15

15.2 Broadcasting and Reducing

When one process has data that are to be communicated to all other processes in a communicator, that process must *broadcast* the data to all other processes. In MPI, this is done with the `MPI_Bcast()` routine, which has the following signature:

```
int MPI_Bcast(
        void*        data,
        int          count,
        MPI_Datatype datatype,
        int          source,
        MPI_Comm     communicator);
```

Similar to what we have seen in the point-to-point communication, this function communicates `count` variables of type `datatype` that are stored in the `data` buffer. The difference here is that now these data are communicated from rank `source` to all other ranks in `communicator`. An example of how this is used in practice can be found in Listing 15.1. The code in Listing 15.1 also demonstrates how multiple data can be transmitted from a `std::vector` on the source rank to `std::vectors` on the receiving ranks. This call works with `std::vector` because the `vector.data()` function is a pointer to the continuously stored data within the vector.

Listing 15.1 Broadcasting data from one rank to all others in the `MPI_COMM_WORLD` communicator. Note how a `std::vector` needs to be handled to work with the `MPI_XXX` routines..

```
#include <iostream>
#include <vector>
#include <algorithm>
#include <string>
#include <mpi.h>

template<typename T>
void print_values(int rank, std::vector<T> values, std::string prefix="") {
  std::cout << prefix << "rank " << rank << ": ";
  for(auto &it : values) {
    std::cout << it << " ";
  }
  std::cout << std::endl;
}

int main(int argc, char* argv[]) {
  // First thing is to initialize the MPI interface.
  MPI_Init(&argc, &argv);
  int rank, size;
  MPI_Comm_rank(MPI_COMM_WORLD, &rank);
  MPI_Comm_size(MPI_COMM_WORLD, &size);

  size_t N = 5;
  std::vector<double> values(N);

  // Assign values only on rank 0
  if(rank==0) {
    std::generate(values.begin(), values.end(), [](){
        static double val{0.0};
        val += 0.1;
        return val;
    });
  }
  MPI_Barrier(MPI_COMM_WORLD);

  print_values(rank,values,"  Before Bcast.  ");

  // Communicate from rank 0 to the rest
  // - note how std::vector data can be communicated with this
  MPI_Bcast(values.data(), values.size(), MPI_DOUBLE, 0, MPI_COMM_WORLD);
  print_values(rank,values,"  After Bcast.  ");

  MPI_Finalize();
  return 0;
}
```

An operation that is in some sense the inverse of a broadcast is a *reduction*. We have already covered reductions in C++ in Section 8.3. A reduction is used when information from all ranks in a communicator need to be aggregated on a single rank. Examples of this include summing data from separate ranks or determining the maximum value of a variable across ranks. Reductions in MPI are performed with the `MPI_Reduce()` routine with the signature:

```
int MPI_Reduce(
      void*        data,
      void*        result,
      int          count,
      MPI_Datatype datatype,
      MPI_Op       operator,
      int          destination,
      MPI_Comm     communicator);
```

`MPI_Reduce()` works on `count` elements of type `datatype` at the `data` location in memory. The variable `result` holds the result of the reduction on the `destination` rank on `communicator`. To obtain a result, an `operator` must be chosen. This is the operation to perform on all the `data` variables across all ranks in `communicator`. Predefined reduction operations are found in Table 15.1; defining your own reduction operations is discussed in Chapter 16.

MPI_Op	Operation
MPI_MAX	maximum
MPI_MIN	minimum
MPI_SUM	sum
MPI_PROD	product
MPI_LAND	logical and
MPI_BAND	bit-wise and
MPI_LOR	logical or
MPI_BOR	bit-wise or
MPI_LXOR	logical exclusive or (xor)
MPI_BXOR	bit-wise exclusive or (xor)
MPI_MAXLOC	max value and location
MPI_MINLOC	min value and location

Table 15.1 Table of predefined `MPI_Op` operators.

Sometimes, the result of a reduction needs to be known to all ranks involved. One way that this could be achieved is by immediately broadcasting the result after the reduction. Because this is a somewhat common requirement of algorithms, MPI provides a separate function to achieve this result: `MPI_Allreduce()`. Again, this dedicated function is provided as a means to help optimize communications behind the scenes. The signature for `MPI_Allreduce()` is

```
int MPI_Allreduce(
      void*        data,
      void*        result,
      int          count,
      MPI_Datatype datatype,
      MPI_Op       operator,
      MPI_Comm     communicator);
```

This is nearly the same as the signature for `MPI_Reduce()`, but it is missing the destination rank where the result would be stored.

15.3 Scattering and gathering

15.3.1 Scattering operations

Continuing in the description of collective communication, when a set of data needs to be split up and distributed across ranks, we make use of *scatter* operations. The scatter operation is a one-to-all communication and is visualized in Figure 15.1. In Figure 15.1, the precondition of a scatter is the top of the figure, and the post-condition is the bottom. Scattering is achieved in MPI with the `MPI_Scatter()` routine, which has the signature

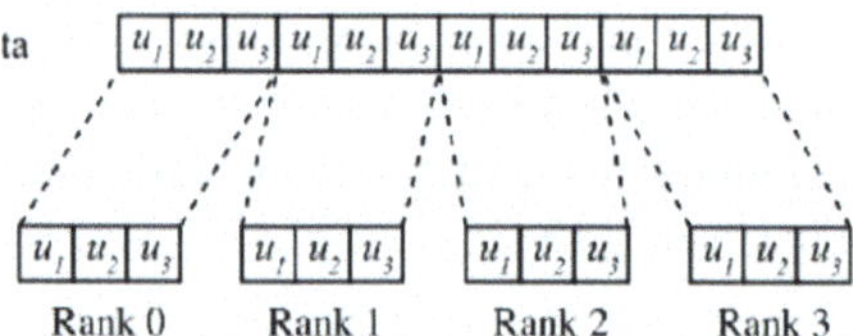

Fig. 15.1 Figure showing gather and scatter operations. `MPI_Scatter()` takes the starting data, all from a single rank (top), and splits them up to many other ranks (bottom). `MPI_Gather()` takes the starting data to be on separate ranks (bottom) and pulls them together on a single rank (top).

```
int MPI_Scatter(
        void*         send_data,
        int           send_count,
        MPI_Datatype  send_datatype,
        void*         recv_data,
        int           recv_count,
        MPI_Datatype  recv_datatype,
        int           source,
        MPI_Comm      communicator);
```

with the following arguments: `send_data` is the buffer of data that exists on the `source` rank of the `communicator` that is to be scattered, and `send_count` is the number of elements of type `send_datatype` to send to each rank. On the receiving end, each rank that is participating in the scatter needs to specify (locally) a `recv_data` buffer to hold `recv_count` elements of type `recv_datatype`. An example usage of `MPI_Scatter()` is given in Listing 15.2.

Listing 15.2 Example code that distributes data using `MPI_Scatter()`.

```
#include <iostream>
#include <vector>
#include <algorithm>
#include <string>
#include <mpi.h>

template<typename T>
void print_values(int rank, std::vector<T> values, std::string prefix="") {
  std::cout << prefix << "rank " << rank << ": ";
  for(auto &it : values) {
    std::cout << it << " ";
  }
  std::cout << std::endl;
}

int main(int argc, char* argv[]) {
  // First thing is to initialize the MPI interface.
  MPI_Init(&argc, &argv);
  int rank, size;
  MPI_Comm_rank(MPI_COMM_WORLD, &rank);
  MPI_Comm_size(MPI_COMM_WORLD, &size);

  size_t N = 3;
  std::vector<double> sendbuf;
  std::vector<double> recvbuf(N);

  // Assign values only on rank 0
  if (rank==0) {
    sendbuf.resize(size*N); // rank 0 holds the entire set
    std::generate(sendbuf.begin(), sendbuf.end(), [](){
        static double val{0.0};
        val += 0.1;
        return val;
    });
  }
  MPI_Barrier(MPI_COMM_WORLD);

  print_values(rank, sendbuf, "  sendbuf before scatter. ");

  // Communicate from rank 0 to the rest
  // - note how std::vector data can be communicated with this
  MPI_Scatter(sendbuf.data(), N, MPI_DOUBLE, recvbuf.data(), N, MPI_DOUBLE,
              0, MPI_COMM_WORLD);
  print_values(rank, recvbuf, "  recvbuf after scatter.  ");

  MPI_Finalize();
  return 0;
}
```

Example output from running Listing 15.2 with 4 processes is

```
sendbuf before scatter. rank 1:
recvbuf after scatter.  rank 1: 0.4 0.5 0.6
sendbuf before scatter. rank 2:
recvbuf after scatter.  rank 2: 0.7 0.8 0.9
sendbuf before scatter. rank 3:
recvbuf after scatter.  rank 3: 1 1.1 1.2
sendbuf before scatter. rank 0: 0.1 0.2 0.3 0.4 0.5 0.6 0.7
  0.8 0.9 1 1.1 1.2
recvbuf after scatter.  rank 0: 0.1 0.2 0.3
```

If data need to be distributed in an uneven manner, then the `MPI_Scatterv()` routine should be used. The signature for `MPI_Scatterv()` is

```
int MPI_Scatterv(
      void*        send_data,
      int*         send_count,
      int*         send_offset,
      MPI_Datatype send_datatype,
      void*        recv_data,
      int          recv_count,
      MPI_Datatype recv_datatype,
      int          source,
      MPI_Comm     communicator);
```

The differences compared to `MPI_Scatter()` are that the `source` rank must also specify two arrays of values for how to communicate with the other ranks: `send_count` is the number of elements to communicate, and `send_offset` is an array of the starting indices of each data subset to be communicated. For this operation, much more setup is generally needed to ensure the participating ranks are dealing with the correct sizes and offsets, as can be seen for even the relatively simple example of Listing 15.3.

Listing 15.3 Example code that distributes data using `MPI_Scatterv()`. The sizes and offsets for the communication take more work to set up, and it is the programmer's responsibility to ensure that these are coherent across the ranks.

```
#include <iostream>
#include <vector>
#include <algorithm>
#include <numeric>
#include <string>
#include <mpi.h>

template<typename T>
void print_values(int rank, std::vector<T> values, std::string prefix="") {
  std::cout << prefix << "rank " << rank << ": ";
  for(auto &it : values) {
    std::cout << it << " ";
  }
  std::cout << std::endl;
}

int main(int argc, char* argv[]) {
  // First thing is to initialize the MPI interface.
  MPI_Init(&argc, &argv);
  int rank, size;
  MPI_Comm_rank(MPI_COMM_WORLD, &rank);
  MPI_Comm_size(MPI_COMM_WORLD, &size);

  std::vector<int>    sizes(size);
  std::vector<int>    offset(size);
  std::vector<double> sendbuf;
  std::vector<double> recvbuf(rank+1);

  // Determine recvbuf size on each rank
  std::generate(sizes.begin(), sizes.end(), [](){
      static int val{0};
      val += 1;
      return val;
  });
  // fill offset
  std::exclusive_scan(sizes.begin(), sizes.end(), offset.begin(), 0);

  // Assign values only on rank 0
  if (rank==0) {
    print_values(rank, sizes, "Sizes to communicate from ");
    sendbuf.resize(size*(size+1)/2);
    std::generate(sendbuf.begin(), sendbuf.end(), [](){
        static double val{0.0};
        val += 0.1;
        return val;
    });
  }
  MPI_Barrier(MPI_COMM_WORLD);

  print_values(rank, sendbuf, "  sendbuf before scatter. ");

  // Communicate from rank 0 to the rest
  // - note how std::vector data can be communicated with this
  MPI_Scatterv(sendbuf.data(), sizes.data(), offset.data(), MPI_DOUBLE,
               recvbuf.data(), sizes[rank], MPI_DOUBLE, 0, MPI_COMM_WORLD);

  print_values(rank, recvbuf, "  recvbuf after scatter.  ");

  MPI_Finalize();
  return 0;
}
```

Example output from running Listing 15.3 with 4 processes looks like

```
Sizes to communicate from rank 0: 1 2 3 4
  sendbuf before scatter. rank 0: 0.1 0.2 0.3 0.4 0.5 0.6 0.7
```

```
  0.8 0.9 1
recvbuf after scatter.  rank 0: 0.1
sendbuf before scatter. rank 1:
recvbuf after scatter.  rank 1: 0.2 0.3
sendbuf before scatter. rank 2:
recvbuf after scatter.  rank 2: 0.4 0.5 0.6
sendbuf before scatter. rank 3:
recvbuf after scatter.  rank 3: 0.7 0.8 0.9 1
```

15.3.2 Gathering operations

In its usual symmetric way, MPI also provides the means to collect the raw data from distributed ranks through a *gather* operation. Gathering is also visualized in Figure 15.1 but now with the precondition that the data exist across many ranks, and the postcondition that all data must exist on a single destination rank. This is an all-to-one communication process, but it is worth noting how this is different from a reduction operation because the raw data are being communicated.

The basic way to gather in MPI is with the `MPI_Gather()` routine, having the signature

```
int MPI_Gather(
        void*        send_data,
        int          send_count,
        MPI_Datatype send_datatype,
        void*        recv_data,
        int          recv_count,
        MPI_Datatype recv_datatype,
        int          destination,
        MPI_Comm     communicator);
```

with arguments as follows: `send_data`, `send_count`, and `send_datatype` specify the characteristics of the local memory that are to be communicated. `recv_data`, `recv_count`, and `recv_datatype` specify the characteristics of the memory that are to be received on rank `destination` of the communicator `communicator`. An example code that performs gathering operations is found in Listing 15.4.

Listing 15.4 Example code that collects data using `MPI_Gather()`.

```
#include <iostream>
#include <vector>
#include <algorithm>
#include <string>
#include <mpi.h>

template<typename T>
void print_values(int rank, std::vector<T> values, std::string prefix="") {
  std::cout << prefix << "rank " << rank << ": ";
  for(auto &it : values) {
    std::cout << it << " ";
  }
  std::cout << std::endl;
}

int main(int argc, char* argv[]) {
  MPI_Init(&argc, &argv);
  int rank, size;
  MPI_Comm_rank(MPI_COMM_WORLD, &rank);
  MPI_Comm_size(MPI_COMM_WORLD, &size);

  size_t N = 3;
  std::vector<double> sendbuf(N);
  std::vector<double> recvbuf;

  // Assign values only on rank 0
  if (rank==0) {
    recvbuf.resize(size*N); // rank 0 holds the entire set
  }
  std::generate(sendbuf.begin(), sendbuf.end(), [rank](){
      static double val{rank};
      val += 0.1;
      return val;
  });

  MPI_Barrier(MPI_COMM_WORLD);

  print_values(rank, sendbuf, "  sendbuf before gather. ");

  // Communicate from rank 0 to the rest, noting the use of std::vector
  MPI_Gather(sendbuf.data(), N, MPI_DOUBLE, recvbuf.data(), N, MPI_DOUBLE,
             0, MPI_COMM_WORLD);
  print_values(rank, recvbuf, "  recvbuf after gather.  ");

  MPI_Finalize();
  return 0;
}
```

Example output from Listing 15.4 looks like

```
sendbuf before gather. rank 1: 1.1 1.2 1.3
recvbuf after gather.  rank 1:
sendbuf before gather. rank 2: 2.1 2.2 2.3
recvbuf after gather.  rank 2:
sendbuf before gather. rank 3: 3.1 3.2 3.3
recvbuf after gather.  rank 3:
sendbuf before gather. rank 0: 0.1 0.2 0.3
recvbuf after gather.  rank 0: 0.1 0.2 0.3 1.1 1.2 1.3 2.1
    2.2 2.3 3.1 3.2 3.3
```

Similar to scattering, there exists a variable-size gather operation in the form of `MPI_Gatherv()`. The signature for `MPI_Gatherv()` is

```
int MPI_Gatherv(
      void*        send_data,
      int          send_count,
      MPI_Datatype send_datatype,
      void*        recv_data,
      int*         recv_count,
      int*         recv_offset,
      MPI_Datatype recv_datatype,
      int          destination,
      MPI_Comm     communicator);
```

The arguments specify the memory characteristics in a similar way as was done for the `MPI_Scatter()` routine, noting that the differences are now in the array of sizes and offsets used in the `destination` rank.

In addition to the previous gathering routines, if all the data are needed on all of the ranks of a communicator, MPI provides `MPI_Allgather()` again to help optimize communication patterns behind the scenes. The signature for `MPI_Allgather()` is

```
int MPI_Allgather(
        void*        send_data,
        int          send_count,
        MPI_Datatype send_datatype,
        void*        recv_data,
        int          recv_count,
        MPI_Datatype recv_datatype,
        MPI_Comm     communicator);
```

We note that `MPI_Allgather()` is an all-to-all communication routine.

15.4 Barriers and synchronization

All the collective communication routines discussed in this chapter introduce *synchronization points*. We discussed synchronization points briefly in Section 14.3 in the context of point-to-point communications, but to quickly reiterate: A synchronization point is a point where multiple processes must arrive before any of them can continue. In Listings 15.2 to 15.4, we see that manual synchronization points can be introduced via the `MPI_Barrier()` routine, with signature

```
int MPI_Barrier(MPI_Comm communicator);
```

Synchronization points generally slow performance, so it is recommended to only use them when absolutely necessary.

Chapter 16
Advanced MPI

This chapter covers the more advanced MPI topics of communicator and group manipulation (for separating processes into different communicators) and the creation of derived datatypes (to facilitate the communication of more complex messages).

16.1 Communicators

MPI communicates exclusively through communicators. We have already seen the use of the default communicator, `MPI_COMM_WORLD`; however, it is possible to create new communicators to enhance program organization. We consider a few different ways of doing this: *grouping* arbitrary sets of ranks from an existing communicator, *splitting* a communicator, or considering a specific pattern of communication that can be handled by creating a *topological* communicator.

16.1.1 Group communicators

Creating new communicators via groups is a three-step process. The first step is to get the group that is associated with an existing communicator using the `MPI_Comm_group()` routine:

```
int MPI_Comm_group(
      MPI_Comm   communicator,
      MPI_Group* group);
```

which takes an existing `communicator` and `group` as input. The second step is to select which ranks of the `communicator` are to be included in the `group`. This is done via the `MPI_Group_incl()` routine:

© The Author(s), under exclusive license to Springer Nature Switzerland AG 2025

R. J. Spiteri and K. Klenk, *Extreme-Scale Computing*, CMS/CAIMS Books in Mathematics 15, https://doi.org/10.1007/978-3-031-89033-8_16

```
int MPI_Group_incl(
      MPI_Group  group,
      int        n,
      const int  ranks[],
      MPI_Group* new_group);
```

which takes as input the existing `group` and selects `n` ranks from `ranks[]` to use in the `new_group`. The third step takes this newly created group (`new_group`) and creates the new communicator using `MPI_Comm_create()`:

```
int MPI_Comm_create(
      MPI_Comm  communicator,
      MPI_Group new_group,
      MPI_Comm* new_communicator);
```

which creates the `new_communicator`.

Thus, the process of splitting `MPI_COMM_WORLD` into two different communicators can be achived by a few lines of code:

```
// 1. Get group from existing communicator
MPI_Group world_group;
MPI_Comm_group(MPI_COMM_WORLD, &world_group);

// 2. Create lists of subsets of ranks (from the original communicator)
std::vector<std::vector<int>> subgroup_ranks {{0,1,2},{1,2,3}}
std::vector<MPI_Group>        subgroup(subgroup_ranks.size());
//    And create the new subgroups
for (int i=0; i < subgroup.size(); ++i) {
  MPI_Group_incl(world_group,
                 subgroup.at(i).size(),
                 subgroup_ranks.at(i).data(),
                 subgroup.at(i).data());
}

// 3. Create the new communicators
std::vector<MPI_Comm> subcomm(subgroup.size());
for (int i=0; i < subgroup.size(); ++i) {
  MPI_Comm_create(MPI_COMM_WORLD,
                  subgroup.at(i),
                  subcomm.at(i).data());

// Clean up the new groups and communicators
for (int i=0; i < subcomm.size(); ++i) {
  if (subcomm.at(i) != MPI_COMM_NULL) {
    MPI_Comm_free(&subcomm.at(i));
    MPI_Group_free(&subgroup.at(i));
  }
}
```

The intent of this code is to create two new communicators that each contain a subset of the ranks in `MPI_COMM_WORLD`. The first new communicator contains the ranks 0, 1, and 2, and the second contains the ranks 1, 2, and 3. We note here that these are the ranks as they are enumerated in the original `MPI_COMM_WORLD` communicator. In the new communicators, these ranks are numbered 0, 1, and 2 for both of them because MPI ranks are assigned by incrementing from 0. At the end of this snippet, there is a vector that contains the two different communicators. We also note the generality of this process: using `MPI_groups` is the most general way of splitting up ranks because it allows you to fully customize which ranks go to new communicators; e.g., ranks 1 and 2 get assigned to both of the new communicators. To split ranks into non-overlapping groups, we use MPI's *split* functionality.

16.1.2 Splitting communicators

The `MPI_Comm_split()` routine splits the ranks of a communicator into non-overlapping groups. Its signature is

```
int MPI_Comm_split(
        MPI_Comm  communicator,
        int       color,
        int       key,
        MPI_Comm* new_communicator);
```

To use this routine, each calling rank in a communicator must provide a *color* (an integer tag for the new communicator to associate with), and a key (rank on the existing communicator). A benefit of splitting over using groups is that it results in a single `new_communicator` that manages the colored communication underneath the hood.

A simple code that splits ranks into odd and even looks like

```
// Assign colors
int world_rank;
MPI_Comm_rank(MPI_COMM_WORLD, &world_rank);
int color = world_rank%2;

// Split the main communicator
MPI_Comm oddeven_comm;
MPI_Comm_split(MPI_COMM_WORLD, color, world_rank, &oddeven_comm);

// Get new communicator ranks
int oddeven_rank;
MPI_Comm_rank(MPI_COMM_WORLD, &oddeven_rank);

// Test the behavior of broadcast from oddeven_rank=0
int value{};
if(oddeven_rank==0) value = 99;
MPI_Bcast(&value, 1, MPI_INT, 0, oddeven_comm);

std::cout << "world_rank = "      << world_rank
          << ", oddeven_rank = " << oddeven_rank
          << ", value = "        << value
          << std::endl;

// Clean up the new communicator
MPI_Comm_free(&oddeven_comm);

}
```

Upon completion of this code snippet, the result should be that whichever group (odd or even) held rank 0 in the `oddeven_comm` communicator will hold the even ranks because MPI ranks are assigned in increasing order. The other members of that group (even) should have the value 99 in them, while the other group (odd) has the default value of 0. This is observed in the output

```
$ mpirun -np 8 ./MPI_split_ex
world_rank = 1, oddeven_rank = 1, value = 0
world_rank = 2, oddeven_rank = 2, value = 99
world_rank = 3, oddeven_rank = 3, value = 0
world_rank = 4, oddeven_rank = 4, value = 99
world_rank = 5, oddeven_rank = 5, value = 0
world_rank = 6, oddeven_rank = 6, value = 99
world_rank = 7, oddeven_rank = 7, value = 0
world_rank = 0, oddeven_rank = 0, value = 99
```

16.1.3 Topological communicators

So far, we have looked at the most general usage of creating a new communicator through using *groups* and the specific case of dividing a communicator into non-overlapping sub-groups through *splitting*. Now, we consider something a little different through the use of *topological* communicators. Topological communicators are useful when there is a specific structure of communication. MPI allows the creation and use of *virtual* topologies to help manage more structured communications.

The two specific types are *Cartesian* communicators, consisting of an N-dimensional ($N \leq 6$) rectangular array of processes with a nearest-neighbor communication pattern, and *graph* communicators, which allow one to specify an arbitrary nearest-neighbor graph of connections. In this chapter, we only consider the Cartesian topological communicator.

16.1.3.1 Cartesian communicator

A *Cartesian topology* in MPI refers to ranks organized in a multidimensional rectangular array up to six dimensions. Defining and working with this topology is useful when the majority of communications is among the nearest neighbors in such a layout.

The `MPI_Cart_create()` routine can be used to create a grid communicator. It has the signature

```
int MPI_Cart_create(
        MPI_Comm  old_communicator,
        int        n_dimensions,
        const int dimensions[],
        const int periodic[],
        int        reorder,
        MPI_Comm* cartesian_communicator);
```

which creates a Cartesian communicator in `n_dimensions`, where each dimension `i` < `n_dimensions` has `dimensions[i]` entries in it. Periodicity in dimension `i` is handled by `periodic[i]` (1 = periodic, 0 = not), `reorder=1` gives MPI the option to reorder neighboring ranks to be physically closer together. Finally, the new communicator is stored in `cartesian_communicator` for future use. A simple program that sets up a Cartesian communicator is found in Listing 16.1 with a picture of the grid communication associated with it in Figure 16.1.

Listing 16.1 Creation of a Cartesian communicator. A rank's Cartesian coordinates can be obtained based on its rank, or conversely, a rank can be identified based on given coordinates. We note also the pattern of `MPI_Barrier`s used here that allows for ordered output. See Figure 16.1

```
#include<iostream>
#include<cassert>
#include<mpi.h>

int main(int argc, char *argv[]) {
  MPI_Init(&argc, &argv);

  int rank, size;
  MPI_Comm_rank(MPI_COMM_WORLD, &rank);
  MPI_Comm_size(MPI_COMM_WORLD, &size);

  // Set the dimensions of the array of ranks (12 ranks, here)
  assert(size == 12);
  int dim[2], period[2], reorder;
  dim[0]=4; dim[1]=3;
  period[0]=1; period[1]=0;
  reorder=1; // allow MPI to reorder based on network layout

  MPI_Comm comm;
  MPI_Cart_create(MPI_COMM_WORLD, 2, dim, period, reorder, &comm);

  // Access Cartesian coordinates from rank
  for (int i = 0; i < size; ++i) {
    if (rank == i) {
      int coord[2];
      MPI_Cart_coords(comm, i, 2, coord);
      std::cout << "Rank " << rank << "'s coordinates are (" << coord[0]
                << "," << coord[1] << ")." << std::endl;
      MPI_Barrier(MPI_COMM_WORLD);
    }
    MPI_Barrier(MPI_COMM_WORLD);
  }

  if(rank==0) std::cout << std::endl;
  MPI_Barrier(MPI_COMM_WORLD);

  // Access rank from Cartesian coordinates
  if (rank==0) {
    for (int i = 0; i < dim[0]; ++i) {
      for (int j = 0; j < dim[1]; ++j) {
        int coord[2], id;
        coord[0]=i; coord[1]=j;
        MPI_Cart_rank(comm, coord, &id);
        std::cout << "Position (" << coord[0] << "," << coord[1] << ")"
                  << " holds rank " << id << "." << std::endl;
      }
    }
  }

  MPI_Finalize();
  return 0;
}
```

The utility of a Cartesian communicator comes from simple specification of the neighbors for any given rank. The neighboring processes can easily be determined by using the `MPI_Cart_shift()` routine, which has the signature

```
int MPI_Cart_shift(
        MPI_Comm communicator,
        int direction,
        int displacement,
        int *rank_source, // out
        int *rank_dest);  // out
```

Here, `direction` is the dimension being looked in, `displacement` is the distance in said direction (>0 for *upwards* shift, <0 for *downwards* shift). On return, `rank_source` is the rank of the process calling this routine, and `rank_dest` is the rank of the process `displacement` steps away in the `direction` direction. If there is no process at the specified displacement away from the calling rank, then `MPI_PROC_NULL` is populated in `rank_dest`.

An example of using this routine is

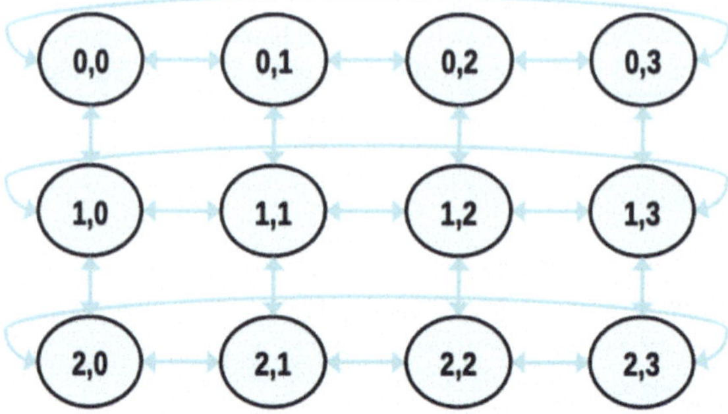

Fig. 16.1 Cartesian topology in two dimensions. The x-direction is periodic. Figure courtesy of codingame.com.

```
// Left and right in the x-direction
int neighbor_right, neighbor_left;
MPI_Cart_shift(comm, 1, 1, &rank, neighbor_right);
MPI_Cart_shift(comm, 1, -1, &rank, neighbor_left);

// Above and below in the y-direction
int neighbor_up, neighbor_down;
MPI_Cart_shift(comm, 0, 1, &rank, neighbor_up);
MPI_Cart_shift(comm, 0, -1, &rank, neighbor_down);
```

where `comm` is a communicator that has been created with `MPI_Cart_create()`.

16.2 Derived datatypes

MPI datatypes allow for the specification of arbitrary data layouts to be transmitted through MPI messages. The act of encoding data such that they can be transmitted through a serial communication framework and recovered elsewhere is known as *serialization*. The act of handling the transmitted data—converting them back into something that has meaning to the program—is known as *deserialization*. Because it is generally useful to have both sides of this procedure, the term *SerDes* is sometimes used to refer to the entirety of this process.

A more widespread example of serialization is found in the Javascript Object Notation (JSON), where objects can be serialized into a particularly formatted text string. Those text strings can then be passed around in whatever way deemed necessary and eventually reconstructed into the objects themselves when needed. Although this procedure does result in serialized data that are human readable (simple ASCII or UTF-8 strings), it is particularly bloated and wasteful for numerical data. To keep message sizes down (recall that networking bandwidth is typically much slower than other components on the motherboard stack), it is generally advisable to work with bits and bytes themselves.

MPI allows for user-defined *derived datatypes* to simplify serialization and deserialization of data. This view of serialization through datatypes allows for simple specification of what needs to be transmitted and allows MPI to effectively automate the serialization and deserialization of messages.

16.2.1 Primitive datatypes

MPI primitive datatypes are what we have been working with so far. These are the datatypes that are equivalent for all C++ native datatypes. Examples include, but are not limited to,[1]

- `MPI_INT` - C++ `int`
- `MPI_DOUBLE` - C++ `double`
- `MPI_CHAR` - C++ `char`

Derived datatypes in MPI rely on previously defined types. In fact, recursive specification of datatypes can be performed, gradually building up more complex datatypes from the existing primitive types. The process for doing this generally goes as follows:

1. Construct a *new* datatype using a derived datatype routine and an *old* datatype.
2. Register the new datatype with MPI with `MPI_Type_commit`.
3. Use the datatype in the program execution.
4. Free the memory associated with the datatype definition when the datatype is no longer needed.

16.2.2 Contiguous datatype

If a program repeatedly needs to communicate fixed-size chunks of data between processes, the organization of the program may benefit from defining a *contiguous* datatype. This is done with the MPI routine

```
int MPI_Type_contiguous(
        int                 count,
        MPI_Datatype        oldType,
        MPI_Datatype*       newType);
```

where `count` is the number of contiguous elements, `oldtype` is an existing datatype, and `*newType` is a pointer to the new datatype. Figure 16.2 (a) shows the contiguous layout. An example of creating, committing, and using a contiguous datatype is found in Listing 16.2.

Listing 16.2 Contiguous datatype example. Some features of note: The contiguous datatype is specified and "committed" before it is used. Broadcasting a single element of datatype `Five_Element` sends five elements of datatype `MPI_INT`. The datatype should be freed when no longer needed.

[1] See Table 13.1 for the complete list.

```
#include <iostream>
#include <array>
#include <algorithm>
#include <mpi.h>

int main(int argc, char* argv[]) {
  // First thing is to initialize the MPI interface.
  MPI_Init(&argc, &argv);

  // each process has an array of this size
  const size_t SIZE=42;
  std::array<int, SIZE> x{0};

  int rank;
  MPI_Comm_rank(MPI_COMM_WORLD, &rank);

  // Create a datatype that uses 5 contiguous elements
  MPI_Datatype Five_Element;
  MPI_Type_contiguous(5, MPI_INT, &Five_Element);
  MPI_Type_commit(&Five_Element);

  if (rank==0) {
    // fill the entirety of rank 0's array
    std::fill(x.begin(), x.end(), 1);
    // Broadcast a single Five_Element contiguous chunk to the rest
    MPI_Bcast(&x[30],1,Five_Element,0,MPI_COMM_WORLD);
  } else {
    // Accept the broadcast
    MPI_Bcast(&x[30],1,Five_Element,0,MPI_COMM_WORLD);
  }

  if(rank==1) {
    std::cout << "Rank: " << rank << "\n";
    for(size_t ii=0; ii<SIZE; ++ii) {
      std::cout << ii << ": " << x[ii] << "\n";
    }
  }

  // Free the created datatype
  MPI_Type_free(&Five_Element);

  // Must shut down the MPI system when you're done.
  MPI_Finalize();

  return 0;
}
```

16.2.3 Strided datatype

Another commonly occurring data access pattern is that of *strided blocks*. In this type, the *block* is a contiguous layout of the old datatype, and the *stride* is the offset between the starting points of the blocks. Figure 16.2 (b) shows what is meant by this, demonstrating a block length of 2 and a stride of 3.

This pattern can be created with the MPI routine

```
int MPI_Type_vector(
      int           count,
      int           blocklen,
      int           stride,
      MPI_Datatype  oldType,
      MPI_Datatype* newType);
```

where `count` is the number of blocks in the datatype, `blocklen` is the size of each contiguous block, `stride` is the offset between the beginning of each block, `oldType` is an existing datatype, and `newType` is a pointer to the new datatype.

An example of creating, committing, and using a contiguous datatype is found in Listing 16.3.

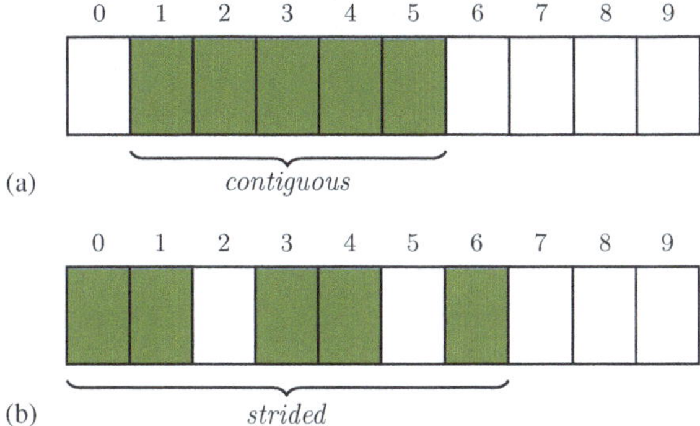

Fig. 16.2 Data layouts for (a) contiguous and (b) strided blocks with block length 2 and stride 3.

Listing 16.3 Strided block datatype example. Feature of note: Broadcasting a single element of datatype `My_Strided` sends six elements (in total) of type `MPI_INT`. The datatype should be freed when no longer needed.

```
#include <iostream>
#include <array>
#include <algorithm>
#include <mpi.h>

int main(int argc, char* argv[]) {
  // First thing is to initialize the MPI interface.
  MPI_Init(&argc, &argv);

  // each process has an array of this size
  const size_t SIZE=42;
  std::array<int, SIZE> x{0};

  int rank;
  MPI_Comm_rank(MPI_COMM_WORLD, &rank);

  // Create a datatype that uses 5 contiguous elements
  MPI_Datatype My_Strided;
  MPI_Type_vector(3, 2, 5, MPI_INT, &My_Strided);
  MPI_Type_commit(&My_Strided);

  if (rank==0) {
    // fill the entirety of rank 0's array
    std::fill(x.begin(), x.end(), 1);
    // Broadcast a single My_Strided datatype
    MPI_Bcast(&x[30],1,My_Strided,0,MPI_COMM_WORLD);
  } else {
    // Accept the broadcast
    MPI_Bcast(&x[30],1,My_Strided,0,MPI_COMM_WORLD);
  }

  if(rank==1) {
    std::cout << "Rank: " << rank << "\n";
    for(size_t ii=0; ii<SIZE; ++ii) {
      std::cout << ii << ": " << x[ii] << "\n";
    }
  }

  // Free the created datatype
  MPI_Type_free(&My_Strided);

  // Must shut down the MPI system when you're done.
  MPI_Finalize();

  return 0;
}
```

A useful application of this pattern is when communicating elements of a multi-dimensional array. This usage is demonstrated by setting up a datatype to represent an entire column of a 2D array in Listing 16.4.

Listing 16.4 Communicating a single column of a two-dimensional array. Feature of note: Broadcasting a single element of datatype `Column` sends five elements of a column to all other processors.

```
#include <iostream>
#include <array>
#include <algorithm>
#include <mpi.h>

int main(int argc, char* argv[]) {
  // First thing is to initialize the MPI interface.
  MPI_Init(&argc, &argv);

  // each process has an array of this size
  const size_t SIZE_1=5;
  const size_t SIZE_2=4;
  std::array<std::array<int, SIZE_2>, SIZE_1> x{0};

  int rank;
  MPI_Comm_rank(MPI_COMM_WORLD, &rank);

  // Create a datatype that uses 5 contiguous elements
  MPI_Datatype Column;
  MPI_Type_vector(SIZE_1, 1, SIZE_2, MPI_INT, &Column);
  MPI_Type_commit(&Column);

  if (rank==0) {
    // fill the entirety of rank 0's array
    for (auto &it : x) {
      std::fill(it.begin(), it.end(), 1);
    }
    // Broadcast a single My_Strided datatype
    MPI_Bcast(&x[0][2],1,Column,0,MPI_COMM_WORLD);
  } else {
    // Accept the broadcast
    MPI_Bcast(&x[0][2],1,Column,0,MPI_COMM_WORLD);
  }

  if (rank==1) {
    std::cout << "Rank: " << rank << "\n";
    for (auto const &it : x) {
      for (auto const &it2 : it) {
        std::cout << it2 << " ";
      }
      std::cout << "\n";
    }
  }

  // Free the created datatype
  MPI_Type_free(&Column);

  // Must shut down the MPI system when you're done.
  MPI_Finalize();

  return 0;
}
```

16.2.4 Indexed datatype

Communicating arbitrary lists of elements can be handled by defining an *indexed block* datatype. An indexed block is defined with the MPI routine

```
int MPI_Type_create_indexed_block(
        int                     count,
        int                     blocklen,
        int*                    displacements,
        MPI_Datatype            oldType,
        MPI_Datatype*           newType);
```

where `count` is the number of blocks in the `displacements` array, `blocklen` is the number of elements in each of the blocks, `oldType` is the MPI datatype of each of the elements, and `newType` is a handle to the newly defined MPI datatype.

For example, defining an MPI datatype that works with the indices 1, 5, and 7 can be done with the following code:

```
std::vector<int> displacements{1,5,7};
MPI_Datatype OneFiveSeven;
MPI_Type_create_indexed_block(displacements.size(), 1, displacements.data(), MPI_DOUBLE, &OneFiveSeven);
```

which can now be used to easily communicate those indices from containers with elements of type `double` by using the type `OneFiveSeven`.

16.2.5 Struct datatype

Similar to C/C++ structs (and classes), the MPI struct datatype allows for multiple different MPI datatypes to be aggregated into a single type. This type can be created with the MPI routine

```
int MPI_Type_create_struct(
        int             count,
        int*            blocklens,
        int*            displacements,
        MPI_Datatype*   types,
        MPI_Datatype*   newtype);
```

where `count` is the number of different blocks to be included in the new datatype, `blocklens` is an array (of size `count`) of the sizes of each block, `displacements` is an array (of size `count`) of offsets for each block, and `types` is an array (of size `count`) of the MPI datatype of each block. Listing 16.5 gives a simple demonstration for how an MPI struct datatype can be defined.

Listing 16.5 Setting up the datatype to communicate a struct. Feature of note: The offsets need to be computed manually from the sizes of the datatypes within the struct.

```
#include <iostream>
#include <array>
#include <algorithm>
#include <mpi.h>

struct Aggregate {
  int  x[2];
  char letter;
};


int main(int argc, char* argv[]) {
  // First thing is to initialize the MPI interface.
  MPI_Init(&argc, &argv);

  // Each process has an Aggregate
  Aggregate agg;

  int rank;
  MPI_Comm_rank(MPI_COMM_WORLD, &rank);

  // Arrays to specify the data layout of the struct
  int lengths[2]{ 2, 1 };
  MPI_Datatype types[2]{ MPI_INT, MPI_CHAR };
  MPI_Aint displacements[2]{ 0, 2*sizeof(int) };

  // Creating the struct datatype
  MPI_Datatype Aggregate_Data;
  MPI_Type_create_struct(2, lengths, displacements, types, &Aggregate_Data);
  MPI_Type_commit(&Aggregate_Data);

  if (rank==0) {
    // set values on process 0
    agg.x[0] = 1;
    agg.x[1] = 2;
    agg.letter = 'b';
    // Brodcast the Aggregate to the rest of the processes
    MPI_Bcast(&agg,1,Aggregate_Data,0,MPI_COMM_WORLD);
  } else {
    // Accept the broadcast
    MPI_Bcast(&agg,1,Aggregate_Data,0,MPI_COMM_WORLD);
  }

  if (rank==1) {
    std::cout << "Rank: " << rank << "\n";
    std::cout << " agg:\n  x[0]: " << agg.x[0] << "\n  x[1]: " << agg.x[1]
              << "\n  letter: " << agg.letter << "\n";
  }

  MPI_Type_free(&Aggregate_Data);
  MPI_Finalize();

  return 0;
}
```

It is worth noting a few slight modifications for using the MPI struct datatype:

1. It can be used with the previous derived datatypes; i.e., if you have an array of structs, you can define a strided datatype for said array.
2. You do not necessarily need to specify the entirety of a struct for the datatype. Only the contents that are intended to be sent through messaging are required.

This struct datatype is only usable like this when everything about a struct or class is known at compile time. If, for example, a struct contains a pointer to some dynamically managed memory, then it is usually not possible to define an offset that points to allocated memory.

16.2.6 C++ objects and deep copying

Suppose we have a class that looks like

```
class Variable {
  std::vector<double> x;
  std::string name;
}
```

Within this class, there are two containers that have variable size. The offsets to the two members of `Variable` are 0 and `sizeof(std::vector<double>)`, but these offsets alone tell us nothing about the size of the contents of the members. Generally, when variables use heap memory (like the `Variable` class above), *deep copies* of these data must be performed to transmit them between processes. A deep copy of data copies the values of the data themselves, rather than just pointers to the data.

This is necessary in MPI because each process works in a different memory space. The value of a pointer (its memory address) for one process is not valid for another process. In other words, the data that are being pointed to must literally be copied.

If we want to pass the entirety of an object of type `Variable` between processes, we need to provide enough data so the object can be recreated on the receiving process. Often, additional code beyond what is provided through MPI datatypes is needed as well. Two possible approaches include:

- Construct a new object of type `Variable` on the receiving process. This requires a constructor that can populate the members of `Variable` on construction as well as multiple messages to transmit the data.
- Populate the members of an existing `Variable` on the receiving process. This requires some means of setting the member variables (either through making the members public or providing public-facing *setter* functionality) and likely still requires multiple messages to transmit the data.

In both of these cases, because of the (potentially) dynamic size of the member variables, at least two messages need to be sent to transmit the data. The first message contains the sizes; the receiving buffer can then use the sizes to allocate appropriately sized buffers to receive the data. The second message contains the actual data. The second message may need to be further subdivided into two messages as well, one for each member, depending on the nature of the problem.

If the sizes of the members do not change through program execution, the messages regarding the size of the members only need to be sent once. Then the buffers allocated for receiving the data can be reused for many subsequent messages; i.e., the program goes through a set-up phase, where information about the future communications in the program is determined, followed by a computational phase, where computations (and messaging of relevant data) take place.

Chapter 17
MPI scaling and recent advanced features

For many scientific researchers, a lot of the point of ESC is to utilize the highest scale of supercomputing to reduce the "time to science". In order to quantify progress toward this goal, a way to measure the time it takes for a program to run *as a function of the resources used* plays an ostensibly critical role. So also is staying abreast of the most recent features of ESC hardware and software.

This chapter begins by discussing the basics of timing MPI programs, as well as the concepts of scalability and efficiency. The chapter finishes with some highlighting of more recently introduced features into the MPI standard: advanced collectives and one-sided communication.

17.1 Scaling and efficiency

Obtaining timing measurements in MPI is straightforward. The MPI routine `MPI_Wtime` gets the current time

```
double t_start = MPI_Wtime();
```

which can then be used at a later point in the program to measure the time taken between two points:

```
double t_final = MPI_Wtime();
double time_taken = t_final -  t_initial;
```

The resolution of the timer (the time between ticks) is obtained from `MPI_Wtick`:

```
double resolution = MPI_Wtick();
```

Typically, it is only possible to obtain reliable timing measurements of events that are greater than a small number of ticks. In practice, measurements should be repeated for the purposes of reproducibility and estimating reliability as well as to mitigate against the effects of competing processes that are inevitably present on any system.

Although we have used `std::chrono::high_resolution_clock` in Listing 9.1, it is recommended to use the `MPI_` timing functions when working with

© The Author(s), under exclusive license to Springer Nature Switzerland AG 2025
R. J. Spiteri and K. Klenk, *Extreme-Scale Computing*, CMS/CAIMS Books in Mathematics 15, https://doi.org/10.1007/978-3-031-89033-8_17

MPI. This is because the clocks used for MPI timing routines are synchronized for all processes within the MPI runtime.

In practice, there are many issues that affect the time it takes for a program to run, and the relative importance of these issues can depend on the hardware (CPU, memory hierarchy, and interconnect), software (compiler, language, and algorithms), and the problem size. It is important to be able to quantify how well a program behaves as a function of the resources allocated to it. Competition for cluster resources is generally stiff; therefore, it is in everyone's interest to allocate an appropriate level of resources to a given program. In particular, it is important not to over allocate resources to programs that do not scale well.

17.1.1 Strong scaling

A natural way to assess the performance of an MPI program (or *how well an MPI program scales*) is to measure the time required for the solution as a function of the number of processes employed. When considering *a problem of fixed size*, a useful measure is the *strong scaling efficiency*, which is defined as

$$\eta_{\text{strong}}(P) = \frac{T_1}{PT_P}, \tag{17.1}$$

where T_P is the time required to solve the problem with P processes.

Perfect strong scaling efficiency $\eta_{\text{strong}}(P) = 1$ is when computation time decreases in exact proportion to the resources thrown at the problem. For example, computation time would be cut in half when using twice as many processes.

Similar to strong scaling efficiency, we can define strong scaling *speedup* as

$$S_{\text{strong}}(P) = P\eta_{\text{strong}} = \frac{T_1}{T_P}. \tag{17.2}$$

Perfect strong scaling speedup would be $S_{\text{strong}}(P) = P$. Strong scaling efficiency is generally useful in practice to identify and avoid diminishing returns, whereas strong scaling speedup is useful to understand Amdahl's law, as discussed in Section 17.1.2.

Although strong scaling may seem to be the most natural way to characterize scaling, it has two inherent limitations. First, strong scaling breaks down in the theoretical limit $P \to \infty$ where there are more processes than operations. In practice, the limit of strong scaling is reached when communication costs are comparable with computation costs per processor. Because communication costs increase with P whereas computation costs decrease with P, this limit is reached fairly easily in practice. Second, strong scaling considers a fixed problem size, and such a scenario is typically not of great interest in practice. However, given that there are times when the desired size of a problem to solve is given, strong scaling provides some insight into an MPI program's ability to operate at that scale. We discuss this topic further at the end of this section.

17.1.2 Amdahl's law

Amdahl's law states: The overall amount of speedup that can be gained by optimizing a single part of a system is limited by the fraction of time that the improved part is actually used. In the context of parallel programming, we can take *optimizing* to mean parallelizing here, but we also want to emphasize the generality of this phenomenon to the reader to keep in mind in all of life's endeavors.

To put Amdahl's law into a mathematical form, we can consider splitting the time taken to run a program into the portion that is being parallelized and that which is restricted to running in serial:

$$T_{\text{Amdahl}} = T = T_{\text{parallel}} + T_{\text{serial}}.$$

If we consider the fraction of execution time $q \in [0, 1]$ to be the fraction of time (T) spent in the (potentially) parallelized section, then we can rewrite this as

$$T_{\text{Amdahl}}(q, P) = T = qT + (1 - q)T.$$

Now, if we assume perfect scaling in the parallelized portion of the code, then we can write the total time as a function of the number of processes (P) thrown at the problem. We thus obtain

$$T_{\text{Amdahl}}(q, P) = T = \frac{q}{P}T + (1 - q)T.$$

Now, when we look at the strong scaling speedup by using equation (17.2) with the total time as a function of the fraction of parallelizable execution time q and the number of processes P, we obtain the expression for *maximum speedup*, known as *Amdahl's law*:

$$S_{\text{Amdahl}}(q, P) = \frac{T_{\text{Amdahl}}(q, 1)}{T_{\text{Amdahl}}(q, P)} = \frac{1}{\frac{q}{P} + (1 - q)} \tag{17.3}$$

equation (17.3) implies a maximal amount of speedup that can be obtained in the situation that only a fraction q of a program can be parallelized:

$$\lim_{P \to \infty} S_{\text{Amdahl}}(q, P) = \frac{1}{1 - q}. \tag{17.4}$$

Figure 17.1 visualizes speedup under the limitations of Amdahl's law. Fortunately for us, most aspects of scientific computations can be parallelized: from the algorithms to the file I/O (discussed more in Part VI). These aspects help to decrease the amount of serial execution, i.e., increase q in equations (17.3) and (17.4).

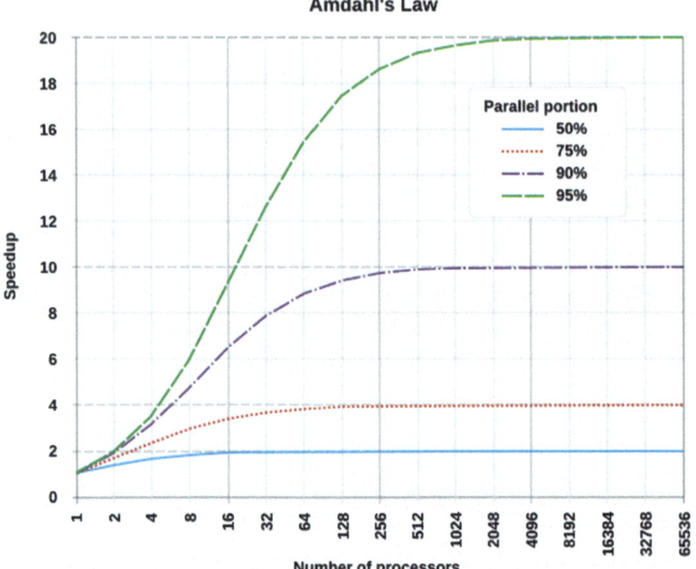

Fig. 17.1 Visualization of Amdahl's law in terms of number of processes. We note the asymptotic approach to the maximum speedup value as the number of processes increases. Courtesy of Wikipedia (`https://en.wikipedia.org//wiki/Amdahl's_law`). As $q \to 1$, Amdahl's law permits larger and larger benefits.

17.1.3 Weak scaling

An alternative measure of scaling to strong scaling efficiency is *weak scaling efficiency*. Weak scaling efficiency measures how long a computation takes to complete when the problem size and computational resources change proportionally. Weak scaling efficiency is defined as

$$\eta_{\text{weak}} = \frac{T_1}{T_P}, \tag{17.5}$$

where T_P is the time required to solve a problem of size KP using P processes *for some fixed constant* K. Perfect weak scaling $\eta_{\text{weak}} = 1$ is when the time to solution remains the same when the problem size and the computational resources are increased by the same factor. For example, computation time remains the same when the problem size and the number of processes are both doubled. Weak scaling is often used as a first pass to evaluate the effectiveness of an MPI program.

For solutions to large problems, we recommend the following two-phase approach to write performant (scalable) code:

1. Solve smaller versions of the desired problem and evaluate the weak scaling of your MPI program as you approach the size of the desired problem.

2. Consider strong scaling down to the smallest number of cores (or nodes) on which your desired problem size can be solved.

This approach ensures that you can evaluate whether your MPI program is running efficiently on multiple cores or nodes. Deficiencies in the performance become readily apparent in the first phase. The first phase can also help determine the *memory scaling* of your program; memory scaling helps to determine the minimum amount of compute resources needed to run the second phase.

17.2 Advanced collectives

This section deals with more advanced collective operations: non-blocking collective communication and neighborhood collective communication.

17.2.1 Non-blocking collective communication

The collective communications described in Chapter 15 are blocking in sense that the ranks in an MPI communicator do not proceed until all ranks in the collective call have completed it. Non-blocking collectives have been introduced to help avoid these (generally costly) synchronization points. Non-blocking collective communications operate similarly to non-blocking point-to-point communications by separating the message sending and the testing for message completion. Non-blocking collectives operate in a similar way to collectives in that they require in-order matching to proceed (rather than using message tags like point-to-point collectives).

The API for non-blocking collectives is the same as that for the collectives, but it uses the routine name `MPI_IXXX` for the non-blocking variant of `MPI_XXX` with a `MPI_Request*` parameter as an additional argument.

For example, the `MPI_Ibcast()` routine looks like

```
int MPI_Ibcast(
      void*         data,
      int           count,
      MPI_Datatype  datatype,
      int           source,
      MPI_Comm      communicator
      MPI_Request*  request);
```

which differs from `MPI_Bcast` (seen in Section 15.2) only in the final argument.

Dealing with the resulting `MPI_Request*` is done just as in Section 14.3, using `MPI_Test()` to test for the completion of a request and `MPI_Wait()` to wait for a message request to be completed. In addition, if a rank needs to wait for more than a single message request to complete, MPI also provides two routines for querying the status of sets of requests. The first is `MPI_Testall()`

```
int MPI_Testall(
        int         count,
        MPI_Request array_of_requests[],
        int*        flag,
        MPI_Status  array_of_statuses[]);
```

which uses `count` to specify the number of entries in the `MPI_Request` and `MPI_Status` arrays, and `flag`, which is either a pointer to `0` or `1` on return depending on whether or not *all* of the requests have been completed. The second is `MPI_Waitall()`

```
int MPI_Waitall(
        int         count,
        MPI_Request array_of_requests[],
        MPI_Status  array_of_statuses[]);
```

which uses `count` for the number of entries in the arrays, `array_of_requests[]` for the set of requests, and `array_of_statuses[]` to hold the status of the corresponding requests on return. `MPI_Waitall()` waits until all of the messages in the list of requests are completed.

17.2.2 Neighborhood collective communication

Neighborhood collectives were introduced in MPI-3 and provide a way to set up implicit sub-communicators that are based on an existing topological communicator.

For collective communications that have the form `MPI_*`, there are coinciding neighborhood collective variants called `MPI_Neighbor_*` that require a *topological* communicator (for example, one from `MPI_Cart_create` or `MPI_Graph_create`).

The conceptual model for neighborhood collectives is as follows. All ranks within a (topological) communicator must call the neighborhood collective, and this includes ranks that have no "neighbors". Despite the fact that all ranks in a communicator call the neighborhood collectives, only those which have neighbors communicate data between them.

An example of a neighborhood collective is the `MPI_Neighbor_allgather()` routine. This routine has the signature

```
int MPI_Neighbor_allgather(
        const void*  send_data,
        int          send_count,
        MPI_Datatype send_datatype,
        void*        recv_data,
        int          recv_count,
        MPI_Datatype recv_datatype,
        MPI_Comm     communicator)
```

We note that the arguments in this routine are the same as for `MPI_Allgather()`, as shown in Section 15.3.2. The only nuance with this neighborhood version is that the `MPI_Comm` in the last argument must be a topological communicator.

17.3 One-sided communication

The schematic difference between one-sided (Put/Get) and two-sided (Send/Recv) communication can be seen in Figure 17.2. Advances in networking hardware have led to the network layer becoming sophisticated enough to orchestrate remote direct memory access (RDMA).

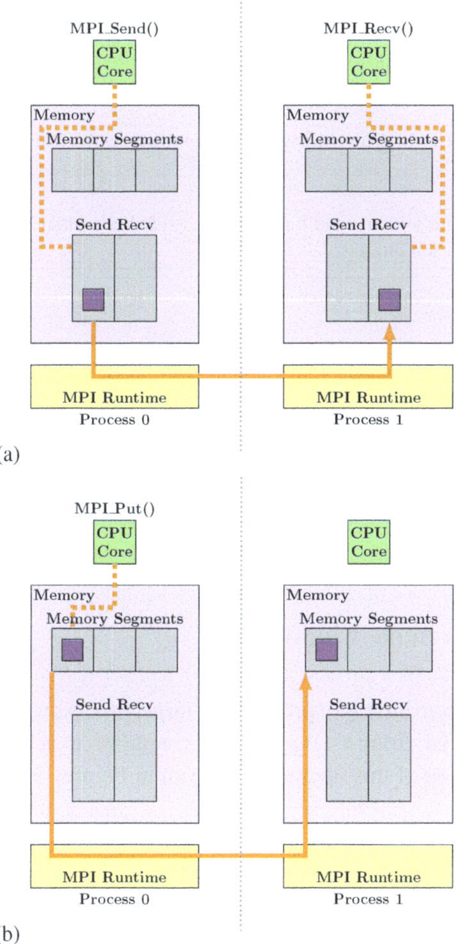

Fig. 17.2 The schematic difference between (a) two-sided and (b) one-sided communication. The CPU of the target process does not need to get involved with one-sided communication.

One-sided communication requires a setup where certain buffers are specified for remote processes to read from and write to. These special buffers are called *windows*.

17.3.1 Creating windows

There are four ways to create windows to hold remotely accessible memory:

1. `MPI_Win_allocate` - Create a window with automatic allocation of a fixed memory region.
2. `MPI_Win_create` - Create a window with explicit allocation of remotely accessible memory.
3. `MPI_Win_create_dynamic` - Create a window with the ability to dynamically add/remove memory to/from it.
4. `MPI_Win_allocate_shared` - Create multiple processes on the same node share a common memory region.

An example of allocating a window for one-sided communications is given in Listing 17.1

Listing 17.1 Allocating a window and buffer for one-sided communication. The buffer contains space for 1000 `int`s on each process.

```
#include <mpi.h>

int main(int argc, char ** argv) {
  int *a;
  MPI_Win win;
  MPI_Init(&argc, &argv);
  /* collectively create remote accessible memory in a window */
  MPI_Win_allocate(1000*sizeof(int), sizeof(int), MPI_INFO_NULL, MPI_COMM_WORLD,
                   &a, &win);
  /* Array a is now accessible from all processes in
   * MPI_COMM_WORLD */
  MPI_Win_free(&win);
  MPI_Finalize();
  return 0;
}
```

17.3.2 Accessing remote windows

Data can be copied from an origin process to a target process window using `MPI_Put`, or they can be copied from a target process window to an origin process using `MPI_Get`. A schematic of this operation is given in Figure 17.3.

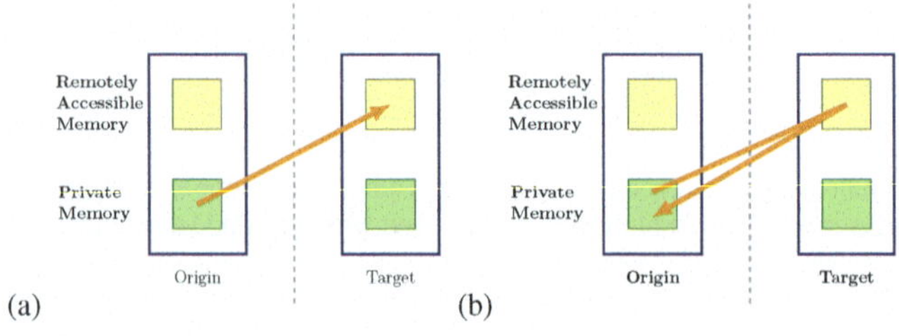

Fig. 17.3 Schematic representation of (a) `MPI_Put` and (b) `MPI_Get`.

The signatures of the routines are

```
int MPI_Put(
        const void*     origin_addr,
        int             origin_count,
        MPI_Datatype    origin_dtype,
        int             target_rank,
        MPI_Aint        target_disp,
        int             target_count,
        MPI_Datatype    target_dtype,
        MPI_Win         win);
```

and

```
int MPI_Get(
        void*           origin_addr,
        int             origin_count,
        MPI_Datatype    origin_dtype,
        int             target_rank,
        MPI_Aint        target_disp,
        int             target_count,
        MPI_Datatype    target_dtype,
        MPI_Win         win);
```

The question of when these operations are permitted is addressed in Section 17.3.4.

17.3.3 Accumulation of remote data

One-sided reductions can be performed into a remote target buffer using the `MPI_Accumulate` routine

```
int MPI_Accumulate(
        const void*  origin_addr,
        int          origin_count,
        MPI_Datatype origin_dtype,
        int          target_rank,
        MPI_Aint     target_disp,
        int          target_count,
        MPI_Datatype target_dtype,
        MPI_Op       op,
        MPI_Win      win);
```

A key difference with the two-sided `MPI_Reduce` operation is that no user-defined reduction operations can be defined for the one-sided `MPI_Accumulate`. Only predefined operations such as `MPI_SUM`, `MPI_PROD`, `MPI_OR`, `MPI_REPLACE`, `MPI_NO_OP`, etc. are permitted.

17.3.4 One-sided memory model

When can remote memory operations occur? MPI provides a few different models, of which we only mention the *fence* model. In the fence model for one-sided communication, all processes participating in a window must call the `MPI_Win_fence` routine to open the fence. With a window opened, `MPI_Put`, `MPI_Get`, and `MPI_Accumulate` operations can be performed on remote memory locations in a one-sided manner. A second call to `MPI_Win_fence` closes the fence. The fences establish an ordering of memory operations to ensure all remote operations from one call are complete before more can occur. This flow is visualized in Figure 17.4.

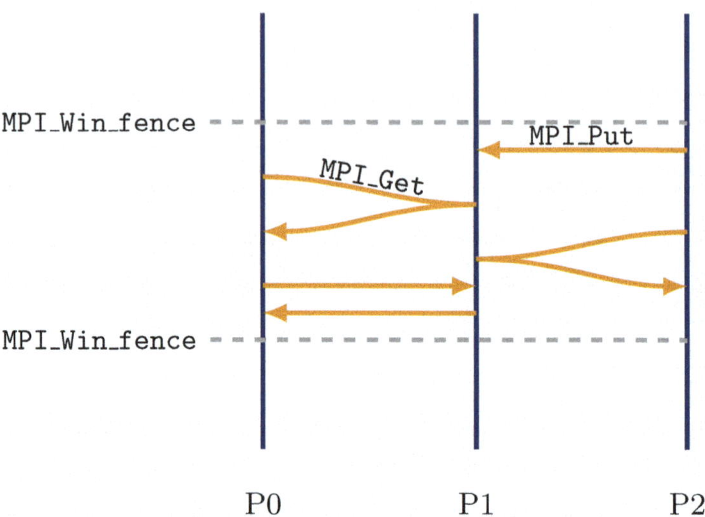

Fig. 17.4 Remote memory operations can only occur between the fences to ensure correct synchronization of the operations. Fences are both opened and closed using the `MPI_Win_fence` routine.

Part V
Shared-memory and accelerator programming

Consisting of nine chapters, this part describes the shared-memory paradigm of programming, that is, when many compute cores have access to the same memory space. We have already seen some exploitation of this sort of environment with the C++ parallel STL invocations in Chapter 9, but now we dive deeper into understanding details of parallel algorithms running in shared memory spaces.

This part has three chapters dedicated to OpenMP multi-core CPU programming, three chapters dedicated to CUDA/GPU programming, and three chapters dedicated to the Kokkos library that provides a higher-level abstraction that can handle multi-core CPU and GPU programming effectively.

Chapter 18
OpenMP I

The *Open Multi-Processing* library (OpenMP) is an open-source library that provides simple extensions to allow multi-threaded programming in C/C++ and Fortran. OpenMP is structured largely around the use of compiler directives to enable, disable, and modify parallel subsections of otherwise serial code. The main difference between OpenMP and MPI is that an OpenMP program runs as a single process on the operating system, and so all the *threads* within the program have access to a single memory space.

This chapter introduces OpenMP, first showing how to include the library and build executables that use it, then continuing with how to use the compiler directives to control and refine the parallel regions, and finishing up with variable scoping and reduction operations.

18.1 Compiling, linking, and running OpenMP programs

GCC, Clang, and Intel C++ compilers can all work with OpenMP. OpenMP functionality can be included in a C++ program by including the omp.h header file

```
#include "omp.h"
```

The compilation and linking of OpenMP programs simply involves specifying certain flags for the compiler. We show the compiler options for GCC and Clang, noting that Intel options can be found in the official OneAPI documentation.[1]

To tell the GCC or Clang compiler to include OpenMP functionality, the option -fopenmp is specified at the command line

```
$ g++ -c hello.cpp -fopenmp
```

[1] https://www.intel.com/content/www/us/en/develop/documentation/oneapi-dpcpp-cpp-compiler-dev-guide-and-reference/top/optimization-and-programming-guide/openmp-support/openmp-library-support/using-the-openmp-libraries.html

© The Author(s), under exclusive license to Springer Nature Switzerland AG 2025

R. J. Spiteri and K. Klenk, *Extreme-Scale Computing*, CMS/CAIMS Books in Mathematics 15, https://doi.org/10.1007/978-3-031-89033-8_18

and to link the resulting object file, the linker needs the `-lgomp` option

```
$ g++ -o hello hello.o -lgomp
```

For simple, single source file codes, GCC both compiles and links to OpenMP properly with a single statement like

```
$ g++ -o hello hello.cpp -fopenmp
```

In addition, if using CMake to generalize the build process, libraries and executables that use OpenMP can be properly linked by using `find_package()` and `target_link_libraries()` in the `CMakeLists.txt` as in

```
add_executable(my_exe)
find_package(OpenMP)
if(OpenMP_FOUND)
  target_link_libraries(my_exe OpenMP::OpenMP_CXX)
endif()
```

In this CMake snippet, `find_package(OpenMP)` sets the variable `OpenMP_FOUND` if a usable version of the OpenMP library is found. If a usable version of the library is available, then linking a target to `OpenMP::OpenMP_CXX` specifies the correct compile and link options for the compiler being used.

When running OpenMP programs from the command line in Linux, the maximum number of threads that are used in a program can be specified by the environment variable `OMP_NUM_THREADS`. For example, the number of threads could be set explicitly before running the program:

```
$ export OMP_NUM_THREADS=4
$ ./hello
```

or by specifying it in the run line with the executable:

```
$ OMP_NUM_THREADS=4 ./hello
```

In a cluster environment, the number of threads used is determined by options provided in your job submission script. For example, in Slurm, specifying the number of threads is handled by using the option

```
#SBATCH --cpus-per-task=4
```

inside the job submission script. The syntax can vary from cluster to cluster, so be sure to consult the documentation provided for your system.

18.2 Parallel sections

In the MPI chapters, the discussion began by introducing ranks, specifically how the number of ranks was determined and how each rank was identified. Here, we do the same, but instead of ranks, OpenMP uses threads. OpenMP programs are run as a single process, and within the program, compiler directives can be used to enable multiple threads in subregions of the code. The most basic form of a parallel region looks like

```
#pragma omp parallel
{
  // Statements to be executed with multiple threads
}
```

where any statements within the curly braces are executed by all threads available to the program.

The number of threads participating in a parallel region can either be taken as the default, which the program takes from the runtime environment (as discussed in Section 18.1), or it can be specified within the code itself. To specify the thread count of the thread pools that OpenMP uses via code, we use the OpenMP routine

```
void omp_set_num_threads(int num_threads);
```

where `num_threads` is the number of threads that is to be used within an `omp parallel` pragma.

To determine how many threads are *currently* running, OpenMP provides the routine

```
int omp_get_num_threads(void);
```

which simply returns the current thread count as an integer. When called outside a parallel region, the function returns the value of 1. To determine each thread's ID, OpenMP provides the routine

```
int omp_get_thread_num(void);
```

which returns an integer for the ID of each thread. The thread IDs returned by this function lie in the interval $[0, \texttt{NTHREADS} - 1]$, and so calling this function outside a parallel region results in a return value of 0.

A simple OpenMP program that shows how parallel execution in OpenMP is used is given in Listing 18.1. We briefly note here the use of the `omp critical` pragma, which tells OpenMP to only allow a single thread at a time to execute the subsequent statement. We discuss `omp critical` in more detail in Section 20.1.2, but its use here helps keep the output of the program readable (try compiling and running the example without that pragma for reference).

Listing 18.1 "Hello, World!": OpenMP edition. This example demonstrates how a pragma statement is used to enable a multi-threaded section. Observe how the thread IDs and number of threads propagate through function calls within the parallel section.

```
// "Hello, World!", OpenMP edition
#include <iostream>
// OpenMP header
#include "omp.h"

void print_thread() {
  auto thread      = omp_get_thread_num();
  auto num_threads = omp_get_num_threads();
  std::cout << "  thread " << thread
      << "/" << num_threads
      << std::endl;
}

int main(int argc, char* argv[]) {

  std::cout << "Hello, World!\n\n";
  std::cout << " Before parallel section:\n\n";
  print_thread();

  std::cout << "\n Inside parallel section:\n\n";
  #pragma omp parallel num_threads(8) // Start parallel region
  {
    #pragma omp critical // Ensure only one thread at a time
    print_thread();
  }
  // End of parallel region
}
```

Sample output from running Listing 18.1 as

```
$ ./hello
```

(i.e., with 8 threads) looks like

```
Hello, World!

 Before parallel section:

  thread 0/1

 Inside parallel section:

  thread 2/8
  thread 6/8
  thread 1/8
  thread 4/8
  thread 7/8
  thread 3/8
  thread 0/8
  thread 5/8
```

Similar to ranks in MPI programs, the order of execution among multiple threads in OpenMP is not deterministic.

18.3 Clauses

The general form of OpenMP pragmas looks like

```
#pragma omp <construct> [clauses]
```

where the mandatory pragma `<construct>` is applied within the subsequent code block and the optional `[clauses]` refines details within the code block. We have already seen the usage of the `parallel` construct in Section 18.2, and now we discuss some clauses that can be used to modify the parallel regions.

18.3.1 The num_threads clause

The `num_threads` clause can be used to specify the number of threads that are used in the parallel section. For example, if we replace the pragma of Listing 18.1 with

```
#pragma omp parallel num_threads(4)
```

then running the code results in output like

```
Hello, World!

 Before parallel section:

  thread 0/1

 Inside parallel section:

  thread 3/4
  thread 0/4
  thread 2/4
  thread 1/4
```

We note here that the `num_threads` clause inside a program can be used to override the number of threads passed from the `OMP_NUM_THREADS` environment variable. Accordingly, however, specifying `num_threads` explicitly inside a code may render it *not performance portable*. That is, a given parallel section that explicitly states its number of threads may either under- or over-subscribe CPU cores when run on different machines. Thus, we recommend that the `num_threads` clause only be used when absolutely necessary.

18.3.2 The shared and private clauses

When inside a parallel region of code, it is often the case that all the different threads may need to access the same variable, and that variable must be the same across all threads. For this, OpenMP provides the `shared` clause. Similarly, sometimes each thread inside a parallel region needs to work on its own copy of a variable that is defined outside the pragma code block. For that case, OpenMP provides the `private` clause to ensure that the variable gets copied into each thread so that it can be worked with by each thread independently. Example usage of the `shared` and `private` clauses can be seen in the example `counter` program in Listing 18.2.

Listing 18.2 Input argument counter. This example takes a list of integers as input, and each thread counts how many times its thread ID shows up in the list, demonstrating both the usage of both the `shared` and `private` clauses.

```
#include <iostream>
#include <vector>
#include "omp.h"

int main(int argc, char* argv[]) {

  int N{argc-1};
  std::vector<int> a(N);
  int count{0};

  for(int i = 0; i < N; ++i) {
    a[i] = atoi(argv[i]);
  }

  #pragma omp parallel num_threads(4) shared(N) private(count)
  {
    auto thread = omp_get_thread_num();
    for(int i = 0; i < N; ++i) {
      count += (a[i] == thread ? 1 : 0);
    }
    #pragma omp critical
    std::cout << "The number " << thread << " appears in the input list "
              << count << " times.\n";
  }
}
```

We note how the `count` variable is specified as `private(count)`. The reason for doing this is because `count` is defined before the pragma, and each thread is to use it independently within the OpenMP code block. The `N` variable is defined as `shared(N)` because it represents the size of the input list and should be known to, but not modified by, all the threads.

Example output of running

```
$ ./counter 5 10 6 1 2 3 4 1 1 2 3 4 4
```

should be

```
The number 0 appears in the input list 1 times.
The number 1 appears in the input list 3 times.
The number 2 appears in the input list 2 times.
The number 3 appears in the input list 2 times.
```

In addition to specifying a variable as `private`, the clause `firstprivate` can be used to explicitly initialize the private copy of a variable on entry to the OpenMP code block. For example, the correct usage in Listing 18.2 would be

```
#pragma omp parallel firstprivate(count)
```

The only reason that the code worked as previously written was that `count` was initialized to zero in the preceding code, and the default initialization value for `ints` is 0. If `count` had any other value going into the OpenMP code block, then the private copies would not have been initialized correctly.

We also take this opportunity to advise against the general use of default behaviors as a best programming practice. Default behaviors can change and thus introduce particularly insidious bugs into code. Moreover, such a practice makes it harder to reason about the code, especially to less experienced (or distracted!) programmers.

18.3.3 The `reduction` clause

Upon exiting an OpenMP pragma block, the variables local to each thread are generally discarded. This is undesirable if the threads are used to partially compute something that must persist beyond the OpenMP block, as is the case for reductions. The OpenMP clause for reductions has the form

```
reduction(<operation>:<variable>)
```

where `<operation>` is the reduction operation to be performed (say, +, -, *, /) and `<variable>` is the variable that holds the reduction. We can modify Listing 18.2 to have `count` used as a reduction variable and persist the total count of all found numbers in the list beyond the end of the OpenMP block. This is seen in Listing 18.3.

Listing 18.3 Input argument counter. This example takes a list of integers as input, and each thread counts how many times its thread ID shows up in the list and reports the results.

```
#include <iostream>
#include <vector>
#include "omp.h"

int main(int argc, char* argv[]) {

  int N{argc-1};
  std::vector<int> a(N);
  int count{0};

  for(int i = 0; i < N; ++i) {
    a[i] = atoi(argv[i]);
  }

  #pragma omp parallel num_threads(4) shared(N) reduction(+:count)
  {
    auto thread = omp_get_thread_num();
    for(int i = 0; i < N; ++i) {
      count += (a[i] == thread ? 1 : 0);
    }
    #pragma omp critical
    std::cout << "The number " << thread << " appears in the input list "
              << count << " times.\n";
  }
  std::cout << "\nTotal of all found numbers: " << count << std::endl;
}
```

The above example is run by providing a list of input integers like

```
$ ./counter_reduce 3 1 2 3 4 1 1 2 3 4 4
```

The output for the above example should be

```
The number 2 appears in the input list 2 times.
The number 3 appears in the input list 3 times.
The number 0 appears in the input list 1 times.
The number 1 appears in the input list 3 times.

Total of all found numbers: 9
```

Chapter 19
OpenMP II

This chapter is concerned with the subdivision of iterations in a `for` loop using OpenMP.

Algorithms implemented by looping over the contents of containers are a common occurrence in computational code. OpenMP provides specific directives that are targeted at subdividing C/C++ and Fortran `for` loops such that multiple threads can handle different iterations. This construct of course assumes that the different iterations of a `for` loop are independent, and care needs to be taken when this is not the case.

19.1 Parallel for loops

The directive that instructs OpenMP to subdivide a `for` loop is `#pragma omp for`, which is used within a `#pragma omp parallel` region as

```
#pragma omp parallel
{
  // some parallel code
  #pragma omp for
  for(... ; ... ; ... ;) {
    // Subdivided parallel loop body
  }
  // some more parallel code
}
```

This code snippet first creates a thread pool as discussed in Chapter 18. Any parallel manipulations can be performed within the `parallel` scope, and then in the single `for` loop inside the `parallel for` scope, the loop iterations are subdivided and split among the threads participating in the pool.

If only the for loop is needed within the parallel section of the code, then the `parallel` and `for` pragmas can be combined:[1]

```
#pragma omp parallel for
for(... ; ... ; ... ;) {
  // Subdivided parallel loop body
}
```

[1] Technically, `parallel for` is a separate construct from either `parallel` and `for`.

© The Author(s), under exclusive license to Springer Nature Switzerland AG 2025
R. J. Spiteri and K. Klenk, *Extreme-Scale Computing*, CMS/CAIMS Books in Mathematics 15, https://doi.org/10.1007/978-3-031-89033-8_19

This code snippet creates the parallel pool of threads and subdivides the for loop iterations among them. *How* the iterations get divided among the threads depends on how the `schedule` for the loop is set; more details are provided in Section 19.2.

Similar to the `parallel` sections, these clauses can be applied to `#pragma omp for`:

`private`:
: Make a thread-local private copy of the parent-scope variable.

`shared`:
: Set all variables to work at the parent scope.

`firstprivate`:
: Like `private`, but with explicit initialization to the parent-scope value.

`reduction`:
: All threads contribute to a variable that persists beyond the pragma.

There are also clauses that are specific to the `for` pragma. These include, but are not limited to,

`schedule`:
: Set a schedule for how indices are to be divided among the threads.

`lastprivate`:
: On exiting a `for` loop, a variable specified with `lastprivate(var)` holds the value it has in the final loop iteration.

`ordered`:
: Ensure strict ordering of loop iterations.

A simple example that uses the `parallel for` pragma is found in Listing 19.1.

Listing 19.1 Example code that computes a dot product of a vector with itself using OpenMP.

```
// for_omp.cpp
#include <vector>
#include <iostream>
#include <algorithm>

int main() {

  // init a vector
  const int N = 10000000;
  std::vector<int> a(N);
  std::fill(a.begin(), a.end(), 4);

  // Perform a dot product of a with itself
  int result{0};
  #pragma omp parallel for reduction(+:result)
  for(int i = 0; i < N; ++i) {
    result += a[i]*a[i];
  }

  std::cout << "result      = " << result << "\n"
            << "true result = " << a[0]*a[0]*N << std::endl;
}
```

The code in Listing 19.1 is computing a dot product of a vector with itself. We see that in addition to using the `parallel for` pragma, we have also used the reduction clause (from Section 18.3.3) to accumulate a sum across all threads into `result`.

19.2 The schedule clause

An important clause that is used with `#pragma omp for` is the `schedule` clause. This clause determines *how* the loop iterations get divided among threads.

`static`:
: Loop indices are divided with fixed block sizes and distributed among participating threads in a pre-determined way.

`dynamic`:
: Loop indices are divided with fixed block sizes and handled as first (thread) come, first (thread) served.

`guided`:
: Loop indices are divided with fixed block sizes and handled by a first (thread) come, first (thread) served. This behavior is similar to `dynamic`, but the block sizes decrease near the end of the loop.

`auto`:
: The schedule is determined by OpenMP automatically through assessment of system conditions such as number of available threads, thread load, and workload per iteration.

`runtime`:
: The schedule is taken from the environment variable `OMP_SCHEDULE` at runtime.

For example, the following snippet divides the `for` loop with a `dynamic` schedule using a block size of 4:

```
#pragma omp parallel for schedule(dynamic,4)
for(... ; ... ; ... ;) {
  // Subdivided parallel loop body
}
```

In this code, the iterations are grouped into blocks with continuous indices of size 4, i.e., $[0, 3], [4, 7]$, etc., and are then handled by the scheduler on a first-come, first-served basis.

OpenMP tries to split up the number of blocked loop iterations as evenly as possible among all the participating threads in the pool. This outcome is of course impossible when the number of blocks does not evenly divide the number of threads; i.e., `nThreads` % `(totalIterations/blockSize)` $\neq 0$. In this case, under the `static` and `dynamic` schedules, we often find that the last block of iterates is smaller in size. In an attempt to avoid such problems by satisfying the hunger of as many threads as possible, the `guided` schedule changes the block size for the final blocks of the iterates. Finally, the `runtime` schedule allows a schedule to be set at runtime of the executable. The `runtime` schedule is handled by the environment variable `OMP_SCHEDULE`. For example,

```
$ OMP_SCHEDULE="guided,8" ./executable
```

runs the executable with a `guided` schedule and a block size 8.

Because the `dynamic` schedule requires threads to make some sort of query to OpenMP to find their next piece of work, it has a performance overhead compared to the `static` schedule. The specific schedule that performs best for a given algorithm running on a given piece of hardware is difficult to predict a priori. The determination of `static` vs `dynamic` comes down to the following question: *Is each iteration of this loop performing approximately the same amount of work?* If *yes*, a `static` schedule that avoids the overhead of `dynamic` scheduling is likely to perform better. If *no*, then a `dynamic` or `guided` schedule can potentially (and even drastically) outperform a `static` schedule, depending on how unbalanced the work of each iteration turns out to be.

Choosing an appropriate block size is even more difficult to ascertain, and the choice for which a given loop will perform best generally needs to be determined by experimenting/profiling different choices. This situation arises because of the complexity of modern CPU architectures, as discussed in Section 10.2.2. As such, the value of the `guided` schedule becomes apparent because it starts with large chunks and automatically adapts the chunk size to handle any load imbalances. The `guided` schedule assigns a block size as follows:

```
let k = user supplied block size
let p = number of threads
let n = number of iterations

q = ceiling(n / p)
assign first thread q iterations

for all remaining threads:
  if (n - q) > (p * k):
    n = n - q
  else:
    n = p * k
  assign next thread n iterations
```

19.3 Single threads

Within a parallel region, there is sometimes the need to have a single thread execute certain instructions. This behavior is achieved via `#pragma omp single` as

```
#pragma omp parallel
{
  // some parallel code
  #pragma omp single
  {
    // Single thread executes this block
  }
  // some more parallel code
}
```

The first thread that reaches a `single` pragma executes it. This behavior is useful for operations such as output. Doing some parallel work followed by dumping some data to a file or a terminal is a common activity: the output can often be managed in such a way that it does not matter *which* thread (in OpenMP) is doing the output.

An example that demonstrates the use of `single` can be found in Listing 19.2. This code starts up an `omp single` parallel region and then performs a number of *global* (non-parallel) iterations. In each iteration, parallel work and singular output are performed. This approach allows threads not participating in the `single` operations to get started on their next iterations.

Listing 19.2 Using `single` to output pieces of data within an `omp parallel region`.

```
// for_omp.cpp
#include <vector>
#include <iostream>
#include <algorithm>

int main() {

  const int N_ITER = 20;

  // init a vector size
  const int N_DATA = 1000;
  std::vector<int> a(N_DATA);

  int result;
  #pragma omp parallel
  {
    for(int i = 0; i < N_ITER; ++i) {

      result = 0;
      // Populate the vector entries
      #pragma omp for
      for(int j = 0; j < N_DATA; ++j) {
        a[j] = (i+1)*(j+1);
      }

      #pragma omp single
      std::cout << "Iteration: " << i << ", a[last]: " << a[N_DATA-1];

      // Sum the vector entries
      #pragma omp for reduction(+:result)
      for(int j = 0; j < N_DATA; ++j) {
        result += a[j];
      }

      #pragma omp single
      std::cout << ", result: " << result << std::endl;

    } // N_ITER

  } // omp parallel

} // main
```

19.4 Thread synchronization

A fundamental construct in OpenMP is to create *worksharing regions*, in which a thread team is created to execute a task. Such regions synchronize threads via an *implicit barrier* at their end. Implicit barriers do not allow any thread proceed past them until all threads participating in the parallel for loop have finished. For example,

```
  #pragma omp parallel
  {
    #pragma omp for
    for(... ; ... ; ... ;) {
    }   // implied barrier to all threads in pool
    #pragma omp for   // starts after all threads finish above
    for(... ; ... ; ... ;) {
    }   // implied barrier
  }
```

Explicit barriers, which are also possible and sometimes necessary, are not required for this example because of the implied barriers that exist at the end of `#pragma omp for` loops.

Implied barriers can be disabled with the `nowait` clause,

```
#pragma omp parallel
{
  #pragma omp for nowait
  for(... ; ... ; ... ;) {
  }  // No implied barrier
  #pragma omp for   // Threads start as soon as they can
  for(... ; ... ; ... ;) {
  }
}
```

In such cases, however, care should be taken to avoid data dependencies across the loops. It is easy to accidentally insert a hidden *race condition* by accessing the same data in the second loop, for example. We discuss more on race conditions in Chapter 20.

If a forced synchronization of all threads is required, the `barrier` pragma can be used to synchronize threads within the pool like so

```
#pragma omp parallel
{
  // some parallel code
  #pragma omp barrier // No thread can pass until all threads are here
  // some more parallel code
}
```

Barriers should generally be avoided when possible, especially in and around the deep loops of expensive compute kernels, because they tend introduce significant performance degradation. A few cases where barriers can be useful are

- for organizing/ordering output among many threads,
- when dealing with a large number of `single` tasks, or
- to explicitly avoid data dependence in a multi-part algorithm.

Chapter 20
OpenMP III

This chapter elaborates on managing the execution of multiple threads. First, we cover how to lock sections of code to avoid *race conditions*, ensuring that only a single thread at a time modifies *critical* data. Then we look at how to manage different threads working on entirely different code paths within a single parallel pragma.

20.1 Race conditions

A *race condition* occurs when multiple threads attempt to update the same data simultaneously— the threads are said to be *racing* to write to the same memory location, which of course is only a problem when data are shared among threads. Race conditions typically occur when, for example,

1. Thread A reads values from the shared data.
2. Thread B writes values to the shared data.
3. Thread A performs computations with the (out of date) data that it has read.

This is problematic when the programmer assumes that threads are always working with the most up-to-date data available.

20.1.1 `atomic` clause

To alleviate race conditions, OpenMP provides two different pragmas that can be used inside parallel regions. The first is the `atomic` pragma, which applies to a single *simple* expression; i.e., it has the general form

```
#pragma omp atomic [atomic-clause]
  single-statement
```

An atomic operation guarantees that the statement is executed as a single operation without interruption. This guarantee is implemented at the hardware level, where the

© The Author(s), under exclusive license to Springer Nature Switzerland AG 2025

R. J. Spiteri and K. Klenk, *Extreme-Scale Computing*, CMS/CAIMS Books in Mathematics 15, https://doi.org/10.1007/978-3-031-89033-8_20

CPU ensures that the operation is completed before another thread can access the data. The `single-statements` allowed depend on the value of the `atomic-clause` specified. Assuming `x` is a shared variable, only the following cases are allowed:

- Assignment of `x` to another variable.
- Assignment of an expression to `x`.
- Update (unary or assignment to binary expression) value of `x`.
- Assignment of an updated version of `x` to another variable.

The `atomic-clause` can be left blank, but in general, it can be specified to add clarity to the code. Possible values for the `atomic-clause` include

`read`:
: A shared variable is being read (and assigned to another variable).

`write`:
: A shared variable is being assigned to.

`update`:
: A shared variable is being read, modified, and assigned to.

`capture`:
: A shared variable is being updated and assigned to another variable.

Further specifics for using the `atomic` clause can be found at `https://www.openmp.org/spec-html/5.0/openmpsu95.html`.

20.1.2 `critical` clause

The second construct for dealing with race conditions is to mark a section of parallel code with the `critical` clause, specifying a `critical` section. `critical` sections ensure that only a single thread operates within these specific sections at a time and have the form

```
#pragma omp critical [(name)]
{
  /* critical statements */
}
```

Compared to `atomic`, the main differences with `critical` are that `critical` sections can contain arbitrary code blocks within them and they can be named.

OpenMP allows for arbitrary blocks of code within `critical` sections by providing thread locks. There are a few different possibilities for how a thread lock can be implemented, e.g., *spinlocks* or mutually exclusive flags (*mutex*es), but OpenMP uses mutexes.

When an OpenMP thread enters a `critical` section, it takes hold of the mutex for that section. Other threads cannot enter the `critical` section until the executing thread releases the mutex. Accordingly, the threads that are waiting must periodically check the state of the mutex to see whether they can proceed. Thus, using `critical` is more computationally expensive than using `atomic` statements — they require the participating CPU cores to do more work for the synchronization.

Naming `critical` sections is important because the mutexes that are used for `critical` sections are on a per-name basis. When names are not used, OpenMP uses a global mutex; i.e., *all* `critical` sections in a code are subject to the state of a single mutex. This situation introduces unnecessary synchronization between potentially unrelated `critical` sections. When a name is specified, OpenMP uses a mutex associated with that name, reducing the degree to which `critical` sections are coupled.

An example use case for naming `critical` sections is to prevent mangled output. Surrounding *all* output code (inside parallel pragmas) like

```
#pragma omp critical (output)  // for non-garbled stdout
      std::cout << "Something to say." << std::endl;
```

ensures that the `stdout` stream never gets mangled from having multiple threads writing to it. In addition, this mutex *only* applies to the use of `critical` with the name `output`.

Further specifics about using the `critical` clause can be found at https://www.openmp.org/spec-html/5.0/openmpsu89.html.

To conclude our discussion on race conditions, we now analyze examples that range from obvious to subtle.

20.1.3 Race condition examples

Example code for a simple type of race condition is found in Listing 20.1.

Listing 20.1 Example code demonstrating a race condition.

```
// race_omp.cpp

#include <vector>
#include <iostream>
#include <numeric>
#include "omp.h"

int main() {

  // init a vector with a sequence from 1 to N
  const int N = 1000;
  std::vector<int> a(N);
  std::iota(a.begin(), a.end(), 1);

  // accumulate the result
  int result{0};
  #pragma omp parallel for
  for(int i = 0; i < N; ++i ) {
    result += a[i];
  }

  // Check the result
  std::cout << "result      = " << result << "\n"
            << "true result = " << N*(N+1)/2 << std::endl;

}
```

The above example tries to sum up the entries of a vector in an `omp parallel for` loop. This simple example sums the sequence {1, 2, . . . , `N` } so that the effect of a race condition can be easily seen by comparing to the known result `N(N+1)/2`. Compiling the example with

```
$ g++ -fopenmp -o race race_omp.cpp
```

and running it in serial with

```
$ OMP_NUM_THREADS=1 ./race
```

produces the correct result:

```
result      = 500500
true result = 500500
```

However, when we run the example with many threads, the final value that gets stored in `result` is different every time. For example,

```
$ OMP_NUM_THREADS=8 ./race
  result      = 69082
  true result = 500500
$ OMP_NUM_THREADS=8 ./race
  result      = 81162
  true result = 500500
```

The reason for this inconsistent output is because `result` is declared outside the OpenMP threading — it is a variable that is shared among all threads once the loop is entered. It is then possible for one thread to load the value of `result` before another thread has stored its new value. So the first thread is working with a *stale* value for `result`, and with many threads, this scenario gets even worse. Race conditions typically manifest as inconsistent results because the ordering in which threads access the data is not guaranteed. Running the same code with a deterministic output but getting different results may be a sign of the existence of a race condition.

For this example, a simple solution is to identify that it is in fact a *reduction* operation (recall Section 18.3.3) that we are performing on `result`. This means the proper pragma to use in Listing 20.1 would be

```
#pragma omp parallel for reduction(+:result)
```

Making this change to the example and trying again, we find the result to be consistently correct:

```
$ OMP_NUM_THREADS=8 ./race2
  result      = 500500
  true result = 500500
$ OMP_NUM_THREADS=8 ./race2
  result      = 500500
  true result = 500500
```

What OpenMP does behind the scenes in this case is create a thread-local version of `result` for each thread, so that each thread does not interfere with another thread's variable. Once the loop is complete, all of the thread local `results` are summed (+) to get the final `result`.

A more sinister issue arises when working with complex objects that are not thread-safe. An example is shown in Listing 20.2.

Listing 20.2 Unsafely adding new key-value pairs to a `std::map`. This code has a bug because `std::map::insert` is not a thread-safe operation.

```
// racemap_omp.cpp
#include <map>
#include <iostream>
#include "omp.h"

int main() {

  std::map<int,int> m;

  #pragma omp parallel shared(m)
  {
    auto tid = omp_get_thread_num();

    // Add a new key-value pair to the map
    m.insert({tid,tid});

    // output each on their own line
    #pragma omp critical (output)
      std::cout << tid << ":" << m.at(tid) << std::endl;
  }
}
```

Really, this situation can occur for any complex operation that is not thread-safe. Even though this example *looks* like the threads are modifying different elements, the underlying implementation of `std::map::insert` is not thread-safe. Often, complex data types (such as containers like `std::map`) have operations that are thread-safe and others that are not. In the case of the standard containers, a good way to start is to think that *accessing* data is thread-safe, whereas *modifying* data is usually not. A corrected modification of Listing 20.2 is shown in Listing 20.3.

Listing 20.3 Safely adding new key-value pairs to a `std::map`.

```
// noracemap_omp.cpp
#include <map>
#include <iostream>
#include "omp.h"

int main(){

  std::map<int,int> m;

  #pragma omp parallel shared(m)
  {
    auto tid = omp_get_thread_num();

    // Add a new key-value pair to the map
    #pragma omp critical (setmap_m) // One thread at a time
      m.insert({tid,tid});          //   calls insert()

    // thread-parallel access is ok
    auto myval = m.at(tid);

    // Output each on their own line
    #pragma omp critical (output)
      std::cout << tid << ":" << myval << std::endl;
  }
}
```

Running Listing 20.3 now results in the following output:

```
$ OMP_NUM_THREADS=8 ./noracemap
0:0
4:4
2:2
6:6
7:7
5:5
3:3
```

```
1:1
```

This output looks much more like what we would expect, whereas Listing 20.2 results in exceptions being thrown randomly when the `m.at(tid)` statement is attempted.

20.2 OpenMP Sections

The previous section discussed how to handle multiple threads attempting to execute the same instruction on the same data. Now we turn to the `sections` construct, allowing different threads to execute different instructions within a parallel region. The basic syntax for using the `sections` construct is as follows:

```
#pragma omp [clauses] parallel
{
  #pragma omp sections
  {
    #pragma omp section
    {
      // A single thread will execute statements here.
    }
    #pragma omp section
    {
      // Likewise, here.
    }
  }
}
```

The `sections` construct is most useful when there are independent tasks that can be performed by different threads. The `clauses` that can apply to OpenMP sections are all ones that we have seen already:

`private`:
List of variables to get thread-private copies.

`shared`:
List of variables to be shared among threads.

`reduction`:
Specify a reduction operation for an existing variable.

`nowait`:
Remove the implicit barrier at the end of the `sections` pragma.

We note here that OpenMP sections also have an implied barrier at the end unless the `nowait` clause is specified.

An example that uses sections with some other C++ functionality previously discussed is found in Listing 20.4.

Listing 20.4 Example of using OpenMP sections. Each section is executed by a single thread. Each thread can operate independently, simultaneously performing completely different tasks from the other threads. Here, one thread prints, one thread sums an array, and one thread gets the time.

```
// sections_omp.cpp
#include <vector>
#include <iostream>
#include <numeric>
#include <execution>
#include <chrono>
#include <ctime>
#include "omp.h"
```

```
int main(){

  #pragma omp parallel sections
  {

    // Just report who does the first section
    #pragma omp section
    {
      auto tid = omp_get_thread_num();
      #pragma omp critical (output)
      std::cout << "Thread " << tid << " ran the first section." << std::endl;
    }

    // Sum a vector with SIMD vectorized instructions in the second section
    #pragma omp section
    {
      auto tid = omp_get_thread_num();
      std::vector<int> a(1000);
      std::iota(a.begin(),a.end(),1);
      auto result = std::reduce(std::execution::par_unseq,
        a.begin(),
        a.end(),
        0,
        std::plus());

      #pragma omp critical (output)
      std::cout << "Thread " << tid << " computed: " << result
                << " (using vectorized instructions)" << std::endl;
    }

    // Get the time in the third section
    # pragma omp section
    {
      auto tid = omp_get_thread_num();
      auto time = std::time(0);

      #pragma omp critical (output)
      std::cout << "Thread " << tid << " got the time: " << std::ctime(&time);
    }
  }
}
```

Recalling Section 9.2, because we are using a parallel execution policy, we need to link Listing 20.4 against the `tbb` library to create an executable. Compiling and running thus looks like

```
$ g++ -fopenmp -o sections sections_omp.cpp -ltbb
$ ./sections
Thread 0 ran the first section.
Thread 9 got the time: aaa bbb dd HH:MM:SS YYYY
Thread 1 computed: 500500 (using vectorized instructions)
```

We note that (by default) this system initialized many more threads than were needed, as seen by thread number 9 being the thread that checked the time.

20.3 OpenMP Tasks

In addition to threading, OpenMP also provides a tasking model to perform *task-based programming*. *Tasks* are blocks of code that can be executed independently and in parallel. To use tasks, we use the construct

```
#pragma omp task
```

Tasks are defined within a *task region*, which consists of all the code to be executed by the task. There are both implicit and explicit tasks, and we have already seen examples of implicit tasks without realizing it. Tasks are created implicitly when a `parallel` construct is encountered. When a thread enters an implicit parallel region or encounters the `parallel` construct, it creates a task and assigns it to a new thread. When the thread that entered the parallel region encounters the implicit barrier at the end of the parallel region, it waits for all tasks to complete before continuing, guaranteeing that all tasks are completed before the thread continues. Explicit tasks, on the other hand, are created by the programmer using the `task` construct shown above. Explicit tasks are usually, but not always, generated by a single thread. Like implicit tasks, explicit tasks execute when the generating thread encounters a barrier.

The benefit of using tasks is that OpenMP can decide on how to schedule them. This is in contrast to using threads, where the programmer has to explicitly specify how to schedule the threads. OpenMP does provide some helpful options for scheduling threads, but tasks provide even more flexibility to the OpenMP runtime.

Using tasks comes down to identifying which portions of code can be executed independently. For example, in the following code snippet, we can identify that the two loops can be executed independently:

```
#pragma omp parallel
{
  #pragma omp single
  {
    #pragma omp task
    for (int i = 0; i < M; ++i) {
      // do something
    }
    #pragma omp task
    for (int i = 0; i < N; ++i) {
      // do something
    }
  }
}
```

As long as the two loops do not depend on each other, they can be correctly defined as tasks. The thread that encounters the `single` construct generates the two tasks, then when it encounters the implicit barrier, the tasks execute. The tasks that are generated can be executed in any order, but they are all guaranteed to be executed before the thread that generated them continues.

Chapter 21
GPU/CUDA programming I

In Section 11.1.2, we provided an introductory explanation of GPUs, highlighting their differences with CPUs along with their increased adoption for general-purpose programming. This chapter kicks off a more systematic treatment of general-purpose GPU (GPGPU) programming, starting with a quick refresher on GPUs, and then progressing to discuss a few first steps with CUDA, NVIDIA's highly popular GPU computing platform and programming model.

21.1 Introduction and GPU refresher

Although originally designed for displaying graphics, the GPU has evolved into a device that is highly suitable for data-parallel processing. This evolution began by realizing that displaying pixels on a monitor is a highly data-parallel operation, analogous to the numerical computations often seen in scientific computing. This good fortune has led to the development of graphics cards such as the NVIDIA P100, V100, and A100 that are solely intended for GPGPU programming. These cards (along with thier AMD and Intel counterparts) can be found in many supercomputing clusters and datacentres around the world and are widely regarded as having made a critical contribution to the achievement of exascale computing.

The computational ability of a GPU can substantially outpace that of a CPU (of a similar generation). However, as introduced in Section 11.1.1, there exists a significant bottleneck when transferring data between the CPU and GPU through PCIe. This bottleneck is important to consider when GPGPU programming because the transfer speed of data can render the use of a GPU ineffective if the computation is not large enough. Therefore, there are two main views one can take when GPGPU programming: "*compute as much as possible on the GPU*," trying to exploit as much computation from the GPU as possible for a given data transfer, or "*balance the computation between the CPU and GPU*," where the GPU receives only a subset of the problem to compute.

© The Author(s), under exclusive license to Springer Nature Switzerland AG 2025
R. J. Spiteri and K. Klenk, *Extreme-Scale Computing*, CMS/CAIMS Books in Mathematics 15, https://doi.org/10.1007/978-3-031-89033-8_21

The most appropriate view depends on a number of factors, including the algorithms being used, their data transfer requirements, and the relative computational power of the CPU and GPU. Programming with both the CPU and GPU is referred to as *hybrid programming* and is especially difficult because it requires careful use of both devices.

Revisiting Figure 11.1, we recall that the architecture of a GPU has many cores that share a control unit and their lowest-level cache, but all cores have access to pools of shared memory in L2 Cache and DRAM. In contrast, each core of a modern CPU tends to have its own control unit and lowest-level cache, with many cores having access to a higher-level cache, and finally all cores having access to the highest level of cache and DRAM.

These architectural differences provide insight into why CPUs are more versatile than GPUs. Having its own control unit gives each core more freedom to operate independently of the other cores. GPUs, on the other hand, have their cores more tightly coupled to the other cores within the same control unit, reducing their flexibility to operate independently but substantially increasing their ability execute the same instruction on many data elements in parallel.

21.2 CUDA basics

CUDA is a proprietary parallel computing platform *and* programming model developed by NVIDIA for GPGPU programming on NVIDIA GPUs. Although alternative platforms and programming models exist for programming the GPUs of other brands (e.g., OneAPI/DPC++ and ROCm) or GPUs in general (e.g., SYCL), we have chosen to work with CUDA for its simplicity and because of NVIDIA's prevalence in the GPU market. Furthermore, learning the abstractions for CUDA should translate to understanding the models of the other platforms.

In CUDA terminology, the CPU is referred to as the *host*, and CUDA-capable GPUs are referred to as *devices*. Devices are connected to the host, and each device has its own separate memory space (RAM and cache). When programming with CUDA, the host is responsible for controlling the execution of the program. To determine where code can be executed, *execution space specifiers* are declared before function names. Functions that use the `__global__` specifier are called from the host and executed on the device; these functions are referred to as *kernels*. When a kernel is executed, we are said to be *launching* the kernel. Functions that use the `__device__` specifier are both called from and executed on the device. To use these functions, they must be called from a kernel. Finally, functions that use the `__host__` specifier or do not have a specifier can only be called from and executed on the host.

We consider a simple example that demonstrates how to launch a kernel and differentiate between the host and device in a "Hello, World!" example.

Listing 21.1 A simple "Hello, World!" example demonstrating how to use the different execution space specifiers.

```
#include <iostream>
#include <string>

__device__ void gpu_hello(){
  printf("Hello, World!: executed on GPU and called from GPU\n");
}

__global__ void cuda_hello(){
  printf("Hello, World!: executed on GPU and called from CPU\n");
  gpu_hello();
}

__host__ void cpu_hello(){
  printf("Hello, World!: executed on CPU and called from CPU\n");
}

int main() {
  // Print Hello, World! from CPU
  cpu_hello();

  // Run kernel on the GPU
  cuda_hello<<<1,1>>>();


  // Check for kernel completion
  {
    cudaError_t cudaerr = cudaDeviceSynchronize();
    if (cudaerr != cudaSuccess) {
      std::cout << "Kernel failed. Error \""
                << cudaGetErrorString(cudaerr) << "\"" << std::endl;
    }
  }

  return 0;
}
```

To compile CUDA code, we need to use the NVIDIA CUDA compiler, `nvcc`. Additionally, CUDA code requires the `.cu` file extension. To compile Listing 21.1, the process is

```
$ nvcc -o cuda_hello cuda_hello.cu
$ ./cuda_hello
Hello, World!: executed on CPU and called from CPU
Hello, World!: executed on GPU and called from CPU
Hello, World!: executed on GPU and called from GPU
```

In this example, we define two functions, `cpu_hello` and `gpu_hello`, and a kernel, `cuda_hello`, that use the `__host__`, `__device__`, and `__global__` specifiers, respectively. From the main function, `cpu_hello` is called first, like a typical C++ function, and demonstrates how to call functions that use the `__host__` specifier. Next, the kernel `cuda_hello` is launched with a special syntax that tells the device how to execute the kernel. We intentionally omit the details of the syntax to be introduced more specifically in the next section. The `cuda_hello` program demonstrates how execution is transferred from the host to the device. Within the `cuda_hello` function, `gpu_hello` is called, showing how separate functions can be called when executing on the device. Finally, the `cudaDeviceSynchronize()` function is called to ensure that the kernel has finished executing before the program terminates.

21.3 CUDA programming model

With the basics of launching a kernel and differentiating between host and device functions introduced, we can begin to explore the CUDA programming model. The CUDA programming model provides a set of abstractions that ease the process of GPGPU programming. There are three main abstractions that the CUDA programming model uses: thread hierarchy, memory hierarchy, and thread synchronization.

The most important abstraction to understand is the thread hierarchy because threads are the fundamental unit of execution on a CUDA device. Similar to a CPU, a thread is what is mapped to a core for execution. The difference with GPUs is that there are many more cores available, and the programmer can execute thousands to millions of threads per kernel launch. Programming so many threads individually is not feasible, and therefore CUDA provides two levels of abstraction above threads to help organize them. These abstractions are *thread blocks* and *grids*.

Threads are organized into thread blocks, and thread blocks are organized into grids. Each kernel launch is executed as a grid, and the programmer specifies the number of thread blocks and threads per block in the kernel launch. The specification of the grid is sometimes called the *execution configuration* and is specified within the `<<<num_block, threads_per_block>>>` syntax after the kernel name but before the arguments to the kernel. How best to define the execution configuration depends on the application, but investigating more thoroughly the relationship between thread execution and the hardware can help us optimize this configuration. Nonetheless, selecting the optimal configuration can sometimes be a trial-and-error process.

NVIDIA GPUs are designed around arrays of Streaming Multiprocessors (SMs). SMs are simply a collection of CUDA cores, and the number of cores within an SM as well as the total number of SMs varies between GPUs. With respect to kernel execution, the grid is separated into thread blocks, and each thread block is mapped to an SM. Therefore, a single thread block executes on a single SM, and other thread blocks execute on remaining SMs. All thread blocks can be imagined to be running in parallel, but in reality they are running concurrently. So in the common case where there are more thread blocks than SMs, the GPU handles the scheduling for us. There is also no distinct maximum number of thread blocks that can compose a grid, but thread blocks and grids can be defined in up to three dimensions (x, y, and z), and there are limits to the number of thread blocks that can be defined in each dimension.

On the other hand, the number of threads per block cannot exceed 1024 and must be a multiple of 32. This maximum limit is imposed by CUDA due to the small amount of shared memory available to thread blocks. Conversely, the multiple-of-32 restriction is a result of how threads are executed on the GPU. Threads are executed in groups of 32 called *warps*, and there are multiple warp schedulers within an SM, with the exact number corresponding to the specific GPU in use. The important thing to note is that threads within a warp are executed at the same time, and to maximize performance, we want to ensure that all threads within a warp are executing the same instruction. That is, we want to avoid situations where some threads within a warp execute different code paths or branches (data dependence).

The memory hierarchy is much simpler than the thread hierarchy. We saw in Figure 11.1 that the GPU has L1 and L2 cache along with DRAM. Each thread has its own private memory, and each thread block has its own shared memory. The DRAM is shared between all threads and thread blocks and can also be accessed by the host. Moreover, there are two read-only memory locations that can be used in specific situations. These are called *constant memory* and *texture memory*. Constant memory is used for data that are read frequently by threads within a warp, and texture memory is optimized memory for graphics processing providing faster access to 2D data.

Finally, the CUDA programming model provides a synchronization mechanism to help coordinate the execution of thread blocks. Synchronization is typically done through the use of barriers, which stop the execution of a thread until all threads within a thread block have reached the barrier. This behavior is achieved via the `__syncthreads()` function.

21.4 CUDA programming environment

To finish this chapter, we explore the CUDA programming environment. We begin with an example of how to obtain information about the available CUDA devices through the *CUDA runtime API*. Then we see how to monitor the activity of the GPU through the `nvidia-smi` command.

21.4.1 CUDA runtime API

The code in Listing 21.2 demonstrates how to obtain information about the available CUDA devices. The first aspect to note is the inclusion of `cuda.h` at the top of the file. This statement is required for the CUDA runtime API, and some functionality may require the additional inclusion of `cuda_runtime.h`. Moreover, to successfully compile this example, we also need to link against the CUDA library. This is accomplished via the `-lcuda` flag in the compilation command:

```
$ nvcc cuda_test.cu -o cuda_test -lcuda
```

To access the CUDA runtime API, we first need to initialize CUDA in a similar way to how we initialize MPI. This is done with the `cudaInit()` function. Next, we can query the number of devices available with the `cudaGetDeviceCount()` function. Then we can iterate over the available devices and query their properties. A full list of properties and other useful CUDA API calls can be found in the developer documentation at `https://docs.nvidia.com/cuda/cuda-runtime-api/index.html`.

Listing 21.2 Obtaining information about available CUDA devices through the CUDA runtime API.cuda_test.cu

```
#include <stdio.h>
#include <iostream>
#include <cuda.h>

int main() {
  // Initialize - Must be called before any other CUDA API calls
  cuInit(0);

  // Get the number of CUDA-enabled GPU devices
  int num_devices = 0;
  cuDeviceGetCount(&num_devices);
  if (num_devices == 0) {
    std::cout << "No CUDA devices found.\n";
    return 0;
  }
  std::cout << "Found " << num_devices << " CUDA device(s).\n";

  // Loop through all of the devices found
  for (auto i = 0; i < num_devices; i++) {
    CUdevice device;
    cuDeviceGet(&device, i);

    char name[100];
    if (cuDeviceGetName(&name[0], sizeof(name), device) == CUDA_SUCCESS) {
      std::cout << "Name: " << name << "\n";
    }

    int cc_major, cc_minor;
    if (cuDeviceGetAttribute(&cc_major, CU_DEVICE_ATTRIBUTE_COMPUTE_CAPABILITY_MAJOR, device) == CUDA_SUCCESS
           &&
        cuDeviceGetAttribute(&cc_minor, CU_DEVICE_ATTRIBUTE_COMPUTE_CAPABILITY_MINOR, device) == CUDA_SUCCESS)
               {
      std::cout << "Compute Capability: " << cc_major << "." << cc_minor << "\n";
    }

    int multi_processors;
    if (cuDeviceGetAttribute(&multi_processors, CU_DEVICE_ATTRIBUTE_MULTIPROCESSOR_COUNT, device) ==
           CUDA_SUCCESS) {
      std::cout << "Multiprocessors: " << multi_processors << "\n";
    }

    int max_threads_per_sm;
    if (cuDeviceGetAttribute(&max_threads_per_sm, CU_DEVICE_ATTRIBUTE_MAX_THREADS_PER_MULTIPROCESSOR, device)
           == CUDA_SUCCESS) {
      std::cout << "Maximum Threads Per Multiprocessor: " << max_threads_per_sm << "\n";
    }

    int max_threads_per_block;
    if (cuDeviceGetAttribute(&max_threads_per_block, CU_DEVICE_ATTRIBUTE_MAX_THREADS_PER_BLOCK, device) ==
           CUDA_SUCCESS) {
      std::cout << "Maximum Threads Per Block: " << max_threads_per_block << "\n";
    }

    int clock_rate;
    if (cuDeviceGetAttribute(&clock_rate, CU_DEVICE_ATTRIBUTE_CLOCK_RATE, device) == CUDA_SUCCESS) {
      std::cout << "GPU Core Clock Rate: " << clock_rate << "KHz\n";
    }

    int memory_clock_rate;
    if (cuDeviceGetAttribute(&memory_clock_rate, CU_DEVICE_ATTRIBUTE_MEMORY_CLOCK_RATE, device) ==
           CUDA_SUCCESS) {
      std::cout << "Memory Clock Rate: " << memory_clock_rate << "KHz\n";
    }

    size_t total_memory;
    if (cuDeviceTotalMem(&total_memory, device) == CUDA_SUCCESS) {
      std::cout << "Total Memory: " << total_memory / (1024 * 1024) << "MiB\n";
    }

    std::cout << "\n\n";
  }

  return 0;
}
```

Some example output from running Listing 21.2 is

```
Found 2 CUDA device(s).
Name: NVIDIA RTX A6000
Compute Capability: 8.6
```

```
Multiprocessors: 84
Maximum Threads Per Multiprocessor: 1536
Maximum Threads Per Block: 1024
GPU Core Clock Rate: 1800000KHz
Memory Clock Rate: 8001000KHz
Total Memory: 48643MiB
```

A notable element in this discussion is the *compute capability* of the device. The compute capability describes the hardware features of the device, which in turn determine the programming abstractions that are available. The compute capability is specified by a version number of the form `major.minor`. The major component of the version number specifies the device's architecture, and the minor component specifies incremental upgrades based on this architecture. From the output of Listing 21.2, we see that the device used is an NVIDIA RTX A6000 and has a compute capability of 8.6, which means it is based on the Ampere architecture (major version 8) with incremental upgrades (minor version 6). Specific details about the compute capabilities of NVIDIA GPUs can be found at `https://docs.nvidia.com/cuda/cuda-c-programming-guide/index.html#compute-capabilities`.

In essence, the CUDA runtime API is a set of functions, all beginning with cuda, that help the programmer manage the execution of the GPU. In later chapters, we see how to leverage the CUDA runtime API to manage GPU memory, transfer data between host and device, and more. In addition to the CUDA runtime API is the CUDA driver API, which offers even more control over GPU execution. In general, however, it is recommended to use the runtime API because it is simpler to use.

21.4.2 Monitoring device activity

Before delving into the finer details of CUDA programming, it is useful to know how to monitor the activity of the GPU. It is helpful to understand whether the system is properly communicating with the GPU as well as how much of the GPU's resources a program is using. The command to monitor GPU activity is `nvidia-smi`, and its output looks like

```
+-----------------------------------------------------------------------------+
| NVIDIA-SMI 520.61.05    Driver Version: 520.61.05    CUDA Version: 11.8     |
|-------------------------------+----------------------+----------------------+
| GPU  Name        Persistence-M| Bus-Id        Disp.A | Volatile Uncorr. ECC |
| Fan  Temp  Perf  Pwr:Usage/Cap|         Memory-Usage | GPU-Util  Compute M. |
|                               |                      |               MIG M. |
|===============================+======================+======================|
|   0  NVIDIA TITAN V      On   | 00000000:21:00.0 Off |                  N/A |
| 28%   38C    P8    25W / 250W |     60MiB / 12288MiB |      0%      Default |
|                               |                      |                  N/A |
+-------------------------------+----------------------+----------------------+

+-----------------------------------------------------------------------------+
| Processes:                                                                  |
|  GPU   GI   CI        PID   Type   Process name                  GPU Memory |
|        ID   ID                                                   Usage      |
|=============================================================================|
```

```
|    0   N/A  N/A      4978      G   /usr/lib/xorg/Xorg                          46MiB |
|    0   N/A  N/A      5127      G   /usr/bin/gnome-shell                        11MiB |
+-----------------------------------------------------------------------------+
```

Kernels that are executing on a device show up under "Processes." The command on its own simply prints the current status and returns. However, the command-line program can also be run in an infinite loop (much like top) by using something like `nvidia-smi --loop=SEC`, where SEC is the number of seconds between updates. This program executes until it is terminated, e.g., by pressing `ctrl+c`.

Chapter 22
CUDA/GPU programming II

This chapter focuses on utilizing the concepts introduced in Chapter 21 to begin writing more practical CUDA programs. In particular, we start to apply concepts surrounding the thread and memory hierarchy. We discuss thread organization, memory management, and error handling.

22.1 Thread management

As introduced in Section 21.3, kernels are always launched as a grid of thread blocks. However, we have not yet discussed how to organize the threads within a block and the blocks within a grid. The key point to consider when GPGPU programming is that all threads within a kernel launch should be viewed as running concurrently in the sense that there are more threads than available cores. The values used for the execution configuration of a kernel launch can be determined statically with hard-coded values or dynamically within the execution of the program.

Figuring out the optimal execution configuration for a kernel is not always straightforward and likely involves some trial and error. However, there are some restrictions that need to be taken into account when determining the execution configuration. The number of threads per block is restricted to a maximum of 1024. Because threads are executed in warps of 32 threads, the number of threads per block should also be defined using multiples of 32. On the other hand, the number of blocks per grid should be large enough to exceed the number of streaming multiprocessors on the GPU. This ensures that the GPU is fully utilized because blocks are scheduled to execute as SMs become available.

© The Author(s), under exclusive license to Springer Nature Switzerland AG 2025
R. J. Spiteri and K. Klenk, *Extreme-Scale Computing*, CMS/CAIMS Books in Mathematics 15, https://doi.org/10.1007/978-3-031-89033-8_22

22.1.1 Indexing and dimensions

Within a grid, each thread is assigned a unique index for the block to which it belongs and for its position within that block. To access these indices, CUDA provides two variables: blockIdx and threadIdx. Both variables have three components (x, y, and z), allowing grids to be defined in up to three dimensions. Furthermore, the dimensionality of the grid and the blocks within the grid can be accessed by each thread using the gridDim and blockDim variables respectively. The gridDim and blockDim variables also have x, y, and z components. To define grids and blocks in dimensions greater than one, the dim3 type must be used. The dim3 type is a vector of up to three dimensions, and any dimension that is unspecified is automatically set to 1, so even in the one 1D case, the dim3 type can be used over a typical integer in the execution configuration of a kernel launch. Listing 22.1 demonstrates how to use the dim3 type and access the indices of all the executed threads.

Listing 22.1 A CUDA program that uses the dim3 type to launch a kernel that prints the index of each thread within the grid and its block to the screen.

```
#include <cuda.h>
#include <cuda_runtime.h>
#include <iostream>

// Simple example of using dim3 to launch a 3D grid of 3D blocks
__global__ void dim3Test() {
    printf("Hello, World! My blockIds are: x = %d, y = %d, z = %d\n"
           "\tMy threadIds are: x = %d, y = %d, z = %d\n",
           threadIdx.x, threadIdx.y, threadIdx.z,
           blockIdx.x, blockIdx.y, blockIdx.z);
}

int main() {

  // Set up a 3D grid of 3D blocks
  dim3 blocksPerGrid(2, 2, 2);
  dim3 threadsPerBlock(2, 2, 2);

  dim3Test<<<blocksPerGrid, threadsPerBlock>>>();
  // Wait for kernel to finish before moving on
  cudaDeviceSynchronize();
  return 0;
}
```

In Listing 22.1, we show how to use the dim3 type to define a 3D grid with thread blocks that are also defined in 3D. Each dimension in blocksPerGrid and threadsPerBlock is set to a size of 2. This gives us a total of 8 blocks within the grid and 8 threads within each block for a total of 64 threads. The kernel dim3Test simply prints the blocks and thread indices to the screen.

22.1.2 Thread synchronization

So far, we have used the cudaDeviceSynchronize() function to synchronize the host with the device. For finer-grained synchronization, the __syncthreads() function can be used to synchronize all threads within a thread block. All threads within a block must reach the synchronization point before any thread can continue. It is important to use caution when synchronizing threads because it can potentially lead to *deadlock*. Deadlock occurs when the code is stuck waiting indefinitely for threads to reach the synchronization point. We discuss in more detail how to avoid deadlock in Chapter 23 but introduce the concept of thread synchronization here because it appears in some examples in the remainder of this chapter.

22.2 Memory management

Memory management is a key component to effective GPGPU programming. Up to this point, it has been difficult to demonstrate any practical examples because we had not yet discussed how to allocate memory on the device and transfer data to it. We address this issue in this section.

22.2.1 CUDA memory basics

As a refresher, CUDA devices have their own memory space that is separate from the host. As a result, space for data must be created in the device's memory. The data must be copied from the host to the device (and the eventually back to the host). One of the goals of GPGPU programming is to minimize the number of times data are copied between the host and device. It is generally advisable to perform larger, less frequent copies than smaller, more frequent ones. Additionally, the cost of data transfers can be hidden by overlapping them with other operations; this topic is covered in Chapter 23.

Similar to host memory, device memory can be allocated both statically and dynamically. Static memory is allocated by the compiler and is available for the duration of a program. To define memory statically, the `__device__` keyword is used before the variable declaration to inform the compiler to allocate space on the GPU. A brief example is given in Listing 22.2.

Listing 22.2 Basic example showing how to use static memory on the GPU.

```
#include"cuda.h"

// Define static array on the GPU
__device__ int single_vector[256];

__global__ void some_kernel() {
  int x = threadIdx.x;
  single_vector[x] = x;
}

int main() {
  // Launch kernel
  some_kernel<<<1,256>>>();
  return 0;
}
```

Although the above example is valid, it is often more practical to allocate memory dynamically. Within the CUDA API are functions that can be used to allocate, copy, and free device memory. When a kernel is launched, it executes outside of the device's memory. There are two ways in which memory can be allocated: as linear memory or as CUDA arrays. Linear memory is the most straightforward to use and behaves much like allocating memory in C, where users explicitly allocate and free memory through specific function calls. CUDA arrays have the potential to offer greater performance and are optimized for what NVIDIA calls *texture fetching*. Texture fetching is used to interact with specific hardware on the device that is fine-tuned for texture objects with 2D and 3D spatial locality. We cover only the use of linear memory, but we do note that CUDA arrays can provide performance benefits in some scientific applications. We feel, however, that they are outside the scope of an introductory text.

The three basic functions for allocating, copying, and deallocating memory are `cudaMalloc`, `cudaMemcpy`, and `cudaFree`, respectively. A simple example showing their usage is shown in Listing 22.3. We note that this example and the ones that follow do not include error checking for brevity. We discuss error handling specifically in Section 22.3.

Listing 22.3 A simple CUDA program with a kernel that adds two integers on a CUDA device demonstrating how to allocate memory on a GPU and copy the result to the host.

```
#include <cuda.h>
#include <cuda_runtime.h>

#include <iostream>

__global__ void add( int a, int b, int* c) {
  *c = a+b;
}

int main(int argc, char** argv) {
  int a{ 1 }, b{ 2 };
  int c { 0 };
  int* dev_c;

  // Allocate space on the device
  cudaMalloc((void**)&dev_c, sizeof(int));

  // Run the kernel
  add<<<1,1>>>(a, b, dev_c);

  // Check for kernel completion
  {
    cudaError_t cudaerr = cudaDeviceSynchronize();
    if (cudaerr != cudaSuccess) {
      std::cout << "Kernel failed. Error \""
                << cudaGetErrorString(cudaerr) << "\"\n";
    }
  }

  // Copy output vector from GPU buffer to host memory.
  cudaMemcpy(&c, dev_c, sizeof(int), cudaMemcpyDeviceToHost);
  std::cout << a << " + " << b << " = " << c << "\n";

  // Clean up device memory
  cudaFree(dev_c);

  return 0;
}
```

Listing 22.3 provides a straightforward example that defines a kernel `add` that adds two integers together. The main function begins by allocating memory on the device with `cudaMalloc`, which takes two arguments: a pointer to the allocated device memory and the number of bytes to allocate. Then, the kernel is launched with a single thread, and completion is verified by calling `cudaDeviceSynchronize` and checking the return value. It is important to note that all kernel launches are non-blocking for the host, so we can use `cudaDeviceSynchronize` to wait for the kernel to complete. Some CUDA functions are synchronous, `cudaMemcpy` being one example, and block the host until the operation is complete. For Listing 22.3, the call to cudaDeviceSynchronize is not necessary because the host will wait at the cudaMemcpy call until the kernel has completed.

The cudaMemcpy function takes four arguments: a pointer to the destination, a pointer to the source, size in bytes of the memory to copy, and the type of transfer. There are many options for the transfer type argument. For this example, we use cudaMemcpyDeviceToHost to copy the result from the device to the host. We could have also used cudaMemcpyDefualt, which would have the CUDA runtime infer the type of transfer based on the pointers provided. Use of the cudaMemcpyDefault argument is preferred when copying memory. Finally, we print the value of c and free the device memory with cudaFree.

When passing data to the GPU, only scalar values and pointers to memory can be used. This is why in Listing 22.3 we did not need to copy the values of a and b to the device. Additionally, classes and structs can be passed to the GPU, but they must have device versions of every member function. When passing data to the GPU, everything is passed by value. In the next example, we show a more realistic scenario that begins to leverage the GPU in parallel to add two arrays together. This example requires that space for each array be allocated on the device and that their values be copied to the device before the kernel is launched.

In Listing 22.4, we demonstrate how to allocate space for three arrays, copy the values of the two input arrays to the device, launch the kernel, copy the result back to the host, and free the device memory. In this example, the kernel is launched using two blocks, each with the number of threads equal to half the size of the output array. This means that there is effectively one thread per array element, and because we are using two blocks, the threads are split across two streaming multiprocessors. Within the kernel addKernel, we use the blockIdx, blockDim, and threadIdx variables to assign each thread a unique index into the array.

Listing 22.4 A CUDA program that adds two arrays together on a device.

```
#include <cuda.h>
#include <cuda_runtime.h>
#include <iostream>

// This kernel assumes it has been launched with a 1D set of blocks
__global__ void addKernel(int* c, const int* a, const int* b, int size) {
  int i = blockIdx.x * blockDim.x + threadIdx.x;
  if (i < size) {
    c[i] = a[i] + b[i];
  }
}

int main(int argc, char** argv) {
  const int arraySize = 5;
  const int a[arraySize] = {  1,  2,  3,  4,  5 };
  const int b[arraySize] = { 10, 20, 30, 40, 50 };
  int c[arraySize] = { 0 };
  int* dev_a, * dev_b, * dev_c;

  // Allocate GPU buffers for three vectors (two input, one output)
  cudaMalloc((void**)&dev_a, arraySize * sizeof(int));
  cudaMalloc((void**)&dev_b, arraySize * sizeof(int));
  cudaMalloc((void**)&dev_c, arraySize * sizeof(int));

  // Copy input vectors from host memory to GPU buffers.
  cudaMemcpy(dev_a, a, arraySize * sizeof(int), cudaMemcpyDefault);
  cudaMemcpy(dev_b, b, arraySize * sizeof(int), cudaMemcpyDefault);

  // Launch kernel with 2 thread blocks with arraySize / 2 threads each
  // giving a total of arraySize threads (1 thread per array element)
  addKernel<<<2, (arraySize + 1) / 2>>>(dev_c, dev_a, dev_b, arraySize);

  cudaError_t cudaerr = cudaDeviceSynchronize();
  if (cudaerr != cudaSuccess) {
    std::cout << "Kernel failed. Error \""
              << cudaGetErrorString(cudaerr) << "\"\n";
  }

  // Copy output vector from GPU buffer to host memory.
  cudaMemcpy(c, dev_c, arraySize * sizeof(int), cudaMemcpyDefault);

  std::cout << "{1, 2, 3, 4, 5} + {10, 20, 30, 40, 50} = {";
  for (int i = 0; i < arraySize; ++i) {
    std::cout << c[i] << ", ";
  }
  std::cout << "}\n";

  cudaFree(dev_a);
  cudaFree(dev_b);
  cudaFree(dev_c);

  return 0;
}
```

22.2.2 Shared memory

In addition to global memory, NVIDIA GPUs also provide a small amount of shared memory that all threads in a block can access. This memory is much faster than global memory, but it is also much smaller. This shared memory is implemented directly onto the SM, and the precise amount is determined by the compute capability of the GPU. Requesting more shared memory than is available results in a failed kernel launch. Shared memory is local to a thread block, meaning that it is not shared between other thread blocks. Declaring this type of shared memory is done by using the `__shared__` keyword before the variable declaration as in Listing 22.5.

Listing 22.5 Example of declaring shared memory

```
__shared__ float static_shared_array[256];
extern __shared__ float dynamic_shared_array[];
```

In Listing 22.5, the first line shows how to declare statically allocated shared memory. However, this is often not a good idea because it restricts the number of threads that can execute within a thread block. The second line shows how to declare shared memory that is allocated at a kernel launch, which is more flexible and allows for the amount of memory to correlate with the number of threads in a block at the time of a kernel launch. A complete example for the use of shared per block memory is shown in Listing 22.6.

In Listing 22.6, we show an example of using shared memory within a kernel to add two matrices together. We start to leverage the thread hierarchy and `dim3` type defined in Section 22.1 to define our execution configuration in 2 dimensions. In this example, we have a total of 4 blocks with 4 threads each for a total of 16 threads. In the `add` kernel, we use the `blockIdx` and `threadIdx` variables to assign each thread a unique index into the matrix. We then sum the matrix and call the `__syncthreads()` function to ensure that all threads within the block have completed their work before we store the result in the output matrix. Finally, after the kernel returns, we copy the result back to the host, print the result, and free the device memory.

Listing 22.6 Example that uses shared memory within blocks. Indexing within the kernel now relies on block indexing, as well as on both local (to the current block) and global indices for correctly handling the local and global arrays.

```
#include <cuda.h>
#include <cuda_runtime.h>
#include <iostream>

const int M = 4;
const int N = 4;
const dim3 blocks(2,2);
const dim3 threads(2,2);
const int a[M][N] = { 1, 2, 3, 4, 5, 6, 7, 8, 9, 10, 11, 12, 13, 14, 15, 16};
const int b[M][N] = { 10, 20, 30, 40, 50, 60, 70, 80, 90, 100, 110, 120, 130, 140, 150, 160};


// Kernel to add two matrices using shared memory
__global__ void add(int *c, const int *a, const int *b, const int N, const int M) {
  // Define shared memory
  extern __shared__ int localc[];

  const int threadsPerBlock = blockDim.x * blockDim.y;

  // Calculate indeces
  int local_ind = threadIdx.x + threadIdx.y * blockDim.x;
  int block_offset = blockIdx.x + blockIdx.y * gridDim.x;
  int global_ind = local_ind + block_offset * threadsPerBlock;

  int global_x = blockIdx.x + threadIdx.x;
  int global_y = blockIdx.y + threadIdx.y;

  // Sum local elements
  if (global_x < M && global_y < N) {
    localc[local_ind] = a[global_ind] + b[global_ind];
  }

  // Synchronize threads in block before moving data to output array
  __syncthreads();
  c[global_ind] = localc[local_ind];
}

int main() {
  int c[M][N] = { 0 };
  int* dev_a, *dev_b, *dev_c;

  // Allocate device memory
  cudaMalloc(&dev_a, M * N * sizeof(int));
  cudaMalloc(&dev_b, M * N * sizeof(int));
  cudaMalloc(&dev_c, M * N * sizeof(int));

  // Copy data from host to device
  cudaMemcpy(dev_a, a, M * N * sizeof(int), cudaMemcpyDefault);
  cudaMemcpy(dev_b, b, M * N * sizeof(int), cudaMemcpyDefault);

  // Launch kernel
  add<<<blocks, threads>>>(dev_c, dev_a, dev_b, M, N);
  cudaError_t cudaerr = cudaDeviceSynchronize();
  if (cudaerr != cudaSuccess) {
    std::cout << "Kernel failed. Error \""
                          << cudaGetErrorString(cudaerr) << "\"\n";
  }

  // Copy results from device to host
  cudaMemcpy(c, dev_c, M * N * sizeof(int), cudaMemcpyDefault);
  // Print result of matrix addition
  for(int i = 0; i < M; ++i) {
    for(int j = 0; j < N; ++j) {
      std::cout << a[i][j] << " ";
    }
    std::cout << "\n";
  }
  std::cout << "\n+\n\n";
  for(int i = 0; i < M; ++i) {
    for(int j = 0; j < N; ++j) {
      std::cout << b[i][j] << " ";
    }
    std::cout << "\n";
  }
  std::cout << "\n=\n\n";
  for(int i = 0; i < M; ++i) {
    for(int j = 0; j < N; ++j) {
      std::cout << c[i][j] << " ";
    }
    std::cout << "\n";
  }

  // Free device memory
  cudaFree(dev_a);
  cudaFree(dev_b);
  cudaFree(dev_c);

  return 0;
}
```

Constant memory is another type of shared memory that can be leveraged within a kernel. Constant memory is read-only memory that exists in a special location on the GPU's main memory. This memory is optimized for read-only access and can provide a performance boost over traditional global memory, but like `__shared__` memory, constant memory is also relatively small. Constant memory exists for the entire lifetime of the kernel and is declared using the `__constant__` keyword during the declaration of a variable. Constant memory can be declared statically or dynamically at runtime through the use of the `cudaMemcpyToSymbol` function. In Listing 22.7, we show an example of a slightly modifed version of Listing 22.6 that uses constant memory.

Listing 22.7 Example that uses constant memory on the GPU. Constant memory is accessed in read-only mode on kernel execution. Space is allocated on a gpu by using a static global variable with the `__device__` keyword, and then populated at runtime (before kernel invocation) by using the `cudaMemcpyToSymbol` routine.

```
#include <cuda.h>
#include <cuda_runtime.h>
#include <iostream>

const int M = 4;
const int N = 4;
const dim3 blocks(2,2);
const dim3 threads(2,2);
const int a[M][N] = { 1, 2, 3, 4, 5, 6, 7, 8, 9, 10, 11, 12, 13, 14, 15, 16};
const int b[M][N] = { 10, 20, 30, 40, 50, 60, 70, 80, 90, 100, 110, 120, 130, 140, 150, 160};

// Use constant dimension sizes on the GPU
__constant__ int dev_M;
__constant__ int dev_N;

// Kernel to add two matrices using shared memory
__global__ void add(int *c, const int *a, const int *b) {
  // Define shared memory
  extern __shared__ int localc[];

  const int threadsPerBlock = blockDim.x * blockDim.y;

  // Calculate indeces
  int local_ind = threadIdx.x + threadIdx.y * blockDim.x;
  int block_offset = blockIdx.x + blockIdx.y * gridDim.x;
  int global_ind = local_ind + block_offset * threadsPerBlock;

  int global_x = blockIdx.x + threadIdx.x;
  int global_y = blockIdx.y + threadIdx.y;

  // Sum local elements
  if (global_x < dev_M && global_y < dev_N) {
    localc[local_ind] = a[global_ind] + b[global_ind];
  }

  // Synchronize threads in block before moving data to output array
  __syncthreads();
  c[global_ind] = localc[local_ind];
}

int main() {
  int c[M][N] = { 0 };
  int* dev_a, *dev_b, *dev_c;

  // Allocate device memory
  cudaMalloc(&dev_a, M * N * sizeof(int));
  cudaMalloc(&dev_b, M * N * sizeof(int));
  cudaMalloc(&dev_c, M * N * sizeof(int));

  // Copy data from host to device
  cudaMemcpy(dev_a, a, M * N * sizeof(int), cudaMemcpyDefault);
  cudaMemcpy(dev_b, b, M * N * sizeof(int), cudaMemcpyDefault);

  // Populate the constant (shader) memory on the GPU
  cudaMemcpyToSymbol(dev_M, &M, sizeof(int));
  cudaMemcpyToSymbol(dev_N, &N, sizeof(int));

  // Launch kernel
  add<<<blocks, threads>>>(dev_c, dev_a, dev_b);
  cudaError_t cudaerr = cudaDeviceSynchronize();
  if (cudaerr != cudaSuccess) {
    std::cout << "Kernel failed. Error \""
```

```
                          << cudaGetErrorString(cudaerr) << "\"\n";
}

// Copy results from device to host
cudaMemcpy(c, dev_c, M * N * sizeof(int), cudaMemcpyDefault);
// Print result of matrix addition
for (int i = 0; i < M; ++i) {
  for (int j = 0; j < N; ++j) {
    std::cout << a[i][j] << " ";
  }
  std::cout << "\n";
}
std::cout << "\n+\n\n";
for (int i = 0; i < M; ++i) {
  for (int j = 0; j < N; ++j) {
    std::cout << b[i][j] << " ";
  }
  std::cout << "\n";
}
std::cout << "\n=\n\n";
for(int i = 0; i < M; ++i) {
  for(int j = 0; j < N; ++j) {
    std::cout << c[i][j] << " ";
  }
  std::cout << "\n";
}

// Free device memory
cudaFree(dev_a);
cudaFree(dev_b);
cudaFree(dev_c);

return 0;
}
```

In Listing 22.7, constant memory is used to set the sizes of the matrices within the kernel. The constant memory is first declared and then populated with `cudaMemcpyToSymbol` before the kernel is launched.

22.3 Error Handling

Admittedly, we have thus far been guilty of performing little to no error checking. This is most definitely not good practice, but it does help reduce the amount of code in the examples for illustrative purposes. All CUDA runtime functions return a `cudaError_t` typed value that can be used to check for errors. In some of the previous examples, we have been using a variable of this type to check whether the kernel completed successfully by comparing the returned value to `cudaSuccess`. There are many types of errors that can be returned by the CUDA runtime system. Each error is associated with an integer that can be used to identify the type of error. The CUDA runtime system provides an `enum cudaError` that maps named errors to the integer values. Success is always mapped to 0. All errors are values above 0, and some examples include

```
cudaErrorInvalidValue = 1
cudaErrorMemoryAllocation = 2
cudaErrorInitializationError = 3
cudaErrorLaunchFailure = 4
...
```

The CUDA kernels themselves do not return an error code, and furthermore they cannot have a return type other than `void`. To check whether the kernel has completed, we use the `cudaDeviceSynchronize` function, which we have already seen in some of our examples. This function blocks the host until the kernel has completed. This blocking is required because the kernel is launched asynchronously, and the host continues to execute code after it has launched the kernel. Alternatively, we can use the `cudaGetLastError()` or `cudaPeekAtLastError()` functions to check whether the kernel encountered any errors pre-launch. These functions also require us to synchronize with the device before checking for errors.

The documentation for all the errors and their associated integer values can be found in the CUDA runtime API documentation at: `https://docs.nvidia.com/cuda/cuda-runtime-api/`.

Chapter 23
CUDA/GPU Programming III

The previous chapters introduced the foundational concepts for getting started with CUDA programming. In this chapter, we continue to build on these concepts, focusing more on performance-oriented topics and leveraging the more advanced features of CUDA. These topics all build toward harnessing the power of multiple GPUs.

23.1 Branching

Just like in CPU programming, branching with `if-else` statements can be used in CUDA kernels with the same C/C++ syntax. We have used branching in some of the previous examples, but it is important to understand the implications of how branches are handled by the GPU because branching can significantly reduce performance and introduce deadlock. In terms of performance, we recall that warps are the unit of execution for groups of at most 32 threads. Our goal is to maximize the number of threads within the warp that are executing simultaneously. When a warp encounters a branch, the threads within the warp may *diverge* on different execution paths within the kernel. When this happens, the GPU is not equipped to handle both execution paths simultaneously and instead executes each path separately. For example, if we have 32 threads encounter an `if-else` statement and 16 threads take the `if` path while the rest take the `else` path, the GPU executes each set of threads separately, meaning that 16 threads become *inactive* when the `if` path is taken, and the other 16 become inactive when the `else` path is taken. If this behavior occurs consistently throughout the kernel, performance will be reduced considerably through the ineffective execution of warps.

The other issue that can arise from branching is the possible introduction of deadlock. CUDA threads can exist in one of three states: *active*, *inactive*, and *exited*. Active threads are threads that are currently executing, inactive threads are threads that are waiting for execution, and exited threads are threads that have finished executing. CUDA handles the management of exited threads automatically, so we typically do not need to worry about them when it comes to synchronization points

© The Author(s), under exclusive license to Springer Nature Switzerland AG 2025

R. J. Spiteri and K. Klenk, *Extreme-Scale Computing*, CMS/CAIMS Books in Mathematics 15, https://doi.org/10.1007/978-3-031-89033-8_23

within the kernel. Inactive threads, on the other hand, can lead to a synchronization point that is never reached, resulting in deadlock. Therefore, it is important to design kernels that avoid branching whenever possible.

23.2 Warp Shuffling

Warp shuffling is a form of communication between threads, sometimes referred to as *lanes* in this context, to exchange data while circumventing the limited amount of shared memory available to a thread block. Warp shuffling allows threads to share their local variables with other threads within the same warp. The benefit is that the exchange of data can happen within a single instruction instead of two, which would otherwise occur when using shared memory. CUDA offers four different shuffle operations shown below:

Listing 23.1 Warp Shuffle Operation

```
T __shfl_sync(unsigned mask, T var, int srcLane, int width=warpSize);
T __shfl_up_sync(unsigned mask, T var, unsigned int delta, int width=warpSize);
T __shfl_down_sync(unsigned mask, T var, unsigned int delta, int width=warpSize);
T __shfl_xor_sync(unsigned mask, T var, int delta, int width=warpSize);
```

Each operation offers a specific way to exchange data between threads. The `__shfl_sync` operation allows a thread to copy data from another thread. The `__shfl_up_sync` and `__shfl_down_sync` operations allow threads to exchange data with threads that are a specified "distance" `delta` away, with `__shfl_up_sync` copying data from `delta`s higher than the current thread, and `__shfl_down_sync` copying data from `delta`s lower than the current thread. Finally, the most intricate of the shuffle operations is the `__shfl_xor_sync` operation, which allows threads to exchange data based on a bitwise XOR of its own thread ID as the `delta` argument.

All shuffle operations use a *mask* to specify which threads are involved in the shuffle operation. The mask is a 32-bit integer where each bit represents which lane participates in the shuffle operation. For example, if we want all threads to participate in the shuffle operation, we can use the mask `0xFFFFFFFF`. If we want only the first 16 threads to participate, we can use the mask `0x0000FFFF`. The second argument is the variable that is being shuffled. The third argument specifies the source lane for the `__shfl_sync` operation, and the delta for the others. Lastly, the optional `width` argument lets us separate the warp into multiple partitions or *subwarps* and perform the shuffle operation only within the subwarp and never across subwarps. In Listing 23.2, we show an example of how to use the above shuffle operations.

Listing 23.2 CUDA program that uses the four shuffle operations introduced in this section.

```
#include <cuda.h>
#include <cuda_runtime.h>
#include <iostream>
#include <vector>

__global__ void shuffleExample(int* result) {
  int laneId = threadIdx.x;
  int value = laneId;

  // Broadcast lane 0's value to all other lanes
  int broadcast = __shfl_sync(0xFFFFFFFF, value, 0);
  result[laneId] = broadcast;

  // Shift all lane values up by one lane
  int shifted_up = __shfl_up_sync(0xFFFFFFFF, value, 1);
  result[laneId + 32] = shifted_up;

  // Use __shfl_down_sync to shift value down by one lane
  int shifted_down = __shfl_down_sync(0xFFFFFFFF, value, 1);
  result[laneId + 64] = shifted_down;

  // Use __shfl_xor_sync to do a butterfly pattern
  int butterfly = __shfl_xor_sync(0xFFFFFFFF, value, 1);
  result[laneId + 96] = butterfly;
}

int main() {
  std::vector<int> host_vec(128);
  int* d_result;

  cudaMalloc(&d_result,  host_vec.size() * sizeof(host_vec[0]));

  shuffleExample<<<1, 32>>>(d_result);

  cudaMemcpy(host_vec.data(), d_result, 128 * sizeof(int), cudaMemcpyDefault);

  for (int i = 0; i < 128; ++i) {
    if (i % 32 == 0)
      std::cout << "\n";
    std::cout << "result[" << i%32 << "] = " << host_vec[i] << "\n";
  }

  cudaFree(d_result);

  return 0;
}
```

The output of the program in Listing 23.2 is shown below:

```
// __shfl_sync
result[0] = 0
result[1] = 0
result[2] = 0
result[3] = 0
...

// __shfl_up_sync
result[0] = 0
result[1] = 0
result[2] = 1
result[3] = 2
...

// __shfl_down_sync
result[0] = 1
result[1] = 2
result[2] = 3
```

```
...
result[30] = 31
result[31] = 31
...

// __shfl_xor_sync
result[0] = 1
result[1] = 0
result[2] = 3
result[3] = 2
result[4] = 5
result[5] = 4
result[6] = 7
result[7] = 6
result[8] = 9
result[9] = 8
result[10] = 11
...
result[30] = 31
result[31] = 30
```

For simplicity in this example, we launch a kernel with 1 block of 32 threads using a single warp to demonstrate the essential behavior of each shuffle operation. For this example, each lane's value is set to its `threadIdx.x` value. The first shuffle operation shown is the `__shfl_sync` operation, where we broadcast the value of lane 0 to all other lanes. In the printed output, we can see that the first set of 32 results are all 0. The next shuffle operation used is the `__shfl_up_sync` operation, where we shift the value of each lane up by 1. The output shows that each lane's value is now one less than its `threadIdx.x` value. We also note that lanes 0 and 1 both have a value of 0; this is because lane 0 has no lane below it to send it a value, and lane 31 does not send its value to lane 0. This behavior may seem counterintuitive. Similarly, the `__shfl_down_sync` operation shifts the value of each lane down by 1. The output shows that each lane's value is now one greater than its `threadIdx.x` except for lane 31, which keeps its value as its `threadIdx.x` value, demonstrating the inverse behavior of the `__shfl_up_sync` operation regarding the first and last lanes. Lastly, the `__shfl_xor_sync` is used to exchange the values of adjacent lanes. Each lane performs a bitwise XOR of its own `threadIdx.x` value with the `delta` argument, resulting it what is referred to as a *butterfly exchange pattern*. The output shows that the values of lanes 0 and 1 are swapped, as are the values of lanes 2 and 3, and so on.

23.3 CUDA Streams

A common goal of GPGPU programming is to keep the GPU as busy as possible performing useful computation; this is referred to as *maximizing occupancy*. We have already stressed the importance being aware of the cost of data transfers between the host and the device. Data transfer is not considered to be useful computation. However, instead of simply minimizing the number of data transfers to maximize occupancy, we can also try to hide this cost by overlapping data transfer with other operations. The way to achieve this is through the use of *CUDA streams*. CUDA streams allow us to queue up operations to be executed in a first-in, first-out order. A major advantage for the use of CUDA streams is that it conveniently unlocks the device's ability to overlap data transfers with computation. Furthermore, multiple streams can be used together to unlock greater parallelism for workloads that contain many tasks that do not necessarily need to take up the entirety of the device's resources. The limit to the number of streams that can be used at once is determined by the device's compute capability and, perhaps more importantly, the type of problem being solved.

The program structure for using CUDA streams is similar to what has been shown in previous chapters. The main difference is in the API calls that need to be used and is best shown through an example. Listing 23.3 shows an example of how to use CUDA streams.

Listing 23.3 CUDA program that uses a stream to asynchronously copy data and add the values of two matrices.

```
#include <cuda.h>
#include <cuda_runtime.h>
#include <iostream>

const int M = 4;
const int N = 4;
const dim3 blocks(2,2);
const dim3 threads(2,2);
const int a_static[M][N] = { 1, 2, 3, 4, 5, 6, 7, 8, 9, 10, 11, 12, 13, 14, 15, 16};
const int b_static[M][N] = { 10, 20, 30, 40, 50, 60, 70, 80, 90, 100, 110, 120, 130, 140, 150, 160};
__device__ static int dev_M;
__device__ static int dev_N;

__global__ void addKernel(int* c, const int* a, const int* b) {
  extern __shared__ int localc[];
  const int threadsPerBlock = blockDim.x * blockDim.y;

  // Calculate indeces
  int local_ind = threadIdx.x + threadIdx.y * blockDim.x;
  int block_offset = blockIdx.x + blockIdx.y * gridDim.x;
  int global_ind = local_ind + block_offset * threadsPerBlock;
  int global_x = blockIdx.x + threadIdx.x;
  int global_y = blockIdx.y + threadIdx.y;

  // sum into local memory
  if (global_x < dev_M && global_y < dev_N)
    localc[local_ind] = a[global_ind] + b[global_ind];

  __syncthreads();
  c[global_ind] = localc[local_ind];
}

int main() {
  int a[M][N];
  int b[M][N];
  int c[M][N];
  int* dev_a, *dev_b, *dev_c;

  // Populate the constant (shader) memory on the GPU
  cudaMemcpyToSymbol(dev_M, &M, sizeof(int));
  cudaMemcpyToSymbol(dev_N, &N, sizeof(int));

  // Allocate memory on the CPU
```

```
    cudaHostAlloc((void**)&a, M * N * sizeof(int), cudaHostAllocDefault);
    cudaHostAlloc((void**)&b, M * N * sizeof(int), cudaHostAllocDefault);
    cudaHostAlloc((void**)&c, M * N * sizeof(int), cudaHostAllocDefault);
    // Populate the host arrays
    for (int i = 0; i < M; i++) {
      for (int j = 0; j < N; j++) {
        a[i][j] = a_static[i][j];
        b[i][j] = b_static[i][j];
      }
    }

    // Allocate memory on the GPU
    cudaMalloc((void**)&dev_a, M * N * sizeof(int));
    cudaMalloc((void**)&dev_b, M * N * sizeof(int));
    cudaMalloc((void**)&dev_c, M * N * sizeof(int));

    // Create a stream
    cudaStream_t stream;
    cudaStreamCreate(&stream);

    // Asynchronously copy data to the GPU
    cudaMemcpyAsync(dev_a, a, M * N * sizeof(int), cudaMemcpyDefault, stream);
    cudaMemcpyAsync(dev_b, b, M * N * sizeof(int), cudaMemcpyDefault, stream);

    // Launch the kernel in the same stream
    addKernel<<<blocks, threads, 0, stream>>>(dev_c, dev_a, dev_b);

    // Asynchronously copy data back to the host
    cudaMemcpyAsync(c, dev_c, M * N * sizeof(int), cudaMemcpyDefault, stream);

    // Wait for all operations in the stream to complete
    cudaStreamSynchronize(stream);

    // Print result of matrix addition
    for(int i = 0; i < M; ++i) {
      for(int j = 0; j < N; ++j) {
        std::cout << a[i][j] << " ";
      }
      std::cout << "\n";
    }
    std::cout << "\n+\n\n";
    for(int i = 0; i < M; ++i) {
      for(int j = 0; j < N; ++j) {
        std::cout << b[i][j] << " ";
      }
      std::cout << "\n";
    }
    std::cout << "\n=\n\n";
    for(int i = 0; i < M; ++i) {
      for(int j = 0; j < N; ++j) {
        std::cout << c[i][j] << " ";
      }
      std::cout << "\n";
    }

    // Clean up
    cudaStreamDestroy(stream);
    cudaFreeHost(a);
    cudaFreeHost(b);
    cudaFreeHost(c);
    cudaFree(dev_a);
    cudaFree(dev_b);
    cudaFree(dev_c);

    return 0;
}
```

Listing 23.3 is a modified version of the matrix addition examples shown in Listing 22.6 and Listing 22.7. The first major difference in Listing 23.3 is the use of `cudaHostAlloc()`, which is used instead of `malloc()` to allocate the host memory. The `cudaHostAlloc()` function allows us to leverage asynchronous memory transfers. After allocating the host memory, we copy the data to the host structures using a statically defined array for simplicity. Once the host data are ready, we allocate memory on the device, then create a CUDA stream using `cudaStream_t stream` to first declare the stream and `cudaStreamCreate(&stream)` to create the stream. To perform the asynchronous memory transfer, we call the `cudaMemcpyAsync()` function for both `dev_a` and `dev_b`, passing in the stream as the last argument. This function queues up the memory transfers to be executed in the order they were called. After the memory transfers are queued up, we call the kernel with two extra arguments in the execution configuration. The first is an option we do not need but have to include in order to specify the second argument, so we pass in the default value of 0 for the shared-memory size. The second is the stream, which is passed in as the final argument. Lastly, we asynchronously copy the result back to the host and synchronize with the stream using `cudaStreamSynchronize(stream)`. This operation waits for all queued operations in the stream to finish before continuing.

23.4 Multiple Devices

As computational problems continue to grow in size, the need to leverage multiple devices becomes increasingly necessary and attractive. There are a few different ways to leverage multiple devices, but the most straightforward way is to use the `cudaSetDevice()` function. If there are multiple devices available on a system, their device IDs can be used to select the desired device by passing it as an argument to the `cudaSetDevice()` function. The device IDs generally start at 0 and increase sequentially for each additional device. However, if the devices required need to meet certain requirements, such as having a certain amount of memory, then the selecting CUDA-capable devices based on their properties can be done using the `cudaDeviceProp` variable. This is shown in Listing 23.4.

Listing 23.4 Choosing a CUDA-capable device based on the amount of memory available.

```
#include <cuda.h>
#include <cuda_runtime.h>
#include <iostream>

// Define a constant for GB
const size_t GB = 1L << 30;
const size_t desired_memory = 40 * GB;
int main(int argc, char*argv[]) {
  cudaDeviceProp prop;
  int dev;

  memset(&prop, 0, sizeof(cudaDeviceProp));

  // Look for a GPU with at least 40GB of global memory
  prop.totalGlobalMem = desired_memory;

  // And choose it
  cudaChooseDevice(&dev, &prop);
  cudaSetDevice(dev);

  // Get the properties of the selected device
  cudaGetDeviceProperties(&prop, dev);

  // Check if the selected device has enough global memory
  if (prop.totalGlobalMem < desired_memory) {
    std::cerr << "No device with enough global memory was found.\n";
    return 1;
  }

  std::cout << "Device " << dev
            << " selected for CUDA computations.\n";

  return 0;
}
```

The example shown in Listing 23.4 demonstrates how to select a device based on the amount of memory available. The `cudaDeviceProp` variable is used to store the desired properties and passed as an argument to `cudaChooseDevice()`. There is a caveat to this approach, however, because the `cudaChooseDevice()` does not guarantee that the device selected meets the requirements. We therefore include an extra check in the code to ensure that it does.

Alternatively, when running an application in an environment with multiple devices, the environment variable CUDA_VISIBLE_DEVICES can be set to limit which devices are available to an application. For example,

```
$ CUDA_VISIBLE_DEVICES=0,2,3 ./cuda_code
```

only makes devices with IDs 0, 2, and 3 visible to the application.

Moving on to using multiple devices within an application, the typical approach is to divide a problem into sub-problems that can be assigned to each device. The benefit to this approach is that existing code can be used with minimal modification, especially in the case where the code is executing within an environment consisting of a single host and only a few devices. Below we show an example of how to divide a problem amongst multiple devices using the cudaSetDevice() function.

The example shown in Listing 23.5 is a simple array addition example that uses multiple devices. The arrays to be added are generated and populated with random values. Then, we get the number of available CUDA devices and allocate an array of pointers to store for each device. Next, we loop through the devices using cudaSetDevice() to select each device and allocate memory to it. We then copy the portion of the array that each device is responsible for before calling the kernel. We do not specify cudaDeviceSynchronize() or cudaMemcpy() after calling the kernel to prevent the host from blocking before launching a kernel to all devices. Instead, we define another for loop outside the kernel loop to copy the results back to the host and free each device's memory. Finally, we verify the results of the device-computed array to within a tolerance of 0.00001.

23.4.1 Unifying Memory

A single address space can be set up for a host and multiple devices. NVIDIA refers to this address space as a *Unified Virtual Address Space* (UVA). To enable UVA, applications must first be compiled as a 64-bit application, which is usually the default for most systems running a 64-bit operating system. Then when allocating memory, the cudaMalloc function must be used over the conventional malloc. The idea behind UVA is to enable applications to use zero-copy memory, which is memory that is pinned in RAM and will not be swapped out to disk or moved somewhere else in RAM. This memory can be allocated using the cudaHostAlloc() function with the cudaHostAllocMapped flag. The advantage of this approach is that the GPUs can now directly access the pinned memory from their kernels without having to go through the host to make a copy of the data explicitly.

For even more convenience, CUDA provides *Unified Memory* (UM) that allows all the hosts and devices within a system to interact with the same memory. Similar to UVA, UM views the entire system as a single memory space, and a single pointer declaration can be used by any host or device, significantly simplifying pointer management. To utilize UM, the `cudaMallocManaged()` function must be used to create a pointer. The magic of UM is that CUDA automatically manages the transfer of data between the host and the devices using the fastest possible route. For example, if the GPUs are connected with NVLink, then the data are automatically transferred directly between the GPUs using NVLink over PCIe. Listing 23.6 shows a modified version of Listing 23.5 that uses UM.

Listing 23.5 Example that uses multiple devices by splitting the problem into sub-problems and assigning each sub-problem to a device.

```
#include <iostream>
#include <cuda.h>
#include <cuda_runtime.h>
#include <cmath>
#include <vector>

__global__ void vectorAdd(const float *A, const float *B, float *C, int numElements) {
  int i = blockDim.x * blockIdx.x + threadIdx.x;
  if (i < numElements) {
    C[i] = A[i] + B[i];
  }
}

int main(int argc, char*argv[]) {
  int numElements = 1e6;
  std::vector<float> h_A(numElements);
  std::vector<float> h_B(numElements);
  std::vector<float> h_C(numElements);

  // Initialize vectors on host
  for (int i = 0; i < numElements; ++i) {
    h_A[i] = rand()/(float)RAND_MAX;
    h_B[i] = rand()/(float)RAND_MAX;
  }

  int deviceCount;
  cudaGetDeviceCount(&deviceCount);
  std::vector<float *> d_A(deviceCount);
  std::vector<float *> d_B(deviceCount);
  std::vector<float *> d_C(deviceCount);

  // Split the problem into smaller problems and solve each on a different GPU
  int elementsPerGPU = numElements / deviceCount;

  for (int dev = 0; dev < deviceCount; ++dev) {
    cudaSetDevice(dev);
    cudaMalloc((void **)&d_A[dev], elementsPerGPU * sizeof(float));
    cudaMalloc((void **)&d_B[dev], elementsPerGPU * sizeof(float));
    cudaMalloc((void **)&d_C[dev], elementsPerGPU * sizeof(float));

    cudaMemcpy(d_A[dev], h_A.data() + dev * elementsPerGPU, elementsPerGPU * sizeof(float), cudaMemcpyDefault)
        ;
    cudaMemcpy(d_B[dev], h_B.data() + dev * elementsPerGPU, elementsPerGPU * sizeof(float), cudaMemcpyDefault)
        ;

    int numBlocks = (elementsPerGPU / 256) + 1;

    vectorAdd<<<numBlocks, 256>>>(d_A[dev], d_B[dev], d_C[dev], elementsPerGPU);
  }

  // Obtain the results from each GPU
  for (int dev = 0; dev < deviceCount; ++dev) {
    cudaSetDevice(dev);
    cudaMemcpy(h_C.data() + dev * elementsPerGPU, d_C[dev], elementsPerGPU * sizeof(float), cudaMemcpyDefault)
        ;
    cudaFree(d_A[dev]);
    cudaFree(d_B[dev]);
    cudaFree(d_C[dev]);
  }

  // Verify that the result vector is correct
  for (int dev = 0; dev < numElements; ++dev) {
    if (std::abs(h_A[dev] + h_B[dev] - h_C[dev]) > 1e-5) {
      std::cout << h_A[dev] << " + " << h_B[dev] << " != " << h_C[dev] << "\n";
      std::cerr << "Result verification failed at element " << dev << "!\n";
      exit(EXIT_FAILURE);
    }
  }
  return 0;
}
```

Listing 23.6 CUDA program that uses Unified Memory

```
#include <iostream>
#include <cuda.h>
#include <cuda_runtime.h>

__global__ void vectorAdd(const float *A, const float *B, float *C, int numElements) {
  int i = blockDim.x * blockIdx.x + threadIdx.x;
  if (i < numElements) {
    C[i] = A[i] + B[i];
  }
}

int main(int argc, char*argv[]) {
  int numElements = 1e6;
  size_t size = numElements * sizeof(float);
  float *h_A, *h_B, *h_C;

  // Allocate Unified Memory - accessible from CPU or GPU
  cudaMallocManaged(&h_A, size);
  cudaMallocManaged(&h_B, size);
  cudaMallocManaged(&h_C, size);

  // Initialize vectors on host
  for (int i = 0; i < numElements; ++i) {
    h_A[i] = rand()/(float)RAND_MAX;
    h_B[i] = rand()/(float)RAND_MAX;
  }

  int deviceCount;
  cudaGetDeviceCount(&deviceCount);
  // Split the problem into smaller problems and solve each on a different GPU
  int elementsPerGPU = numElements / deviceCount;

  for (int dev = 0; dev < deviceCount; ++dev) {
    cudaSetDevice(dev);
    int offset = dev * elementsPerGPU;
    int numElementsPerGPU = elementsPerGPU;
    if (dev == deviceCount - 1) {
      numElementsPerGPU = numElements - offset;
    }
    vectorAdd<<<(numElementsPerGPU / 256) + 1, 256>>>(h_A + offset, h_B + offset, h_C + offset,
           numElementsPerGPU);
  }

  // Wait for the kernel to finish
  cudaDeviceSynchronize();

  // Verify that the result vector is correct
  for (int dev = 0; dev < numElements; ++dev) {
    if (std::abs(h_A[dev] + h_B[dev] - h_C[dev]) > 1e-5) {
      std::cout << h_A[dev] << " + " << h_B[dev] << " != " << h_C[dev] << "\n";
      std::cerr << "Result verification failed at element " << dev << "!\n";
      exit(EXIT_FAILURE);
    }
  }

  // Free memory
  cudaFree(h_A);
  cudaFree(h_B);
  cudaFree(h_C);

  return 0;
}
```

Listing 23.6 is based on the same array addition example as Listing 23.5. The main difference is that instead of allocating memory for each device, we allocate a single pointer for each array using `cudaMallocManaged()`. This allows us to use the same pointer for both the host and the device, cleaning up the memory management portion of the code. Like Listing 23.5, we still specify which device we are assigning work to using `cudaSetDevice()`, but instead of copying the data to the device explicitly, we simply calculate the portion of the array for which each device is responsible. We then call `cudaDeviceSynchronize()` to ensure that all kernels have finished and check our results before freeing the only three pointers we allocated.

Chapter 24
Kokkos I

So far, we have considered CPU and GPU programming separately. In these next three chapters, we consider the Kokkos programming model, which aims to connect the hardware and software through a unified abstraction. The primary motivation for Kokkos is to promote *performance portability* of code across different hardware architectures, all the while reducing the need for implementation-specific versions of the same code.

24.1 Kokkos overview

The Kokkos programming model is part of a larger *Kokkos Ecosystem*[1] that includes the Kokkos core library, the Kokkos Kernels Math Libraries, and tools for both profiling and debugging. Kokkos is open-source and was developed within the US Department of Energy's Exascale Computing Project. Kokkos is a C++ programming model, but efforts are increasing to provide interoperability with other languages such as Fortran and Python. In this textbook, we focus largely on the Kokkos core C++ library, and we begin by introducing the steps to get started with Kokkos.

24.1.1 What is Kokkos?

Kokkos is a C++ programming model for performance portability that aims to be *descriptive* rather than *prescriptive*. It is implemented as a template library on top of CUDA, OpenMP, and other high-performance libraries. Kokkos hides the specific implementation details of parallelization from the user and instead encourages users to describe how their computations should be parallelized more generally. The Kokkos library then maps these *descriptions* to the target hardware. The key princi-

[1] https://kokkos.org/

© The Author(s), under exclusive license to Springer Nature Switzerland AG 2025

R. J. Spiteri and K. Klenk, *Extreme-Scale Computing*, CMS/CAIMS Books in Mathematics 15, https://doi.org/10.1007/978-3-031-89033-8_24

ple of Kokkos is to offer *single-source, performance-portable code*. That is, Kokkos aims to provide the software tools that allow programs to execute efficiently on the variety of super-computing architectures that are available today and in the future.

This idea of being descriptive rather than prescriptive should sound familiar because it is similar to the idea of *parallel execution policies* introduced in Chapter 9. Recalling parallel execution policies, algorithms are implemented on iterators over standard containers, and this implementation allows (some) algorithms to be parallelized efficiently over CPU threads and vector registers. The main difference in Kokkos is that it provides additional abstractions for how containers can be organized, and so a parallel execution policy can generalize to run efficiently on both CPUs and GPUs.

The idea of separating the algorithm from the implementation is a fundamentally good idea. It allows expressive high-level code to be written, incorporating the best-performing implementations under the hood for the target hardware on which the algorithm is running.

24.1.2 Obtaining and installing Kokkos

Kokkos can be obtained and installed using the instructions from the official Kokkos GitHub repository at `https://github.com/kokkos/kokkos`. The easiest way to get started with Kokkos in your own project is to use CMake. Kokkos can be linked to a project by adding the following lines to your `CMakeLists.txt` file:

```
find_package(Kokkos REQUIRED)
target_link_libraries(<my_target> Kokkos::kokkos)
```

24.2 Basic data parallel patterns

Prior to delving into basic data parallel algorithms, it is important to first introduce some fundamental Kokkos terminology. At the core of Kokkos are *parallel dispatch operations*, which are used to execute code in parallel. These operations are made up of three components: *a parallel pattern*, *an execution policy*, and *a computational body*. The parallel pattern describes the structure of the parallel computation to be performed. Examples of some Kokkos parallel patterns are `Kokkos::parallel_for`, `Kokkos::parallel_reduce`, `Kokkos::parallel_scan`, and `Kokkos::fence`. The execution policy describes *how* the computation is to be performed. Examples of Kokkos execution policies include `Kokkos::RangePolicy`, `Kokkos::MDRangePolicy`, and `Kokkos::TeamPolicy`. The computational body is the code that is to be executed for each iteration of the parallel pattern.

With these basic concepts in mind, we now give an example of how to use `Kokkos::parallel_for` and `Kokkos::parallel_reduce` to populate an array with data and compute the sum of the array in parallel.

Listing 24.1 An example to demonstrate how to define and use functors within parallel constructs.

```
#include <Kokkos_Core.hpp>

// Define a functor for the parallel_for operation
struct ForFunctor {
  double* data;

  ForFunctor(double* data): data(data) {}

  KOKKOS_INLINE_FUNCTION
  void operator() (const int i) const {
    data[i] = i;  // Initialize each element with its index
  }
};

// Define a functor for the parallel_reduce operation
struct ReduceFunctor {
  const double* data;

  ReduceFunctor(const double* data): data(data) {}

  KOKKOS_INLINE_FUNCTION
  void operator() (const int i, double& sum) const {
    sum += data[i];  // Add each element to the sum
  }
};

int main(int argc, char* argv[]) {
  Kokkos::initialize(argc, argv);
  {
  const int N = 100;
  double* data = (double*) std::malloc(N * sizeof(double));

  // Perform a parallel_for operation
  Kokkos::parallel_for("for_loop", N, ForFunctor(data));

  // Perform a parallel_reduce operation
  double sum = 0.0;
  Kokkos::parallel_reduce("reduce", N, ReduceFunctor(data), sum);

  printf("Sum of array elements: %f\n", sum);

  std::free(data);
  }
  Kokkos::finalize();
  return 0;
}
```

In Listing 24.1, there are a few initial details to note before discussing the `main` function. The first is that the `Kokkos_Core.hpp` header file must be included to use Kokkos. Second, Kokkos requires parallel dispatch operations to operate on *functors*. Functors (first introduced in Section 7.3.2) are passed as the computational body to the parallel dispatch operations. In Listing 24.1, two functors are defined: a `ForFunctor` that populates an array `data` using `parallel_for` and a `ReduceFunctor` that computes the sum of the array, `data`, with `parallel_reduce`.

The `main` function begins by initializing the Kokkos environment with the `Kokkos::initialize` command. Just like with MPI, Kokkos requires that all Kokkos calls be placed between `Kokkos::initialize` and `Kokkos::finalize`. Additionally, the code between `Kokkos::initialize` and `Kokkos::finalize` should be contained within curly braces to ensure proper scoping of the Kokkos environment. After initializing Kokkos, the array `data` of size `N` is allocated with the traditional `malloc` function. Then a `parallel_for` is called to populate the array using `ForFunctor` as the computational body and a `parallel_reduce` is called using `ReduceFunctor` as the computational body to compute the sum of the array. In each parallel operation, a name is specified as the first argument, and it is recommended to name each parallel operation because it can aid in debugging and clarify the location of errors raised by Kokkos at runtime. The second argument is the execution policy, and in both cases, `N` automatically defaults to the

`Kokkos::RangePolicy`, which specifies a continuous range over which to iterate. The third argument is the functor that is to be executed in parallel. Lastly, the result is printed, the array `data` is deallocated, and the Kokkos environment is finalized.

When using Kokkos, the parallel operations that are defined are usually called *kernels*. Under the hood, Kokkos takes care of how the parallelization is performed for every kernel launch. The method that Kokkos uses to parallelize the kernel depends on the available backends with which Kokkos was originally compiled. If multiple backends are present, there is an order of precedence that Kokkos follows. Most simply, if a GPU backend (e.g., CUDA) and CPU (e.g., OpenMP) backend are both present, Kokkos defaults to the GPU backend. This is problematic for our example in Listing 24.1 because the `std::malloc` operation fails for a GPU backend. We discuss how to rectify this with a more in-depth treatment of memory spaces in Section 24.4.

24.2.1 Lambda functions

Using functors can sometimes be too verbose. Fortunately, lambda functions (introduced in Section 7.3.3) can be used to simplify Kokkos code. In fact, lambda functions are the preferred method for defining the computational body of Kokkos kernels. In the example below, we have replaced the functors in Listing 24.1 with lambda functions.

Listing 24.2 An example that shows how to express functors as lambda functions.

```
#include <Kokkos_Core.hpp>

int main(int argc, char* argv[]) {
  Kokkos::initialize(argc, argv);
  {
  const int N = 100;
  double* data = (double*) std::malloc(N * sizeof(double));

  // Perform a parallel_for operation
  Kokkos::parallel_for("for_loop", N, KOKKOS_LAMBDA(const int i) {
    data[i] = i;  // Initialize each element with its index
  });

  // Perform a parallel_reduce operation
  double sum = 0.0;
  Kokkos::parallel_reduce("reduce", N, KOKKOS_LAMBDA(const int i, double& sum) {
    sum += data[i];  // Add each element to the sum
  }, sum);

  printf("Sum of array elements: %f\n", sum);

  std::free(data);
  }
  Kokkos::finalize();

  return 0;
}
```

In Listing 24.2, each functor has been replaced with a `KOKKOS_LAMBDA` to define the computational body. Using lambda functions significantly reduces the verbosity of the code. Moreover, normal lambda functions, like those used in Section 7.3.3, can be used. However, `KOKKOS_LAMBDA` is preferred because it promotes portability. Depending on the backend(s) present, `KOKKOS_LAMBDA` will be replaced with the appropriate backend-specific code, e.g., marking the lambda function as `__device__` for CUDA. Otherwise, the programmer would have to manually ensure that the lambda functions are marked with the appropriate backend-specific qualifiers.

24.2.2 Equivalence to STL and OpenMP

The `Kokkos::parallel_for` can be used to perform a variety of STL algorithms depending on the data lifetimes involved in the computation kernel. The difference with `Kokkos::parallel_for` is that it iterates through an *iteration index* rather than through an *iterator object*. The difference is subtle, but it is where the power of the `Kokkos::parallel_for` lies. Using an iteration index allows `Kokkos::parallel_for` to remain portable across different hardware because simple for-loop iteration structures that use a index rather than an iterator object can easily be mapped to compute units of a GPU or CPU.

Furthermore, Kokkos aims to provide a style similar to OpenMP for defining parallel loops. The advantage is more concise and portable syntax. With OpenMP, parallelization is done with pragmas around the `for` loop, whereas parallelization in Kokkos is done through function calls like `parallel_for`. Although pragmas are still relatively concise, they still require increased attention when reading the loops to ensure the programmer can follow what is going on, and of course, OpenMP pragmas are not all portable across different hardware.

24.3 Execution Spaces

Execution spaces in Kokkos are the places (devices) and manner (algorithms) in which parallel code is executed. They execute on a set of homogeneous cores. Similar to the use of GPUs, there is a separation between the intended execution spaces of code. For CUDA, this separation was between host and device. For Kokkos, it is host code and parallel code. Host code runs on the CPU as part of the main process. Anything that is part of a parallel construct is parallel code. Parallel code is what is run in a specific execution space (CPU or GPU). To adjust where the body of a parallel construct is run, we must change the execution space. This is achieved by compiling the Kokkos library with the appropriate backend. For example, to run on NVIDIA GPUs, we must compile Kokkos with the CUDA backend enabled. In the case where there are multiple backends enabled, we can specify the execution space of a parallel construct through its *execution policy*. So far, we have only set the policy to a singular value, which invokes the `Kokkos::RangePolicy` specifically. If we want to change the execution space, we must be explicit in defining our policy and its execution space.

Additionally, to make use of anything that is not a CPU, we need to ensure our parallel constructs are marked with a macro. We have already seen the macros used for lambda functions in Section 24.2.1 as a way to make the code increasingly portable, letting the compiler decide the appropriate qualifiers to use. With traditional functors, we can use the #KOKKOS_FUNCTION or #KOKKOS_INLINE_FUNCTION macros to achieve the same effect as KOKKOS_LAMBDA.

Turning back to Listings 24.1 and 24.2, we can now set the execution space to overcome the problem of illegal memory access that occurs when Kokkos defaults to the CUDA backend. In Listing 24.3, we specify the execution space to Kokkos::OpenMP inside the execution policy argument of both our parallel_for and parallel_reduce operations.

Listing 24.3 An example of specifying the execution space to use OpenMP.

```
#include <Kokkos_Core.hpp>

int main(int argc, char* argv[]) {
  Kokkos::initialize(argc, argv);

  const int N = 100;
  double* data = (double*) std::malloc(N * sizeof(double));

  // Perform a parallel_for operation
  Kokkos::parallel_for("for_loop",
      Kokkos::RangePolicy<Kokkos::OpenMP>(0, N),
      KOKKOS_LAMBDA(const int i) {
    data[i] = i;  // Initialize each element with its index
  });

  // Perform a parallel_reduce operation
  double sum = 0.0;
  Kokkos::parallel_reduce("reduce",
      Kokkos::RangePolicy<Kokkos::OpenMP>(0, N),
      KOKKOS_LAMBDA(const int i, double& sum) {
    sum += data[i];  // Add each element to the sum
  }, sum);

  printf("Sum of array elements: %f\n", sum);

  std::free(data);
  Kokkos::finalize();

  return 0;
}
```

In the next section, we discuss a different kind of space called a *memory space* that allows us to utilize the GPU instead of working around it.

24.4 Memory Spaces

A memory space is a place within a system where data are stored and managed. This can be CPU RAM, GPU RAM, or even disk drive storage space. Typically, the default memory space is the same as the default execution space. However, because we have been allocating memory using std::malloc in the previous examples, we have *not* been allocating memory to the default memory space in situations where the default execution space is a GPU. To rectify this, we can use a Kokkos-specific function to allocate and deallocate memory, as we illustrate in the following example.

Listing 24.4 An example that shows the usage of Kokkos C-style memory management.

```
#include <Kokkos_Core.hpp>

int main(int argc, char* argv[]) {
  Kokkos::initialize(argc, argv);
```

```
  const int N = 100;
  double* data = (double*) Kokkos::kokkos_malloc(N * sizeof(double));

  // Perform a parallel_for operation
  Kokkos::parallel_for("for_loop", N, KOKKOS_LAMBDA(const int i) {
    data[i] = i;  // Initialize each element with its index
  });

  // Perform a parallel_reduce operation
  double sum = 0.0;
  Kokkos::parallel_reduce("reduce", N, KOKKOS_LAMBDA(const int i, double& sum) {
    sum += data[i];  // Add each element to the sum
  }, sum);

  printf("Sum of array elements: %f\n", sum);

  Kokkos::kokkos_free(data);
  Kokkos::finalize();

  return 0;
}
```

In Listing 24.4, the calls to `std::malloc` and `std::free` are changed to `Kokkos::kokkos_malloc` and `Kokkos::kokkos_free` functions, respectively. These functions let Kokkos decide where to allocate memory based on the execution space. Additionally, these Kokkos-specific functions should be used over `std::malloc` and `std::free` because they are more portable and work with any execution space. We omitted their use in the previous sections to clearly distinguish between how execution spaces and memory spaces behave within Kokkos.

When using `Kokkos::kokkos_malloc` and `Kokkos::kokkos_free`, a specific memory space can be specified if desired. This is done by passing the memory space as a template parameter to the `Kokkos::kokkos_malloc` function as `Kokkos::kokkos_malloc<MemorySpace>`. The `MemorySpace` template parameter can be *any valid space*, meaning either an execution space or a memory space can be specified. This specificaiton is possible because each execution space has a corresponding default memory space.

Now with a basic understanding of the fundamental concepts of Kokkos, we introduce an even better way to manage memory in the next chapter using *views*.

Chapter 25
Kokkos II

In the previous chapter, we covered the fundamentals of using Kokkos. We learned how to write basic kernels and how execution and memory spaces work. We kept the discussion in the framework of C-style memory management with the use of `std::malloc` and `std::free` and eventually introduced the Kokkos equivalents `Kokkos::kokkos_malloc` and `Kokkos::kokkos_free`. In this chapter, we discuss the preferred way to manage memory in Kokkos, i.e., via the `Kokkos::View` class.

25.1 Views

The most important aspect about writing performance-portable code is the memory-access pattern. In Section 6.1, we demonstrated how the order in which we access memory can have a significant impact on the performance of an application. To further complicate matters, the optimal way in which data should be laid out in memory is different for different hardware architectures. Luckily, Kokkos provides *views* as a way to ease the burden of managing memory between different architectures.

Views are C++ classes that are used to represent multidimensional arrays. They are intended to be lightweight and contain a metadata component that describes the contents of the view and a data component that points to the actual data stored as the aforementioned multidimensional array. The dimensions of a view can be specified at either run or compile time. The template specification for a `Kokkos::View` is

```
template <class DataType [, class LayoutType] [, class MemorySpace] [, class MemoryTraits]>
class View;
```

The *only* required argument to the creation of a view is the `DataType`, which is not only used to specify the type of the data but also the dimensionality of the underlying array. The `LayoutType` is covered in Section 25.1.1, the `MemorySpace` behaves as discussed in Section 24.4, and the `MemoryTraits` argument is discussed in Section 25.1.2. Some examples of view creation include

```
Kokkos::View<double*>     view1("one", N); // 1D double, run-time size
Kokkos::View<const int*>  view2("two", N); // 1D int, run-time size, const data
```

© The Author(s), under exclusive license to Springer Nature Switzerland AG 2025

R. J. Spiteri and K. Klenk, *Extreme-Scale Computing*, CMS/CAIMS Books in Mathematics 15, https://doi.org/10.1007/978-3-031-89033-8_25

```
Kokkos::View<double[4][4]> view3("three"); // 2D double, 2 compile-time sizes
Kokkos::View<Foo**>        view4("four", N, N); // 2D class, 2 run-time sizes
```

Views can only consist of a single data type, and the data within a view can be of type `const` (read-only). Just like when defining a kernel, the `Kokkos::View` constructor can include a string name. The elements of a view are accessed using the `(...)` operator instead of the `[...]` operator that is used for `std::vector` or `std::array`. To offer a more complete example, we show how to utilize views to perform parallel matrix multiplication in Listing 25.1.

Listing 25.1 Matrix multiplication with Kokkos views

```
#include <Kokkos_Core.hpp>

int main(int argc, char* argv[]) {

  Kokkos::initialize(argc, argv);
  {
  int N = 10000; // Matrix size

  // Create 2D views for the matrices
  Kokkos::View<double**> A("A", N, N);
  Kokkos::View<double**> B("B", N, N);
  Kokkos::View<double**> C("C", N, N);

  // Initialize A and B matrices with some values
  Kokkos::parallel_for("Init", Kokkos::RangePolicy<>(0, N), KOKKOS_LAMBDA(int i) {
    for (int j = 0; j < N; ++j) {
      A(i, j) = 1.0 * i + j;
      B(i, j) = 1.0 * i - j;
    }
  });

  // Perform matrix multiplication C = A * B
  Kokkos::parallel_for("MatMul", Kokkos::RangePolicy<>(0, N), KOKKOS_LAMBDA(int i) {
    for (int j = 0; j < N; ++j) {
      double sum = 0;
      for (int k = 0; k < N; ++k) {
        sum += A(i, k) * B(k, j);
      }
      C(i, j) = sum;
    }
  });

  // Print the first element of the result matrix
  Kokkos::parallel_for("Print", Kokkos::RangePolicy<>(0, 1), KOKKOS_LAMBDA(int i) {
    printf("C(0,0) = %f\n", C(50, 50));
  });
  }
  Kokkos::finalize();
  return 0;
}
```

In Listing 25.1, we have defined three views, `A`, `B`, and `C`. The `A` and `B` views are initialized with some data to be multiplied, and `C` is filled with the results. We simply name each view and then specify each dimension of the view as part of the construction. In this case specifically, we are using a 2D view, so we specify two dimensions of size `N` as part of our constructor calls. We note that we do not need to allocate or deallocate memory anywhere in the example. Views behave much like a `std::shared_ptr`, and so memory is allocated when the view is constructed, and reference counting handles the automatic deallocation when the view goes out of scope. If we needed to re-specify the sizes of the view, we could use either the `Kokkos::realloc` or `Kokkos::resize` functions. The actual matrix multiplication is nothing we have not seen before, except that we were explicit in specifying an execution policy for the `parallel_for` loop. In this case, we are using the `Kokkos::RangePolicy`, which is what was used in the previous example. Finally, we print the elements of the `C` view to verify that the matrix multiplication was performed correctly.

25.1.1 Layouts

Turing our attention back to the `Kokkos::View` template, we see that there are three optional template parameters: `LayoutType`, `MemorySpace`, and `MemoryTraits`. We focus on `LayoutType` here and save the discussion of `MemorySpace` and `MemoryTraits` for the next chapter.

The notion of *row-major* and *column-major* ordering of memory was first introduced in Section 6.1. This ordering is important when considering the performance of a program. When programming specifically for a CPU, row-major ordering is generally preferred because it plays to the strengths of the CPU cache. However, when programming for a GPU, column-major ordering is generally preferred, due to the absence of a similar cache-per core architecture.

Within Kokkos, the `LayoutType` template parameter specifies the memory layout of the view. There are three options for the `LayoutType` template parameter: `Kokkos::LayoutLeft`, `Kokkos::LayoutRight`, and `Kokkos::LayoutStride`. The `Kokkos::LayoutLeft` is column-major ordering and is ideal for coalesced memory access on GPUs. Coalesced memory access refers to a memory access pattern of a program such that thread t accesses memory location i and thread $t + 1$ accesses memory location $i + 1$. This pattern is ideal for GPUs that contain large blocks of threads. On the other hand, `Kokkos::LayoutRight` is row-major ordering and is ideal for cached memory access on CPUs. Cached memory access is preferable for the pattern where thread t accesses memory location i and $i + n$ to take advantage of the additional elements that are in the CPUs cache. Finally, `Kokkos::LayoutStride` is a custom layout that is generally useful for taking non-contiguous subviews. If left unspecified, Kokkos attempts to choose the layout that is optimal, a topic touched upon in the next chapter.

25.1.2 Memory Traits

Memory traits can be specified to a `Kokkos::View` to specify specific properties of how the memory should behave when accessed. To specify memory traits for a view, we use the `Kokkos::MemoryTraits` template parameter as in

```
double* data = new double[N];
Kokkos::View<double*, Kokkos::HostSpace, Kokkos::MemoryTraits<Kokkos::desired_trait> view(data, N);
```

where `Kokkos::desired_trait` is the desired memory trait. There are four arguments that can be used to specify the memory traits of a view: `Kokkos::Unmanaged`, `Kokkos::RandomAccess`, `Kokkos::Atomic`, and `Kokkos::Restrict`.

The `Kokkos::Unmanaged` trait is used to wrap a view around an existing data structure. This is useful for trying to interface Kokkos with existing code. With an unmanaged view, we lose some convenience of the typical managed view, such as reference counting, automatic deallocation, and most importantly, no check for the correct memory space. To create an unmanaged view, we simply pass a pointer to the data around which we want to wrap the view as the first argument to the view constructor along with its dimensions.

The `Kokkos::RandomAccess` trait is used to specify that the view is to be accessed with a random access pattern. This is a useful declaration for a view because it informs the compiler that the view is not to be accessed uniformly or iteratively. With this information, the compiler can make the necessary optimizations to the code depending on the specific memory space in which the view is accessed.

The `Kokkos::Atomic` trait is used to specify that the view can only be read from or written to using atomic operations. Usually, when data are read from or written to memory, there are multiple instructions that the hardware executes to perform a read or write. This can be problematic because the data can suddenly change while the hardware is executing the specific set of instructions. This problem can be circumvented by using locks around the read or write operation, but such a solution is neither thread scalable nor portable because GPUs do not have good support for locks. Instead, the solution is to use atomic operations, where the hardware executes reads and writes using special atomic instructions that are just a single instruction and guaranteed not to be interrupted.

The final memory trait is `Kokkos::Restrict`, which is used to specify that a view is not referenced by any other view or that parts of the view are not contained in any other view. This is useful because it can allow for the compiler to make certain optimizations to the code that it could not make otherwise. However, the compiler cannot help determine whether a view with the `Kokkos::Restrict` trait is being aliased. It is still up to the developer to ensure that a view with the `Kokkos::Restrict` trait is not aliased.

Below, we provide a complete example of how to use memory traits to specify an unmanaged view.

Listing 25.2 Example of using an unmanaged view.

```
#include <Kokkos_Core.hpp>
#include <iostream>

int main() {

  // Crate array outside of Kokkos scope
  const int size = 100;
  double* data = new double[size];

  Kokkos::initialize();
  {

  // Create a Kokkos view with the Unmanaged trait
  // This view won't be reference counted, and you manage its memory explicitly
  Kokkos::View<double*, Kokkos::MemoryTraits<Kokkos::Unmanaged>> my_view(data, size);

  // Initialize the view (for demonstration purposes)
  Kokkos::parallel_for("populate_array", size, KOKKOS_LAMBDA(const int i) {
      my_view(i) = i;
  });

  // Access the data
  double sum = 0.0;
  Kokkos::parallel_reduce("sum_array", size, KOKKOS_LAMBDA(const int i, double& sum) {
      sum += my_view(i);
  }, sum);

  std::cout << "Sum of elements: " << sum << std::endl;
  // Clean up: delete the data array
  }
  // Finalize Kokkos
  Kokkos::finalize();

  // Access array data outside of Kokkos scope
  for (int i = 0; i < 10; ++i) {
```

```
    std::cout << data[i] << std::endl;
  }

  // Delete Outside of Kokkos Scope
  delete[] data;

  return 0;
}
```

In Listing 25.2, we initialize an array outside the Kokkos environment. We then create an unmanaged view that wraps around the array. We then populate the array with some data and sum the elements using a parallel reduction. To demonstrate that the views contents are not deallocated when the view goes out of scope, we print some elements of the array after the `Kokkos::finalize` call.

25.2 Mirrors

A *mirror* is a Kokkos tool that allows us to create a view in one memory space that is a copy of a view in another memory space. Mirrors are the preferred solution to the problem of needing to access data in one memory space from another. The reason for using mirrors should not be surprising: it is again for performance-portability. For example, if we were to utilize CUDA's Unified Virtual Memory (UVM) to access data from the host, we would incur some cost to the CUDA runtime system managing the data between the host and device.

Instead, we can just create a mirror view, which is a view in the host memory that is the same as the view (including the data) in another memory space. That is, if we have a view in another memory space, and we create a mirror of it, it always allocates to the host. In the case that our existing mirror already exists on the host, then no data are copied, and the mirror just points to the same data as the original view. Listing 25.3 provides an example of how to set up a mirror view.

Listing 25.3 Example of setting up a mirror view.

```
#include <Kokkos_Core.hpp>

int main(int argc, char* argv[]) {
  Kokkos::initialize(argc, argv);
  {

  int size = 10000;  // Size of the matrices

  // Create some views for matrix multiplication
  Kokkos::View<double**> A("A", size, size);
  Kokkos::View<double**> B("B", size, size);
  Kokkos::View<double**> C("C", size, size);

  // Create the host mirrors to simulate the data read from files
  Kokkos::View<double**>::HostMirror hostViewA = Kokkos::create_mirror_view(A);
  Kokkos::View<double**>::HostMirror hostViewB = Kokkos::create_mirror_view(B);
  Kokkos::View<double**>::HostMirror hostViewC = Kokkos::create_mirror_view(C);
  // Simulate Reading of a matrix from a file
  // Kokkos::parallel_for("Read A", Kokkos::RangePolicy<>(0, size), KOKKOS_LAMBDA(int i) {
  for (int i = 0; i < size; i++) {
    for(int j = 0; j < size; j++) {
      hostViewA(i, j) = i + j;  // Initialize each element with its index
      hostViewB(i, j) = i - j;  // Initialize each element with its index
    }
  };

  // Copy the data to the device
  Kokkos::deep_copy(A, hostViewA);
  Kokkos::deep_copy(B, hostViewB);
```

```
    // Perform the matrix multiplication on the device
    // Perform matrix multiplication C = A * B
    Kokkos::parallel_for("MatMul", size, KOKKOS_LAMBDA(int i) {
      for (int j = 0; j < size; ++j) {
        double sum = 0;
        for (int k = 0; k < size; ++k) {
          sum += A(i, k) * B(k, j);
        }
        C(i, j) = sum;
      }
    });

    // Copy the result back to the host
    Kokkos::deep_copy(hostViewC, C);

    // Print the first element of the result matrix
    printf("C(0,0) = %f\n", hostViewC(0, 0));

  }
  Kokkos::finalize();
  return 0;
}
```

In Listing 25.3, we simulate a common situation where a device is used to perform some computation, but the data are initially stored in a file. In this case, the device cannot be used to perform file I/O, and the host is required for this operation. In Listing 25.3, we begin by creating our views `A`, `B`, and `C` in the default memory space, which is CUDA, assuming that the CUDA backend is enabled. We then use the `Kokkos::create_mirror_view` function to create a mirror of each view. We then simulate the reading of data from a file with a nested `for` loop to populated the host views, `hostViewA` and `hostViewB`. Then just before we perform the computation, which is a simple matrix multiplication, on the device, we complete a deep copy of the host views to the device views. The computation is then performed on the device, the result is copied back to the host, and the first element is printed to stdout.

25.2.1 Dual View

Performing many deep copies between memory spaces is expensive. If two references to the same data are needed between the host and device for an extended period of time, then a dual view might be an appropriate solution. A dual view is similar to a mirror view, but it is more appropriately used for situations where data or views need to be managed in two separate memory spaces. A dual view takes two views and allows them to appear as a single data structure. The dual view can then be used to keep track of when and where data are being modified. It is up to the programmer, however, to ensure that the appropriate flags are updated to reflect when a view's data have been modified. Listing 25.4 shows how to set up and use a dual view.

Listing 25.4 Example of setting up a dual view.

```
#include <Kokkos_Core.hpp>
#include <Kokkos_DualView.hpp>

int main(int argc, char* argv[]) {
  Kokkos::initialize(argc, argv);
  {

  // Create a DualView
  Kokkos::DualView<double*> data("data", 10);

  // Modify the DualView on the host
  for (int i = 0; i < 10; i++) {
    data.h_view(i) = i * 1.0;
```

```
  }
  data.modify_host();

  // Print the DualView data on the host
  for (int i = 0; i < 10; i++) {
    printf("data(%d) = %f\n", i, data.h_view(i));
  }

  // Modify the DualView on the device
  data.sync_device();
  Kokkos::parallel_for("modifyOnDevice", Kokkos::RangePolicy<>(0, 10), KOKKOS_LAMBDA(int i) {
    data.d_view(i) += 1.0;
  });
  data.modify_device();

  data.sync_host();
  for (int i = 0; i < 10; i++) {
    printf("data(%d) = %f\n", i, data.h_view(i));
  }

  }
  Kokkos::finalize();
  return 0;
}
```

The example shown in Listing 25.4 is a straightforward example to demonstrate the functionality of a dual view. First, an additional header file is required to utilize the dual view, `Kokkos_DualView.hpp`. Next, the Kokkos environment is initialized, and a dual view of a one-dimensional array of size 10 is created. We then populate the dual view with some data from the host using a regular `for` loop. We note that we access the host side of the dual view with the `h_view(i)` method. After filling the dual view with some data, we mark the host view as modified with the `modify_host()` method. If the `modify_host()` method is not called, then a subsequent call to synchronize the data will not behave as expected. After marking the host view as modified, we then print out the array to ensure that we can observe the changes we are about to perform. Before we pass the dual view to a kernel that will add 1 to each element using a device, we synchronize the data with the `sync_device()` method. After the kernel is called, we then mark the data as modified by the device with the `modify_device()` method. Finally, we synchronize the data back to the host with the `sync_host()` method to print the modified array from the host and verify the changes made by the device.

These steps may seem laborious, but they are necessary to ensure that the data are managed correctly between the host and device. Generally, these steps are contained at the beginning and end of a function that is known to compute on the device or host. One may ask why Kokkos does not handle the tracking of data modifications by device and host for us. The answer is that it could, in principle, but the overhead in performing the house-keeping would be a significant performance bottleneck.

25.3 Subviews

To finish off this chapter and our discussion of views, we cover the use of subviews. It is common that only a portion of a view is needed for a particular computation. In this case, we can use a subview to extract the portion of the view that is needed. Subviews are motivated by the same *slicing* functionality that exists in other languages like Python or Fortran. We now walk through an example of how to use subviews.

Listing 25.5 Example of setting up a subview.

```
#include <Kokkos_Core.hpp>

int main(int argc, char* argv[]) {
  Kokkos::initialize(argc, argv);
  {
  int init_size = 5;


  // Create a 2D Kokkos::View
  Kokkos::View<double*> data_vector("data_vector", init_size);

  // Fill the view with data using a parallel_for
  Kokkos::parallel_for("fillData_vector", init_size, KOKKOS_LAMBDA(int i) {
    data_vector(i) = i;
  });

  auto subview = Kokkos::subview(data_vector, Kokkos::pair<int,int>(1,4));

  // Print the subview data using a parallel_for
  Kokkos::parallel_for("printData", subview.extent(0) , KOKKOS_LAMBDA(int i) {
    printf("subview(%d) = %f\n", i, subview(i));
  });


  }
  Kokkos::finalize();
  return 0;
}
```

In Listing 25.5, we have a simple example of how to define a subview. We begin by creating a one-dimensional view of size 10 and populate it with some data. We then define a subview using `Kokkos::subview`, specifying the original view as the first argument and the slice we want as the subview using `Kokkos::pair`. With `Kokkos::pair`, we simply define the start and end of the slice of the dimension we want. In this case, we only have one dimension; if we had additional dimensions, we would specify a `Kokkos::pair` for each dimension. Alternatively, we could use `Kokkos::ALL` to specify that we want all elements in that dimension; this syntax is identical to using `:` in Python or Fortran. Finally, we print the contents of the subview. In the policy, we use `subview.extent(0)` to get the size of the subview. We note that `extent` is a method that is used to get the size of a view or subview dimension as specified by the argument.

Chapter 26
Kokkos III

In this final chapter on Kokkos, we look deeper into extracting more parallelism from our Kokkos code. So far, we have seen how to use Kokkos for high-level parallelism at the level of loops. However, when loops are nested within other loops, for example, we can potentially extract even more parallelism if the hardware on which we are running our code has the capability to execute greater degrees of parallelism and obtain more performance. Accordingly, we look into hierarchical parallelism and the methods that Kokkos provides for extracting additional parallelism from our code.

26.1 Hierarchical parallelism

Hierarchical parallelism is the idea that parallelism can be implemented at different levels of granularity. The canonical example is that of nested loops, where each level of loop can be parallelized. It is becoming increasingly common for hardware, both CPUs and GPUs, to have support for hierarchical parallelism. For example, a CPU may have a number of cores that can execute instructions in parallel, and furthermore, the CPU core may have the ability to execute vector instructions, where an array of data can have the same instruction performed on it in a single operation for an extra level of parallelism.

Kokkos provides the support for appropriately mapping hierarchical parallelism to the hardware on which the code is running. However, this mapping is not always as straightforward as the mapping of parallelism was with using single calls to `parallel_for`, `parallel_reduce`, or `parallel_scan`, etc. In some cases, each Kokkos kernel may need to be set up differently to accommodate the correct mapping of parallelism. This difference is because the correct way in which to map hierarchical parallelism can be algorithm-dependent. According to the documentation,[1] Kokkos identifies three opportunities for hierarchical parallelism, namely,

---

[1] https://kokkos.org/kokkos-core-wiki/ProgrammingGuide/HierarchicalParallelism.html#nested-parallelism

© The Author(s), under exclusive license to Springer Nature Switzerland AG 2025

R. J. Spiteri and K. Klenk, *Extreme-Scale Computing*, CMS/CAIMS Books in Mathematics 15, https://doi.org/10.1007/978-3-031-89033-8_26

- loops that are *not* considered to be *tightly nested*. More specifically, inner loops that are not dependent on the outer loops;
- situations in which an inner loop gathers data for an iteration of an outer loop; and
- situations where cache blocking (maximizing cache data re-use) can improve the performance of the code.

In the rest of this section, we look into three specific ways that Kokkos provides support for hierarchical parallelism, starting with thread teams as the foundational element for hierarchical parallelism.

26.1.1 Thread teams

A thread team is simply a grouping of threads. In the previous chapters, we became familiar with a one-dimensional mapping of threads to the hardware. This worked well on the parallel loops that we have seen so far, where each thread could work independently on its own iteration. Thread teams, on the other hand, use a two-dimensional mapping of threads to the hardware. The two-dimensional mapping is made up of a *league size*, which is the number of teams, and a *team size*, which is the number of threads within a team. Each thread team is identified by a *league rank*, and each thread within a team is indexed by a *team rank*.

There is no limit to the number of teams that can form a league, but the number of threads that can form a team is limited by the hardware. Synchronization can only be performed within a team and not between teams. Importantly, thread teams behave concurrently, similar to the way that thread blocks behave in CUDA. When a kernel is launched, it appears as if all the thread teams are running, but in reality it may be that only a portion of them are actually running at a given time, and the rest are waiting for their opportunity to run while the execution units of the kernel are busy with other thread teams.

To use thread teams, the `TeamPolicy` execution policy needs to be used. Listing 26.1 is an example of how to set up and use thread teams.

Listing 26.1 Example of using a `TeamPolicy` to launch a kernel with thread teams.

```
#include <Kokkos_Core.hpp>

int main(int argc, char* argv[]) {
  Kokkos::initialize(argc, argv);
  {

  // Define the team policy
  int team_size = 2;
  int league_size = 2;
  Kokkos::TeamPolicy<> policy(league_size, team_size);

  // Create a parallel_for with a team policy
  Kokkos::parallel_for("TeamIDs", policy, KOKKOS_LAMBDA(
                        const Kokkos::TeamPolicy<>::member_type& team) {
    int l_r = team.league_rank();
    int t_r = team.team_rank();
    int t_s = team.team_size();
    printf("League Rank = %d :: Team Rank = %d :: Team_Size = %d\n",
           l_r, t_r, t_s);
  });
```

```
  }
  Kokkos::finalize();
  return 0;
}
```

We begin Listing 26.1 by defining an instance of a team policy. This is a required prerequisite for launching a kernel that uses thread teams. The `TeamPolicy` takes in the `league_size` and the `team_size` as arguments. The specific execution space can be defined within <> brackets. In this example, we keep things small with a league of 2 teams, each with 2 threads. In our kernel, we pass our policy `policy`, and within the `KOKKOS_LAMBDA`, we need to specify the `member_type` of the `TeamPolicy` as the first argument. This is a special type that is used to access our teams within the kernel. Finally, we show how we can access some properties of the ranks of the teams. Specifically, we access the league rank and the team rank along with its size. The output of Listing 26.1 is as follows:

```
League Rank = 1 :: Team Rank = 0 :: Team_Size = 2
League Rank = 1 :: Team Rank = 1 :: Team_Size = 2
League Rank = 0 :: Team Rank = 0 :: Team_Size = 2
League Rank = 0 :: Team Rank = 1 :: Team_Size = 2
```

Selecting the team size can be difficult, and the optimal value can depend on the hardware on which the code is running. To alleviate this dependency, Kokkos can attempt to automatically select the proper team size based on the hardware. This automatic selection is invoked by setting the team size in the `TeamPolicy` instance to `Kokkos::AUTO`. This setting is the recommended way to set the team size unless using specific values is justified through increased performance.

26.1.2 TeamThreadRange

Now that we have seen how to define thread teams, we can look into how to utilize hierarchical parallelism within a kernel. There are three levels of parallelism that can be used within a kernel. The first is what we have seen in the previous chapters, where we use a single call to a parallel operation like `parallel_for`, `parallel_reduce`, etc. The second level of parallelism is where we can begin to fully realize the potential of thread teams, and that is through the use of `TeamThreadRange`. The third level of parallelism (covered in Section 26.1.3) is vector-level parallelism.

The execution policy `TeamThreadRange` is used within a parallel construct to split an iteration space over the threads within a team. Each thread team is then guaranteed to be running concurrently. Listing 26.2 is an example of how to use the `TeamThreadRange` execution policy.

Listing 26.2 Example of using the `TeamThreadRange` to launch a kernel.

```
#include <Kokkos_Core.hpp>

int main(int argc, char* argv[]) {
  Kokkos::initialize(argc, argv);

  // Define the team policy
  int team_size = 4;
```

```
  int league_size = 10;
  Kokkos::TeamPolicy<> policy(league_size, team_size);

  // Create a parallel_for with a team policy
  Kokkos::parallel_for("useTeamThreadRange", policy, KOKKOS_LAMBDA(const Kokkos::TeamPolicy<>::member_type&
       team) {
    // Each team performs multiple parallel_fors
    for (int n = 0; n < 3; n++) {
      Kokkos::parallel_for(Kokkos::TeamThreadRange(team, team_size), [&](int i) {
        printf("Team %d, Thread %d, Iteration %d\n", team.league_rank(), i, n);
      });
      team.team_barrier();
    }
  });

  // Ensure all kernels have completed
  Kokkos::fence();

  Kokkos::finalize();
  return 0;
}
```

In Listing 26.2, we compute the pairwise distances between a set of points. First, we define a struct `Point` to represent a point in 3D space. We then define two views, a set of points `points` and a matrix `distances` to store the pairwise distances. We then define a `TeamPolicy` instance `policy` with a league size of `numPoints` and a team size set by Kokkos using `Kokkos::AUTO`. Here, setting the league size to the number of points is a good choice because we effectively have each team filling in the rows of the matrix `distances`. In this specific example, if we specified the league size to be less than the number of points, then we would have missing rows in the matrix `distances`. We then define the kernel, which first gets the league rank to be used as the row index for the matrix `distances`. We then use the `TeamThreadRange` to fill in the row of the matrix `distances` with the pairwise distances.

Some aspects of using `TeamThreadRange` to be aware of are that the mapping of the parallelism to the hardware is still dependent on the hardware, meaning that Kokkos tries to optimize how the parallelism is performed based on the target hardware on which the code is running. This effort includes trying to optimize the layout of the data according to the hardware. Additionally, when nesting parallel constructs, the inner parallel constructs do not need the `KOKKOS_LAMBDA` because the compiler can determine which architecture the lambda is for based on the first occurrence of `KOKKOS_LAMBDA`. Finally, the team size does not need to be a multiple of the number of iterations for which a thread team is responsible.

26.1.3 Vector-level parallelism

The third level in the hierarchy of parallelism is vector-level parallelism, which can be used with the `ThreadVectorRange` execution policy. This level of parallelism is used for Kokkos to allow the compiler to use vectorization where possible. The use of `ThreadVectorRange` is almost identical to the use of `TeamThreadRange`, the difference being that the `ThreadVectorRange` must contain work that is vectorizable; i.e., the work must be able to utilize vector operations (SIMD instructions). Listing 26.3 is a modified version of Listing 26.2 to use `ThreadVectorRange`.

Listing 26.3 Example of using the `ThreadVectorRange` to launch a kernel.

```
#include <Kokkos_Core.hpp>
#include <cstdio>

struct Point {
  double x, y, z;
};

int main(int argc, char* argv[]) {
  Kokkos::initialize(argc, argv);
  {

  int numPoints = 1000; // Matrix size

  Kokkos::View<Point*> points("points", numPoints);
  Kokkos::View<double**> distances("distances", numPoints, numPoints);

  Kokkos::TeamPolicy<> policy(numPoints, Kokkos::AUTO());
  Kokkos::parallel_for("compute_distances", policy, KOKKOS_LAMBDA(
              const Kokkos::TeamPolicy<>::member_type& team) {
    int i = team.league_rank();

    Kokkos::parallel_for(Kokkos::ThreadVectorRange(team, numPoints), [=] (int j) {
      // printf("i = %d, j = %d\n", i, j);
      double dx = points(i).x - points(j).x;
      double dy = points(i).y - points(j).y;
      double dz = points(i).z - points(j).z;
      distances(i, j) = sqrt(dx*dx + dy*dy + dz*dz);
    });
  });


  }
  Kokkos::finalize();
  return 0;
}
```

We have modified the internal loop of the kernel to use `ThreadVectorRange` instead of `TeamThreadRange`, allowing the compiler to make any optimizations for vectorization that it can. For this particular example, especially if we are running the code on older hardware, we may not see any performance improvements from using `ThreadVectorRange` because the square-root function is an expensive operation that may not be vectorizable.

26.1.4 Scratch pads

Scratch pads are memory spaces that are available to use within a thread team. The concept of a scratch pad is similar to the shared memory available to threads within a thread block in CUDA. A scratch pad, when created, exists for the duration of the team. The point of a scratch pad is to reduce the stress on global memory by allowing thread teams to load data to their scratch pad and reduce contention in the global memory space. This feature is especially useful in situations where a team is accessing or operating on the same data many times. It allows the data to stay close to the thread team and allows Kokkos to take advantage of any special hardware like the aforementioned shared memory available to thread blocks in CUDA.

Scratch pads are defined through the `TeamPolicy`, which sets the size of the scratch pad for each team with the `set_scratch_size` method. Views can also be created that belong to the scratch pad by passing in `Kokkos::ScratchMemorySpace` as a template parameter. Below is an example of how to use a scratch pad.

Listing 26.4 An example of how to use a scratch pad with a view

```
#include <Kokkos_Core.hpp>
```

```
int main(int argc, char* argv[]) {
  Kokkos::initialize(argc, argv);

  // Define the team policy
  int team_size = 4;
  int league_size = 10;
  Kokkos::TeamPolicy<> policy(league_size, team_size);

  // Define the scratch pad size
  int scratch_size = team_size * sizeof(double);

  // Create a parallel_for with a team policy
  Kokkos::parallel_for("useScratchPad",
                       policy.set_scratch_size(0, Kokkos::PerTeam(scratch_size)),
                       KOKKOS_LAMBDA(const Kokkos::TeamPolicy<>::member_type& team) {
    // Each team gets a scratch pad
    Kokkos::View<double*, Kokkos::ScratchMemorySpace<Kokkos::DefaultExecutionSpace>> scratch(team.team_scratch
        (0), team_size);

    // Each thread in the team writes to the scratch pad
    Kokkos::parallel_for(Kokkos::TeamThreadRange(team, team_size), [&](int i) {
      scratch(i) = i * 1.0;
    });

    // The first thread in each team prints the scratch pad data
    if (team.team_rank() == 0) {
      for (int i = 0; i < team_size; i++) {
        printf("scratch(%d) = %f\n", i, scratch(i));
      }
    }
  });

  // Ensure all kernels have completed
  Kokkos::fence();

  Kokkos::finalize();
  return 0;
}
```

In Listing 26.4, we demonstrate basic usage of a scratch pad. We first define a size `scratch_size` for the scratch pad. There are multiple levels of scratch pads that can be defined, and each level has a maximum size that can be defined depending on the hardware. The first level (0) is the smallest because this is the scratch pad that corresponds with L1 cache. The second level (1) corresponds with *high-bandwidth memory*, and the final level (2) corresponds with RAM and storage devices. The level of scratch memory is defined as the first argument to the `set_scratch_size` function. In this example, we use the first level. We then specify the second argument as the size of the scratch pad for the team. We could define an additional third argument to specify the size of the scratch pad for each thread. Next, we define a view `scratch` to represent the scratch pad for each team. The first argument to the view `team.team_scratch` creates a handle to the scratch pad for each team at level 0. The second argument is the runtime dimensions of the scratch pad. It is best to just use the team size as the dimensions of the scratch pad. We follow this by having each thread write to the scratch pad and have the first thread in each team print out the contents of its team's scratch pad.

26.2 Tasking

The final topic that we cover for Kokkos is *tasking*. Tasking or *task-based programming* is an alternative method of writing parallel code in terms of tasks, as opposed to in terms of threads or processes. Tasking in Kokkos specifically is just another way to express parallelism. There is a caveat to tasking in Kokkos, namely that the problem types are restricted to ones that consist of large numbers of small data parallel tasks with arbitrary dependencies. Because of this, the tasking interface in Kokkos may not be suitable for other problems that can be represented using a task-based approach. Everything that we have learned about Kokkos concepts so far still holds true; for example, tasks still follow the pattern, policy, and functor structure that the other parallel constructs follow. The main difference is in how tasks are structured, and this difference includes having distinct patterns and policies separate from the ones used with the parallel constructs introduced in previous chapters.

Like the computational body of a parallel construct, tasks are still defined within a functor. At the time of writing, tasks cannot be defined with lambda functions. The first statement inside a task's functor needs to be its return type. This condition is satisfied by defining the alias `value_type` within the functor. When it comes to defining a functor for a task, the first argument to the () operator must be a `TeamMember`, which for simplicity can be provided as a template parameter to the functor. The value that gets returned from the completion of the task must be provided as the last parameter to the () operator. For launching tasks, the process that is used is different from what we have seen in previous chapters. This is a result of having to specify a `scheduler` for the tasks. The `scheduler` is another Kokkos concept that is used within an `execution_space` to manage the execution of tasks. It is important to use the Kokkos-defined schedulers to allow tasking code to remain portable across different hardware. Demonstrating the setup involved in tasking is best done through an example, which is shown in Listing 26.5.

Listing 26.5 Example of using tasking in Kokkos.

```
#include <Kokkos_Core.hpp>
#include <iostream>

template <class Scheduler>
struct SumArray {
  using value_type = double; // Task Return type

  double *array;
  int start;
  int end;
  Kokkos::BasicFuture<double, Scheduler> left;
  Kokkos::BasicFuture<double, Scheduler> right;


  KOKKOS_INLINE_FUNCTION
  SumArray(double *array, int start, int end):
      array(array), start(start), end(end) {}

  template <class TeamMember>
  KOKKOS_INLINE_FUNCTION
  void operator()(TeamMember& member, double &result) {
    auto& scheduler = member.scheduler();
    if (end - start <= 10) { // Compute the sum
      int sum = 0;
      for (int i = start; i < end; i++) {
        sum += array[i];
      }
      result = sum;

    // Check if the left and right futures are ready
    } else if (!left.is_null() && !right.is_null()) {
      result = left.get() + right.get();
    } else {
      int mid = (start + end) / 2;
      left = Kokkos::task_spawn(Kokkos::TaskSingle(scheduler),
                                SumArray{array, start, mid});

      right = Kokkos::task_spawn(Kokkos::TaskSingle(scheduler),
                                 SumArray{array, mid, end});

      Kokkos::BasicFuture<void, Scheduler> sum_array[] = { left, right };
```

```
      auto array_all = scheduler.when_all(sum_array, 2);
      printf("Respawning task\n");
      Kokkos::respawn(this, array_all);
    }

  }
};



int main(int argc, char* argv[]) {
  Kokkos::initialize(argc, argv);
  {
    // This is a very manual process to defining the scheduler and memory pool
    using scheduler_type = Kokkos::TaskScheduler<Kokkos::DefaultExecutionSpace>;
    using memory_space = typename scheduler_type::memory_space;
    using memory_pool = typename scheduler_type::memory_pool;
    size_t memory_pool_size = 1 << 22;

    auto mpool = memory_pool(memory_space{}, memory_pool_size);
    auto scheduler = scheduler_type(mpool);

    Kokkos::BasicFuture<double, scheduler_type> result;

    int n = 100;
    int answer = 0;
    double* array = new double[n];
    for (int i = 0; i < n; i++) {
      array[i] = i;
      answer += i;
    }

    result = Kokkos::host_spawn(Kokkos::TaskSingle(scheduler),
                                SumArray<scheduler_type>{array, 0,n}
                );
    Kokkos::wait(scheduler);
    std::cout << "Sum from tasks is: " << result.get() << std::endl;
    std::cout << "Sum from direct computation is: " << answer << std::endl;

  }
  Kokkos::finalize();
}
```

In Listing 26.5, we attempt to use tasking to sum the elements of a vector. Generally, this problem would be solved using `parallel_reduce`; however, this problem still provides a good opportunity to demonstrate the use of tasking. We begin by defining a functor `SumArray` and set its `value_type` alias as `double` for the return type. Member variables are then defined including two of type `Kokkos::BasicFuture`. A `Kokkos::BasicFuture` is used as a handle to the result of a task for when it completes and allows the spawning of additional tasks when used within the functor. More simply, it is a place where tasks can store their results sometime in the future. Next, we define the constructor for the functor, which takes in an array, a start index, and an end index for the array. We follow the constructor by specifying the `()` operator, which takes in a `TeamMember` specified through a template parameter and a double reference to store the result of the task. Our strategy for Listing 26.5 is to split the array into smaller portions that are summed then accumulated and finally returned. The example looks similar to a recursive function. We first define the case where we have an array small enough to be summed in the `if` block. In the `if_else` block, we check the status of the results. This check is important for when we use `Kokkos::respawn` to respawn the task. If the results are not ready, then we execute the `else` block, which separates the array into two parts and spawns two additional tasks for each part. The next few lines in the `else` block are used to *aggregate* the results of the two tasks into another future and inform the scheduler to respawn the tasks when the results of `left` and `right` become available. We do this because there is no way to explicitly wait for the results of another task within a task. Instead, we specify for the task to respawn when the results are ready, which happens when the both `left` and `right` futures get populated. This approach then guarantees the tasks to respawn when the results are ready and retrieve them as part of the `if_else` block of the functor. To summarize, the task initially breaks itself down into smaller sub-tasks until the `if` condition is true, and then the `Kokkos::respawn` method is used to predicate the task on the results of the sub-tasks to sum and collect the results and return them.

Shifting focus now to the `main` function in Listing 26.5, we first define a `Kokkos::TaskScheduler` instance for our scheduler type. After defining the scheduler type, a pool of memory is allocated to the scheduler for the execution of tasks. Unfortunately, this process is manual because it is difficult to determine the amount of memory that a user's particular problem will need. In Listing 26.5, we use a rough default based on the Kokkos lecture series.[2] With this amount, we can go a long way, but if a different amount is needed, it can be adjusted accordingly. Before spawning our tasks, we need to define a `Kokkos::BasicFuture` to store the final result of the task. We then spawn our task using `Kokkos::host_spawn()`, which consists of the *task policy*, `Kokkos::TaskSingle`, and a call to our functor. Instead of using `Kokkos::TaskSingle`, we could also use `Kokkos::TaskTeam` to launch a team of tasks. We note that `Kokkos::host_spawn()` is used outside a task functor, whereas `Kokkos::task_spawn()` is used inside a task functor. Finally, we are allowed to wait for the result from the host code, unlike the task code, with `Kokkos::wait(scheduler)`. We then get and print the result of the task.

[2] https://www.youtube.com/watch?v=xEAyOod57-c&t=5427s

Part VI
Parallel file systems and parallel input/output

This part consists of three chapters and introduces software abstractions that allow for portable high-performance usage of filesystems. Each chapter discusses a single library that is commonly used in extreme-scale computing.

Chapter 27 is on HDF5, the *Hierarchical Data Format* library, which provides an abstract means of dealing with binary files in a portable manner.

Chapter 28 is on NetCDF, the *Network Common Data Format*, which specifies a portable, self-describing data format for working with large-scale array data.

Chapter 29 is on VTK, the *Visualization Toolkit*, which is a C++ library that is designed to work well with 3D visualization.

Chapter 27
HDF5

27.1 Introduction

HDF5 is a file format designed to store and organize large amounts of data. It is portable and flexible, making it suitable for a wide range of applications on many systems. HDF5 offers efficient and performant I/O in a style similar to XML that enables users to define the structure and relationships within the data in a self-describing format. In other words, the information required to interpret an HDF5 file is contained within the file itself. HDF5 includes a *data model* for organizing and providing efficient access to data from an application and *software* for working with HDF5 and storing data in the HDF5 file format with the `.h5` extension. The HDF5 file format provides a flexible abstraction for storing and representing data in a hierarchical structure based on the data model.

27.2 Data Model

27.2.1 Groups and Datasets

The foundation of HDF5 is the *HDF5 Data Model*; it outlines how data are to be specified and organized into the HDF5 file format. The HDF5 data model follows a hierarchical structure for organizing data into two primary objects: *groups* and *datasets*.

Groups are a powerful categorization tool used to assemble the contents of an HDF5 file. They are versatile containers used to hold both datasets and other groups, allowing users to curate the HDF5 file format to meet their needs. Groups are arranged much like directories in a Linux filesystem, with '/' denoting the *root group*. Identification of both groups and datasets within an HDF5 file is achieved by specifying their group path starting from the root group.

© The Author(s), under exclusive license to Springer Nature Switzerland AG 2025

R. J. Spiteri and K. Klenk, *Extreme-Scale Computing*, CMS/CAIMS Books in Mathematics 15, https://doi.org/10.1007/978-3-031-89033-8_27

Datasets are the storage elements used for composing collections of application data into a format that can be stored and retrieved from an HDF5 file. They are multidimensional arrays of data elements paired with supporting *metadata*. The metadata are what make the HDF5 file format self-describing. The metadata include the dataset's datatype, dataspace, properties, and (optionally) attributes. Listing 27.1 demonstrates how to create an HDF5 file, add a group, and add two datasets to the created group.

Listing 27.1 HDF5 introductory example that creates a group and two datasets within that group.

```
#include <iostream>
#include <string>
#include "H5Cpp.h"

using namespace H5;

// Define our dataset dimensions
const int          X    = 6;
const int          Y    = 12;
const int          RANK = 2;

int main(void) {
  // Create a new file and specify to overwrite any existing file if one exists.
  H5File file("hdf5_basics.h5", H5F_ACC_TRUNC);

  // Create a group in the file.
  Group group = file.createGroup("our_first_group");

  // Create the data space for the dataset.
  hsize_t dataset_dimensions[2];
  dataset_dimensions[0] = X;
  dataset_dimensions[1] = Y;
  DataSpace dataspace(RANK, dataset_dimensions);

  // Create the dataset with an absolute path.
  DataSet dataset = file.createDataSet("/our_first_group/our_first_dataset", PredType::STD_I32BE, dataspace);
  // Create the dataset with a relative path.
  DataSet dataset2 = group.createDataSet("our_second_dataset", PredType::STD_I32BE, dataspace);

  // Create dummy data to write to the dataset.
  int data[X][Y];
  int data2[X][Y];
  for (int i = 0; i < X; i++) {
    for (int j = 0; j < Y; j++) {
      data[i][j] = i * j - j;
      data2[i][j] = i * j + j;
    }
  }

  // Write the data to the dataset
  dataset.write(data, PredType::NATIVE_INT);
  dataset2.write(data2, PredType::NATIVE_INT);
}
```

In Listing 27.1, we begin by creating the HDF5 file `hdf5_basics.h5`. In addition to the file name, we also need to specify the desired access mode. In this case, we use the predefined flag `H5F_ACC_TRUNC` to create a new file and overwrite any existing files with the same name. Upon successful creation of the HDF5 file, we obtain an `H5File` object with which we can interact with to create our group and datasets. By calling the `createGroup` method on the file object, we create a group called `our_first_group`. Our new group is situated under the root group and has the path '`/our_first_group/`'.

In order to create the dataset, we need to specify its desired layout. The mechanism we use to achieve this is a *dataspace*. We create a dataspace object by passing it a *rank* and an array that specifies each of the dimensions of our dataset. The rank is simply the dimensionality of our dataset, and the array of dimensions denotes the length of each dimension. We then pass the dataspace object as the final parameter in the calls to the `createDataSet` method. The second parameter, `PredType::STD_I32BE`, is used to specify the type of the elements within the dataset. In the example, we specify the use of 32-bit big-endian signed integers. We postpone the explanation of the first parameter because we demonstrate two ways in which we can create a dataset within a group. For `our_first_dataset`, we call `createDataSet` on the file object and specify an absolute path for the dataset to be contained within `our_first_group`. For `our_second_dataset`, we call `createDataSet` on the group object returned from the creation of our group and only need to specify the name of the newly created dataset. We then populate both datasets. Again, we need to specify the type; we use `PredType::NATIVE_INT` to indicate the use of the system's integer value. This value is converted to `PredType::STD_I32BE` by HDF5 when it is written. One final note about the HDF5 C++ API is that files do not need to be closed explicitly; they are automatically closed when the objects go out of scope.

27.2.2 Dataspaces

As previously alluded to, multidimensional arrays are used to store data and metadata in an HDF5 dataset. We briefly introduced dataspaces in Listing 27.1 as a way to describe the dimensionality of storage arrays. Here, we provide a more detailed explanation of dataspaces.

A simple dataspace is a rectangular N-dimensional array. The dimensions of a dataspace are specified by a rank (number of arrays) and an array of dimensions (length of each array). Dataspaces may be of fixed dimension, they may be of unlimited dimension, or they may describe only a subset of a dataset. An unlimited dimension can be updated with additional elements, expanding the length of the dimension over the course of the dataset's lifetime. When using a dataspace to describe a subset of a dataset—called a *hyperslab*—we are specifying the portion of the dataset with which we want to interact. Using subsets is helpful when we do not need the entire contents of a file. Subsetting allows the dataspace to only set up buffers with the size we need. This reduces memory usage and can increase the efficiency of an application's interaction with HDF5.

27.2.3 Datatypes

HDF5 uses both pre-defined and derived datatypes. Pre-defined datatypes are managed by HDF5 and are not guaranteed to be the same from one HDF5 session to the next. This is because HDF5 uses a handle to open and close the datatype, but the underlying value may change when opening the same HDF5 file on another platform or with a different version of HDF5.

There are two pre-defined datatypes in HDF5: standard and native. Standard datatypes are consistent across all platforms and follow a specified standard such as IEEE-754. Native datatypes are specific to the platform and are not guaranteed to be the same across different platforms. On the other hand, derived datatypes are specified using combinations of the pre-defined types, similar to MPI's derived datatypes. An example of a derived datatype is the *compound datatype*. A compound datatype is a collection of other datatypes in a similar way that a C `struct` is a collection of C types. Listing 27.2 demonstrates how to create a compound datatype and store it within an unlimited dimension.

Listing 27.2 HDF5 example that creates and saves a compound datatype using an unlimited dimension.

```
#include "H5Cpp.h"
using namespace H5;

// define a struct
typedef struct our_compound_type {
  int a;
  float b;
  double c;
} our_compound_type;

const int RANK = 1;

void append_to_dimension(DataSet& dataset, const our_compound_type& data) {
  // Get the current size of the dataset
  hsize_t current_size[1];
  dataset.getSpace().getSimpleExtentDims(current_size, NULL);

  // Get the number of elements to append
  hsize_t data_size[1] = {1};

  hsize_t new_size[1] = {current_size[0] + data_size[0]};

  dataset.extend(new_size);
  // Select the hyperslab that corresponds to the appended data
  DataSpace fileSpace = dataset.getSpace();
  fileSpace.selectHyperslab(H5S_SELECT_SET, data_size, current_size);

  // Create a memory space for the appended data
  DataSpace memSpace(1, data_size, NULL);
  // Write the data to the dataset
  dataset.write(&data, dataset.getDataType(), memSpace, fileSpace);

}

int main (void) {
  // Create the HDF5 file
  H5File file ("compound_type.h5", H5F_ACC_TRUNC);

  // create a compound datatype from the struct
  CompType mtype (sizeof (our_compound_type));
  mtype.insertMember ("a_name", HOFFSET (our_compound_type, a), PredType::NATIVE_INT);
  mtype.insertMember ("b_name", HOFFSET (our_compound_type, b), PredType::NATIVE_FLOAT);
  mtype.insertMember ("c_name", HOFFSET (our_compound_type, c), PredType::NATIVE_DOUBLE);

  // create dataspace
  hsize_t init_dim_size [RANK] = { 0 };
  hsize_t max_dims [RANK] = { H5S_UNLIMITED };
  DataSpace space (RANK, init_dim_size, max_dims);

  // enable chunking for this dataset
  H5::DSetCreatPropList dataset_properties;
  hsize_t chunk_dims [RANK] = { 1 };
  dataset_properties.setChunk (RANK, chunk_dims);

  // create a dataset
  DataSet dataset = file.createDataSet("compound_example", mtype, space, dataset_properties);

  // Store the compound type in the dataset
  our_compound_type comp_data_1 = { 1, 2.0, 3.0 };
  append_to_dimension(dataset, comp_data_1);
  our_compound_type comp_data_2 = { 4, 5.0, 6.0 };
  append_to_dimension(dataset, comp_data_2);

  return 0;
}
```

In Listing 27.2, we create a compound type `mtype`. We start by defining a C struct that is used to store our compound type. In `main`, we create a new file and compound type. We pass the size of our struct to the constructor of the compound type so that HDF5 knows how much storage to use. We then specify each of the members along with their offsets within the structure. Next, we create a dataspace with rank 1, an initial dimension size of 0, and unlimited dimension size. We also have to enable chunking in order to use an unlimited dimension. The chunksize is set to the size of a single element of our compound type. The dataset is then created by passing it a name, the compound type, the dataspace, and the chunking properties. We then turn our attention to the function `append_to_dimension()`. We defined this function to append data to our unlimited dimension. In the function, we get the current size of the dataset and the number of elements we wish to append. We use these values to determine the amount by which to extend the dataset. We then use the `selectHyperslab()` method on our filespace to select the appropriate space we want to write to our dataset. The function finishes by creating the memory space for the data and writing them. We call our function twice to write two structures to `compound_type.h5`.

27.2.4 Properties and Attributes

The final parts of the HDF5 Data Model are properties and attributes. A property is a characteristic of an HDF5 object. For example, a dataset has a data storage layout property that can change the ordering of the physical data within the file. This data storage layout is set to *contiguous* by default. However, to increase performance, the data storage layout can be changed to *chunked* or *chunked and compressed*. We demonstrated creating a property characteristic in Listing 27.2 for the `compound_example` dataset. In that example, compression could have been added with the line

```
dataset_properties.setDeflate(6);
```

for level-6 compression.

In addition to properties, HDF5 also supports attributes. Attributes are optional components of an HDF5 object made up of a name and a value. They are used to store metadata about the HDF5 object. For example, attributes can be used for storing the units of the data within an HDF5 object or the date and time the data were created. In Listing 27.3, we demonstrate how to create and write attributes to a dataset. In addition, we also demonstrate how to read data from a dataset.

Listing 27.3 HDF5 example showing how to create and write attributes to a dataset.

```
#include <iostream>
#include <string>
#include <vector>
#include "H5Cpp.h"

using namespace H5;

// Define our dataset dimensions
const int         X     = 6;
const int         Y     = 12;
const int         RANK  = 2;

int main(void) {
  // Open an existing file allowing read and write access.
  H5File file("hdf5_basics.h5", H5F_ACC_RDWR);
```

```
  DataSet our_first_dataset = file.openDataSet("/our_first_group/our_first_dataset");

  // Create a vector of strings we want to write as attributes
  std::vector<std::string> str_vec;
  str_vec.push_back("X_Axis = Expirement Number");
  str_vec.push_back("Y_Axis = Tempurature (deg C)");

  // create a vector of char pointers
  std::vector<const char*> ptr_vec;
  for (size_t i = 0; i < str_vec.size (); i++) {
    ptr_vec.push_back (str_vec[i].c_str ());
  }

  // create a dataspace for the dataset for our attribute
  hsize_t dim = ptr_vec.size ();
  int attr_rank = 1;
  DataSpace dataspace(attr_rank, &dim);

  // create a variable length string datatype
  StrType str_type (0, H5T_VARIABLE);

  // create an attribute
  Attribute attribute = our_first_dataset.createAttribute("Units", str_type, dataspace);
  // write the vector of char pointers to the attribute
  attribute.write(str_type, (void*)&ptr_vec[0]);

  int data_out[X][Y];
  our_first_dataset.read(data_out, PredType::NATIVE_INT);

  for (int i = 0; i < X; i++) {
    for (int j = 0; j < Y; j++) {
      std::cout << data_out[i][j] << " ";
    }
    std::cout << std::endl;
  }

  return 0; // successfully terminated
}
```

In Listing 27.3, we re-use the HDF5 file to populate `our_first_dataset` with attributes. There is some setup involved in order to store strings as the attribute type. Suppose we now want to add an attribute that explains what each axis represents to the 2D array `our_first_dataset`. We start by opening the file with the `H5F_ACC_RDWR` flag. This flag specifies that we want to open `hdf5_basics.h5` with read-write access. Next, we create a vector of strings and convert them to `char` pointers. Attributes, just like datasets, need a dataspace to specify their size. We create the dataspace with rank 1 and dimension 2. Just before creating the attribute, we define a variable length string type to be used for it. After specifying the new type, we create the attribute by calling the `createAttribute()` method on `our_first_dataset`. We then call the `write()` method on the created attribute to write it to an HDF5 file. For the second part of the example, we read in the data saved in Listing 27.1. Reading in data is straightforward and only requires the use of the already-open `our_first_dataset` dataset object. To read in the data, we call the `read()` method on `our_first_dataset` to fill `data_out` with the contents of the dataset.

27.3 Working with HDF5

To conclude our discussion of HDF5, we discuss some practical considerations and tools for working with HDF5. HDF5 can be used in many of the popular programming languages. We have introduced basic examples centered around the C++ API. However, HDF5 can also be used with C, Fortran, Java, and Python. When working with HDF5 in C and Fortran, it is important to handle interactions with the

HDF5 objects correctly. Interactions with HDF5 objects typically follow three steps: i) open the object, ii) access/manipulate the object, and iii) close the object. It is important to close the object to prevent memory leaks, undefined behavior, and the unnecessary consumption of system resources.

It is also useful to understand the particularities of prefixes in the HDF5 C and C++ APIs. Different parts of the HDF5 interface use prefixes that start with "`H5`" to distinguish them from each other. For example, the `H5A` prefix is used for dealing with attributes, the `H5F` prefix is used for dealing with files, the `H5D` prefix is used for dealing with datasets, and the `H5S` prefix is used for dealing with dataspaces. In our examples, Listing 27.1, Listing 27.2, and Listing 27.3, we saw the `H5F` prefix when specifying the appropriate flag to the H5File constructor.

Much of our discussion of HDF5 has been centered around the HDF5 API. However, HDF5 also provides a number of command-line tools to help increase productivity. One of the most useful tools is `h5dump`. The `h5dump` tool prints the contents of an HDF5 file in ASCII text to the terminal. There are many options that can be provided to `h5dump` to control the output. For example, `-d` can be used to specify a dataset to print, `-H` can be used to print the header information of the file, and `-a` can be used to print a specific attribute. Using `h5dump` is helpful to quickly determine the contents of an HDF5 file. HDF5 also provides compiler wrappers for compiling and linking applications that use HDF5. These wrappers work in the same way as the MPI wrappers `mpicc` and `mpic++` by using existing compilers on the system and specifying additional include and link flags to enable compiling and linking to an installed HDF5 library. The HDF5 wrappers are `h5cc`, `h5c++`, and `h5fc` for C, C++, and Fortran, respectively. They can be used to automatically link the HDF5 library to an application.

Lastly, HDF5 can be built to support parallel I/O. Parallel HDF5 is built on top of MPI-IO. This means that the MPI-2 standard is required to use parallel HDF5. Parallel HDF5 files are compatible with serial HDF5 files. When using parallel HDF5, a single file image is shared by all processes. A single file image is used instead of separate files for each process to reduce the resources consumed by the processes. When using parallel HDF5, the HDF5 file is opened with an MPI communicator. All processes in the communicator can access the opened file. A limitation of the parallel implementation of HDF5, however, is that processes can only open a single HDF5 file at a time. This means that either i) the parallel file with which the processes of a communicator are working must be closed before a new parallel HDF5 file can be opened, or ii) a different communicator (with absolutely no overlap in physical ranks from the first) must be used to open a second parallel file. HDF5 does not support shared-memory programming paradigms and is strictly designed for use with MPI.

Unfortunately, at the time of writing, there is no C++ API for parallel HDF5. If one wants to make use of parallel API from a C++ code, the C API must be used.

27.4 Reference material

- High-level introductory material: `https://support.hdfgroup.org/HDF5/Tutor/HDF5Intro.pdf`
- More details on the basics: `https://support.hdfgroup.org/HDF5/Tutor/introductory.html`

Chapter 28
NetCDF

In this chapter, we introduce the netCDF data format. NetCDF is a popular data format used in scientific computing and is particularly popular within the earth sciences. NetCDF is more formally known as the *network Common Data Form*. NetCDF is a set of software libraries specialized for the reading and writing of array-oriented scientific data. Its attractive feature is the ability to create datasets that are self-describing and machine independent. NetCDF is built using the C programming language but provides interfaces for C++, Fortran, and Java. Since the release of netCDF 4.0, netCDF is built on top of HDF5 and offers some interoperability with HDF5. HDF5 was covered in detail in Chapter 27.

28.1 NetCDF data model

NetCDF consists of two data models since the release of version 4.0. The original data model is called the *classic data model*, and the more recent data model is called the *enhanced data model*. It might be more appropriate to consider the enhanced data model as an extension to the classic data model because it is fully backwards compatible. With this perspective in mind, we cover the classic data model first and then introduce the new features that were added in the enhanced data model.

28.1.1 Classic Data Model

The classic data model is made up of three main components that serve as the foundation for netCDF datasets: dimensions, variables, and attributes.

The dimension component describes the sizes of the variables within a netCDF file. A good way to think about dimensions is to consider them as the lengths of the axes in a multidimensional array. Within a netCDF dataset, only one dimension is

© The Author(s), under exclusive license to Springer Nature Switzerland AG 2025

R. J. Spiteri and K. Klenk, *Extreme-Scale Computing*, CMS/CAIMS Books in Mathematics 15, https://doi.org/10.1007/978-3-031-89033-8_28

allowed to be the *unlimited dimension*. An unlimited dimension is a dimension that can grow in size over the lifetime of a netCDF dataset.

Variables are used to store data as N-dimensional arrays, where each dimension of the variable has to match an existing dimension defined within the netCDF dataset. Variables within the classic data model can only be one of the following types: `byte`, `char`, `short`, `int`, `float`, or `double`.

Finally, attributes are metadata that can be associated with a variable or with the dataset itself. Attributes help describe aspects of the dataset, such as the units of the data or the author of the dataset. Attributes can only be a single value or a one-dimensional array, with the latter being ideal for storing strings of characters. When an attribute is associated with a dataset, it is considered a *global attribute*.

Now that we have described the basic components of the classic data model, we can look at an example of how to create a netCDF file and populate it with data using the C++ interface.

Listing 28.1 A basic netCDF example. Here we demonstrate how to create a file, add dimensions, add a variable, and populate the variable with data.

```
#include <iostream>
#include <netcdf>

// Define the dimensions for our netCDF file
#define xDimSize 12
#define yDimSize 6

// Return this in event of a problem
constexpr int nc_err = 2;

int main() {
  // Create a netCDF file. Replace parameter tells netCDF to overwrite
  // this file, if it already exists.
  try {
    netCDF::NcFile dataFile("netcdf_basics.nc", netCDF::NcFile::replace);

    // Define the dimensions
    auto xDim = dataFile.addDim("x", xDimSize);
    auto yDim = dataFile.addDim("y", yDimSize);

    // Define an unlimted dimension
    auto timeDim = dataFile.addDim("time");

    // Define units for the time dimension
    auto timeUnits = dataFile.putAtt("time units", "seconds since 1970-01-01 00:00:00");

    // Define a variable
    auto data = dataFile.addVar("our_first_variable", netCDF::ncInt, {timeDim, xDim, yDim});

    // Populate the variable in a loop with some data
    for (size_t i = 0; i < 10; i++) {
      std::vector<size_t> start_vec = {i, 0, 0}; // starting index for variable dimensions
      std::vector<size_t> count_vec = {1, xDimSize, yDimSize}; // number of elements to write for each
            dimension

      // Create vector of data to write
      int data_vec[xDimSize][yDimSize];
      for (int j = 0; j < xDimSize; j++) {
        for (int k = 0; k < yDimSize; k++) {
          data_vec[j][k] = i * j * k;
        }
      }

      data.putVar(start_vec, count_vec, &data_vec);
    }

    return 0;

  } catch (netCDF::exceptions::NcException &e) {
    std::cout << e.what() << std::endl;
    return nc_err;
  }
}
```

Analyzing the example in Listing 28.1, we first point out the try-catch block. The netCDF documentation recommends that you use a try-catch block when using the netCDF C++ interface. All netCDF exceptions are derived from the `NcException` class, which is used to catch any errors that may occur when using the netCDF C++ interface. Inside the try block, we first create a netCDF file. Upon successful creation of the file, a `NcFile` *object*, referred to as `dataFile`, is returned. We then use the newly created `dataFile` object to call the `addDim()` method. We create three dimensions in this example, x and y dimensions with different sizes, and a time dimension to demonstrate how to create an unlimited dimension. Unlimited dimensions are created by omitting the second argument to the `addDim()` method. After the creation of the time dimension, we create a *global attribute* that describes the units for the time data. Lastly, we show how to create a variable using the `addVar()` method, then populate it with data, and write it to the netCDF file with the `putVar()` method. For the `putVar()` method, we provide a `start_vec` and a `count_vec`; more detail on these arguments is provided in Section 28.2. In short, the `start_vec` indicates where to start writing within the netCDF dataset, and the `count_vec` indicates how many elements to write. The result of Listing 28.1 is a netCDF file *netcdf_basics.nc* that contains three dimensions and the variable `our_first_variable`. It is worth noting that when using netCDF in C, the netCDF file must be closed explicitly. In C++, however, the netCDF file is closed automatically once the `dataFile` object goes out of scope.

For our next example, we look at how to read and inspect the netCDF file we have just created.

Listing 28.2 How to open a netCDF file and inspect some of its contents.

```cpp
#include <iostream>
#include <netcdf>

// Return this in event of a problem
constexpr int nc_err = 2;

int main() {
  try {
    // Open the NetCDF file for reading
    netCDF::NcFile dataFile("netcdf_basics.nc", netCDF::NcFile::read);

    // Get the time dimension and check its size
    auto timeDim = dataFile.getDim("time");
    std::cout << "time dimension size: " << timeDim.getSize() << std::endl;

    // Get the time attribute we set in the previous example
    std::string output;
    auto timeUnits = dataFile.getAtt("time units");
    timeUnits.getValues(output);
    std::cout << "time units: " << output << std::endl;

    // Get our_first_variable
    auto data = dataFile.getVar("our_first_variable");
    std::cout << "variable name: " << data.getName() << std::endl;
    std::cout << "variable dimensions: " << data.getDimCount() << std::endl;
    std::cout << "dimension 0: " << data.getDim(0).getName() << std::endl;
    std::cout << "dimension 1: " << data.getDim(1).getName() << std::endl;
    auto xDimSize = data.getDim(1).getSize();
    std::cout << "dimension 2: " << data.getDim(2).getName() << std::endl;
    auto yDimSize = data.getDim(2).getSize();

    // Read the first time step
    std::vector<size_t> start_vec = {2, 0, 0}; // starting index for variable dimensions
    std::vector<size_t> count_vec = {1, xDimSize, yDimSize}; // number of elements to write for each dimension
    int data_vec[xDimSize][yDimSize];
    data.getVar(start_vec, count_vec, &data_vec);
    std::cout << "value of data_vec[2][3] " << data_vec[2][3] << std::endl;

    return 0;

  } catch (netCDF::exceptions::NcException &e) {
```

```
    std::cout << e.what() << std::endl;
    return nc_err;
  }
}
```

In Listing 28.2, we open the file in a similar way to how it was created in the previous example. The difference here is the argument we pass to the `NcFile` constructor. We pass `NcFile::read` to indicate that we want to open the file in read-only mode. We then use the `getDim()` method to get the `time` dimension and check its size. Additionally, we inspect the global attribute to see what units of time are being used. We then use the `getVar()` method to get the variable `our_first_variable` and check some of its properties. Finally, we read the data from the variable. We create a vector of starting indices to indicate where in the file we want to start the read for each dimension. We then specify a count vector that says how many data are to be read from each dimension and output one of the values within the variable to the screen. The output of Listing 28.2 is

```
time dimension size: 10
time units: seconds since 1970-01-01 00:00:00
variable name: our_first_variable
variable dimensions: 3
dimension 0: time
dimension 1: x
dimension 2: y
value of data_vec[2][3] 12
```

28.1.2 Enhanced Data Model

With the introduction of netCDF version 4.0 (netCDF-4), the classic data model was extended to include *groups* and *user-defined data types*. Dimensions, variables, and attributes are still at the core of the enhanced data model and are used in the same way as described in Section 28.1.1. However, now they can be grouped together and created to form hierarchical structures within a netCDF dataset. Each group in a netCDF dataset is much like a dataset from the classic data model. One way to conceptualize how the enhanced data model works is that each group is a file from the classic data model. When using groups, each group can have its own dimensions, but subgroups can inherit dimensions from their parent group. Because groups act as their own datasets, they can also contain variables with the same name as other groups. We now look at an example of how to use groups in netCDF.

Listing 28.3 A netCDF example that introduces how to use groups.

```
#include <iostream>
#include <netcdf>

// Define the dimensions for our netCDF file
#define xDimSize 12
#define yDimSize 6
#define zDimSize 10

// Define an error code for netCDF
constexpr int nc_err = 2;
```

```
int main() {
  // Create a netCDF file.
  // This will be our root group
  try {
    netCDF::NcFile rootGroup("netcdf_groups.nc", netCDF::NcFile::replace);

    // Define the dimensions
    auto xDim = rootGroup.addDim("x", xDimSize);
    auto yDim = rootGroup.addDim("y", yDimSize);

    // Define a variable in the root group
    auto variable_1 = rootGroup.addVar("variable_1", netCDF::ncInt, {xDim, yDim});

    // Create a group
    auto group1 = rootGroup.addGroup("group_1");

    // Create a new dimension in the group
    auto zDim = group1.addDim("z", zDimSize);

    // Define a variable in the group
    // notice groups inherit dimensions from parents (xDim, yDim)
    auto variable_2 = group1.addVar("variable_2", netCDF::ncInt, {xDim, yDim, zDim});

  } catch (netCDF::exceptions::NcException &e) {
      std::cout << e.what() << std::endl;
      return nc_err;
  }
}
```

In Listing 28.3, we start by creating a netCDF file in exactly the same way we did for the classic data model; i.e., there is no difference in the way a file is created in the enhanced data model. When the file is created, it is implicitly considered the *root group*. We then add dimensions and a variable within the root group. To create a subgroup, we call the `addGroup()` method on our root group object, returning the new subgroup, `group1`. We then add a dimension and a variable to the subgroup. We note that a created variable can use the same dimensions as the root group.

The other extension to the enhanced data model is the introduction of user-defined types. NetCDF supports three versions of user-defined types. The first version of user-defined type is the *compound type*. Compound types are similar in structure to a C `struct`. They are a collection of arbitrary data of different types that can include other compound data types. In Listing 28.4, we provide an example of how to create a compound type and store some data in it.

Listing 28.4 An example of how to create a compound data type. When using compound data types we need a corresponding `struct` that we use to define the structure of the compound data type.

```
#include <iostream>
#include <netcdf>

// Define the dimensions for our netCDF file
#define teamsDimSize 8
#define rosterDimSize 20
#define maxTeamNameLength 20

// Return this in event of a problem
constexpr int nc_err = 2;

struct team {
  std::string team_name;
  int team_id;
  double roster_ids[rosterDimSize];
};

int main() {
  try {
    netCDF::NcFile dataFile("user_defined_types.nc", netCDF::NcFile::replace);

    // We are creating a computing type of a team_name, team_id, and roster_ids
    // To create the compound type we need to know the size of all the members

    // Define a compund type
    netCDF::NcCompoundType compoundType = dataFile.addCompoundType("my_compound_type", sizeof(team));
```

```
    // Add the members to the compound type
    compoundType.addMember("team_name", netCDF::ncString, 0);
    compoundType.addMember("team_id", netCDF::ncInt, offsetof(team, team_id));

    // The shape of the array we are adding, each index is a dimension of the array we want to add
    std::vector<int> roster_ids_shape = {rosterDimSize};
    compoundType.addMember("roster_ids", netCDF::ncDouble, offsetof(team, roster_ids), roster_ids_shape);

    // Define the dimensions
    auto teamsDim = dataFile.addDim("numTeams", teamsDimSize);
    auto allTeamsVar = dataFile.addVar("allTeams", compoundType, {teamsDim});

    team teams[teamsDimSize];
    for (int i = 0; i < teamsDimSize; i++) {
      teams[i].team_name = "Team " + std::to_string(i);
      teams[i].team_id = i;
      for (int j = 0; j < rosterDimSize; j++) {
        teams[i].roster_ids[j] = i + j + (i * 19);
      }
    }

    // Write the teams to a file
    allTeamsVar.putVar(&teams);


  } catch (netCDF::exceptions::NcException &e) {
    std::cout << e.what() << std::endl;
    return nc_err;
  }
}
```

In Listing 28.4, we create a compound data type called "`my_compound_type`". When creating a compound data type, we need a corresponding C struct to communicate the size of the compound data type to the `addCompoundType()` method. We then add the data members to our compound data type using the `addMember()` method. When adding members, we need to specifically state the name of the member, the type, the offset, and (if we are using arrays) the shape of the array. To then store data in the compound data type, we simply create the C struct and pass it to the `putVar()` method.

The second version of user-defined types is the *opaque type*. Opaque types are used to store data that are not explicitly understood by netCDF; however, netCDF can still store the bits of data that make up the opaque type to be retrieved later. Opaque types behave similarly to void pointers in C.

The third version of user-defined types is the *variable length (VLEN) array*, which consists of arrays of the basic data types. At each save point, the VLEN array can be a different length. In Listing 28.5, we provide an example of how to create both opaque and VLEN types.

Listing 28.5 An example of how to store an opaque type and a VLEN array.

```
#include <iostream>
#include <netcdf>

// Define the dimensions for our netCDF file
#define teamsDimSize 8
#define rosterDimSize 20
#define maxTeamNameLength 20

// Return this in event of a problem
constexpr int nc_err = 2;

class example_class {
public:
  int a;
  int b;
  int c;

  example_class(int a, int b, int c) {
    this->a = a;
    this->b = b;
    this->c = c;
```

```
  }
};

int main() {
  try {
    netCDF::NcFile dataFile("user_defined_types_2.nc", netCDF::NcFile::replace);

    // ****************************
    // Opaque Type Example
    // ****************************
    // Create an opaque type
    example_class example(1, 2, 3);
    netCDF::NcOpaqueType opaqueType = dataFile.addOpaqueType("my_opaque_type", sizeof(example_class));

    // store the example class in a variable
    auto exampleVar = dataFile.addVar("example", opaqueType);

    // Write the example class to a file
    exampleVar.putVar(&example);

    // ****************************
    // VLEN Type Example
    // ****************************
    // Create a vlen type
    netCDF::NcVlenType vlenType = dataFile.addVlenType("my_vlen_type", netCDF::ncInt);

    // Define the dimensions
    auto timeDim = dataFile.addDim("timeDim", 2);

    // Create a variable that uses the vlen type
    auto vlenVar = dataFile.addVar("vlenVar", vlenType, {timeDim});

    // Create some data to write to the file
    nc_vlen_t vlenData[3];
    int data1[3] = {1, 2, 3};
    int data2[4] = {4, 5, 6, 7};
    vlenData[0].p = data1;
    vlenData[0].len = 3;
    vlenData[1].p = data2;
    vlenData[1].len = 4;

    // Write the data to the file
    vlenVar.putVar(&vlenData);

  } catch (netCDF::exceptions::NcException &e) {
    std::cout << e.what() << std::endl;
    return nc_err;
  }
}
```

In Listing 28.5, we create an opaque type called `my_opaque_type` and show how to store a C++ object in it even though netCDF does not know what the object is. We then create a VLEN type called `"my_vlen_type"` and use it to create the `vlenVar` variable. We then store two arrays of different sizes in the `vlenVar` variable.

This section should have sounded familiar to you because these extensions in the enhanced model of netCDF map closely to the structure of HDF5 data as discussed in Chapter 27.

28.2 Subsetting

Sometimes we may not want to read the entire dataset from a netCDF file. This can be due to various reasons such as the dataset is too large to fit in memory or only a subset of the data is needed. NetCDF provides a way to subset data from a netCDF file by specifying *start* and *count* arguments to the `getVar()` and `putVar()` methods. In Listing 28.1 and Listing 28.2, these arguments were passed to the `getVar()` and `putVar()` methods. These arguments are used to specify the starting index and the number of elements to read or write from the variable. Each argument needs to match the shape of the variable we are reading or writing. For example, if we have a variable with dimensions `time`, `x`, and `y`, then we would need to specify a start argument as a vector with three elements specifying where in the variable we are starting to write to or read from. The count argument follows the same rule and needs to be of size three. The count argument specifies how many elements are to be read or written as measured from the starting index. Data that are read in this subset form are thus hyper-rectangular in geometry.

Another note on reading and writing is that netCDF automatically handles the conversion of the data types in both directions. For example, if we have a netCDF variable of type `float` and we want to read it to a C++ variable of type `double`, netCDF automatically handles the data conversion for us. The automatic data conversion also works in the other direction: If we have a C++ variable of type `double` and we want to write it to a netCDF variable of type `float`, netCDF again automatically handles the data conversion for us.

28.3 NetCDF Command-Line Tools

The developers of netCDF and others have created a number of command-line tools for working with netCDF datasets. These tools can be used to inspect the contents of a netCDF file, convert between different netCDF file formats, and even compress the data within a file. Some basic command-line tools provided by the netCDF developers are `ncdump`, `ncgen`, and `nccopy`.

As a brief introduction to these tools, `ncdump` is used to inspect the contents of a netCDF file. It can be used to see how many dimensions exist as well as variables, and attributes. It can even be used to output the data of variables to the command line. The `ncgen` tool can be used to create netCDF files from ASCII files that are stored in the network Common Data form Language (CDL) format. The `nccopy` tool is used to copy netCDF files. The `nccopy` tool may be of particular interest because it can be used to chunk data within a file.

Chunking data within a large netCDF file can have a massive effect on performance because it allows you to only parse through the data you need when opening the file. When chunking data, you specify the dimensions you want to chunk by and the size of the chunks. For example, if we have a variable with dimensions `time`, `x`, and `y` with respective sizes of 1000, 100, and 100, we can chunk the data by specifying the `time` dimension and a chunk size of 100. This would result in 10 chunks of data. With the order of the dimensions of the variable as `time` : `x` : `y`, the command to chunk these data would look like

```
nccopy -c 4 ourVariable:100,100,100 input.nc output.nc
```

Chunking netCDF datasets to improve performance of reading or writing nc files is often system-dependent. For a given cluster, particularly those with network-based storage, there should be recommendations of chunk sizes and dimensions that work well with the networking architecture available and their capabilities.

More information on the above command-line tools can be found at `https://docs.unidata.ucar.edu/netcdf-c/current/netcdf_working_with_netcdf_files.html`.

There are also a number of third-party tools that can be used to work with netCDF files. NCO (netCDF Command-Line Operators) is a popular third-party tool that can be used to manipulate and inspect netCDF files. We list some of the more useful tools and their links below:

- NCO: `https://nco.sourceforge.net/`
- UDUNITS: `https://www.unidata.ucar.edu/software/udunits/`
- IDV: `https://www.unidata.ucar.edu/software/idv/`
- NCL: `https://www.ncl.ucar.edu/`
- GrADS: `http://cola.gmu.edu/grads/`

28.4 Parallel access

Unlike the HDF5 library, the C++ interface for netCDF (`netcdf-cxx`) allows for parallel interaction with netCDF-4 files, as long as the underlying `netcdf-c` and `hdf5` libraries have been built with parallel support. More details on this setup can be found at `https://www2.nrel.colostate.edu/projects/irc/public/Documents/Software/netCDF/cpp4/html/page_parallel.html`.

28.5 Additional reference material

- The NetCDF data model: `https://www.unidata.ucar.edu/software/netcdf/docs/netcdf_data_model.html`
- Introductory user guide: `https://docs.unidata.ucar.edu/nug/current/index.html`
- NetCDF Tutorial: `https://docs.unidata.ucar.edu/netcdf-c/current/tutorial_8dox.html`

Chapter 29
VTK

29.1 Introduction

Visualization is the act of depicting data in graphical form. Learning to visualize data can help you understand and communicate the story behind the data more effectively. The Visualization Toolkit (VTK) is a powerful library for 3D visualization, especially of scientific data. Although the underlying architecture may seem complex at first, the goal of VTK is to be easy to use and learn.

VTK provides an object-oriented approach to visualization. VTK is implemented as a C++ class library, but it also allows user access through several higher-level language interface layers including Python, TCL, and Java. VTK is a frequently used I/O tool in common finite-element libraries such as `deal.II`, `FEniCS`, and `FreeFEM` for solution visualization, and it is a key feature of many other applications including biomedical image analysis, sound engineering, and geoscience research.

29.2 Using VTK

Installing VTK is straightforward. Many Linux distributions contain VTK (and development versions) in their repositories, although they may not be the most recent version nor be compiled with all the components that VTK has to offer. Fortunately, building VTK from source is not difficult and is well documented on the VTK website, `https://vtk.org/documentation/`, and wiki pages, `https://vtk.org/Wiki/VTK`. In short, all we need is to download the source code and configure and build VTK using CMake.

After installing VTK, we can use it in our own programs by further leveraging CMake to simplify the compilation process. To do so, we can define a `CMakeLists.txt` file that contains the `find_pacakage(VTK [OPTIONS])` command. This command is used to locate the installation of VTK. To link VTK to the executable <`exe`>, we can use the CMake command

© The Author(s), under exclusive license to Springer Nature Switzerland AG 2025

R. J. Spiteri and K. Klenk, *Extreme-Scale Computing*, CMS/CAIMS Books in Mathematics 15, https://doi.org/10.1007/978-3-031-89033-8_29

target_link_libraries(<exe> ${VTK LIBRARIES}).We then run CMake to configure our project, followed by make to build, with CMake having set up the necessary includes and linking. An example CMakeLists.txt file that builds two examples in this chapter is given in Listing 29.1.

Listing 29.1 A CMakeLists.txt file that builds two examples in this chapter.

```
cmake_minimum_required(VERSION 3.10)
project(test)

find_package(VTK REQUIRED)
include(${VTK_USE_FILE})

add_executable(vtk_smart_pointer vtk_smart_pointer.cpp)
target_link_libraries(vtk_smart_pointer ${VTK_LIBRARIES})

add_executable(vtk_process_objects vtk_process_objects.cpp)
target_link_libraries(vtk_process_objects ${VTK_LIBRARIES})
```

VTK is designed using the object-oriented programming paradigm. Most VTK classes inherit from the vtkObject superclass. At times, some classes inherit from the vtkObjectBase superclass. When instantiating objects in VTK, we must explicitly use the class's New() method. When we are finished with the object, we must also explicitly call the object's Delete() method. Failure to do so results in memory leaks, which can lead to undefined behavior or the program crashing outright. VTK classes should therefore be treated as pointers in their use.

Luckily, VTK has implemented their own reference-counted smart pointer. VTK's smart pointers are called vtkSmartPointer<>, and they can be included with the vtkSmartPointer.h header file. We should recognize that the behavior of the vtkSmartPointer<> is similar to the C++ std::shared_ptr<> introduced in Chapter 7. When we use the vtkSmartPointer<>, calls to the object's Delete() method are automatically handled when the object goes out of scope, and the reference count of the object is 0, thus preventing memory leaks within our program.

An example of how to use the vtkSmartPointer<> for a vtkPolyData dataset is found in Listing 29.2.

Listing 29.2 Using vtkSmartPointer<> to manage a vtkPolyData dataset

```
#include "vtkPolyData.h"
#include "vtkSmartPointer.h"

int main(int argc, char* argv[]) {
  // Create a dataset
  vtkSmartPointer<vtkPolyData> myObject = vtkSmartPointer<vtkPolyData>::New();

  // Do something with the dataset

  // The dataset will be deleted automatically when it goes out of scope
  return EXIT_SUCCESS;
}
```

29.3 VTK Objects and the Visualization Pipeline

29.3.1 Data Objects

The most abstract representation of an object in VTK is the *data object*. Data objects are simply a collection of data, e.g., scalars, vectors, or tensors, without any specific representation of what the data are or what form they should take. Often data objects are built on the superclass `vtkDataArray`. The `vtkDataArray`superclass provides methods to manipulate the data in the objects; no direct access should occur. These methods can get characteristics about the data such as minimum or maximum values.

The internal structures of data objects can differ from each other. In other words, different data object have different methods and different effects on performance, storage, and the process objects with which they interact. Depending on the needs of the user, different data objects can be used to represent the same data.

29.3.2 Datasets

Data objects on their own provide little value in visualization because we cannot operate on them with visualization algorithms. Accordingly, we need to define data objects in a formal structure in order for visualization algorithms to operate on them. We can do this by organizing data objects into *datasets* (class `vtkDataSet`). Datasets consist of geometric data (such as points), topological data (such as connections among points), and attribute data (such as fields defined on the points). Dataset objects store most of this information in subclasses of the `vtkDataArray` type, such as `vtkDataSet`. Similar to `vtkDataArray`, the class `vtkDataSet` also has methods to access information about a given dataset's structure and associated data. VTK distinguishes geometry from topology, defining topology as the set of properties that remain the same after applying geometric transformations, whereas geometry is considered the application of the topology into a structured object. In other words, geometry is always simply a list of coordinates, and topology represents the connectedness of the geometry.

For example, a VTK dataset may have many points defined within it. The dataset may also have a triangle that is defined by a list of the three point IDs that make up its corners. For that triangle, the list of point IDs represents its topology, and the point locations make up the geometry. In practice, a dataset may have many such triangles defined from the list of points; this is in fact how a `vtkUnstructuredGrid` is created.

29.3.2.1 VTK Dataset Types

There are six primary dataset types used by VTK: *polygonal data* (vtkPolyData) for representing arbitrary polygons, *image data* (vtkImageData) for representing square-spaced (*raster*) grids, *rectilinear grid* (vtkRectilinearGrid) for representing rectangular grids that may have variable widths between edges, *structured grid* (vtkStructuredGrid) for representing rectangular array data with particular geometrical transformation, *unstructured grid* (vtkUnstructuredGrid) for representing grids of arbitrary polygons, and lastly *unstructured points* (vtkPolyData or vtkUnstructuredGrid) for representing arbitrary distributed points. For unstructured points, there is no corresponding dataset type like the other five types have; however, the vtkPolyData and vtkUnstructuredGrid types can be used to represent unstructured points. These types can be categorized into two groups: *structured* and *unstructured*.

29.3.2.2 Points

A dataset stores point data as a Points element. This element consists of a vtkDataArray, where each holds three values corresponding to each point's x, y, and z coordinates. Coordinates data may also be stored as a Coordinates element. Adding points to a VTK dataset looks like

```
// collection of points
vtkSmartPointer<vtkPoints> points = vtkSmartPointer<vtkPoints>::New();
// loop over points to add
points->InsertNextPoint(x,y,z);
// Add collection of points to the dataset
data_set->SetPoints(points);
```

where data_set is one of the dataset types described in the previous subsection.

29.3.2.3 Cells

The information about a dataset's cells are often stored as a Cell element. This includes point connectivity, the offset in the connectivity array that represents the end of each cell, and the cell type. A few examples of linear cell types include VTK_VERTEX, VTK_LINE, VTK_TRIANGLE, and VTK_PIXEL, and a few examples of non-linear cell types are VTK_QUADRATIC_TRIANGLE and VTK_QUADRATIC_TETRA. Each cell type is encoded by a number that can easily be stored in a dataset's cell array. Cells may also be defined explicitly as Verts, Lines, Strips, or Polys elements, which do not contain additional information about cell type.

29.3.2.4 Indexing within Datasets

VTK uses its own type for indexing. Instead of using unsigned int, it is preferable to use a vtkIdType object. This comes in handy because VTK is used for datasets ranging from small to large and can thus be configured to use 64-bits for storing indices. For example, if you wished to loop over all cells of a dataset and get the list of points held by each cell, you would need to use `vtkIdTypes` rather than an `unsigned int` as the loop index:

Listing 29.3 Indexing example from the VTK Users' Guide.

```
for (vtkIdType cellId = 0; cellId < numCells && !abort; ++cellId) {
  // Get the list of points for this cell.
  input->GetCellPoints(cellId, ptIds);
  vtkIdType numIds = ptIds->GetNumberOfIds();
}
```

29.3.2.5 Attribute Data

Attribute data are data objects (e.g., scalars, vectors, tensors) stored in the form of a `vtkDataArray` that are associated with points and cells. Attribute data associated with points are called *point attribute data*, and attribute data associated with cells are called *cell attribute data*. Both point and cell attribute data are subclasses of the `vtkFieldData` class. Therefore, field data are the generic representation of data that can be separated into cell or point data depending on what the object requires. `FieldData`, `CellData`, and `PointData` are dataset objects that store attribute data alongside the geometry and topology of a dataset. The `AddArray(vtkDataArray *)` routine can be used to associate a VTK array with a dataset object.

29.3.3 Process Objects

Data objects are connected with process objects, also called *filters* or *algorithms*, to create the visualization pipeline. Process objects operate on input data to generate transformed output data. Input to process objects includes one or more data objects and local parameters to control their operation. A variety of supporting abstract superclasses are available to derive new objects including data objects and filters. The visualization pipeline is designed to convert data into graphical primitives via *mappers*. Mappers are a type of process object that are the sinks of the pipeline and interface to `vtkActors`. `vtkActors` are the objects that represent the data that are on screen. They are not to be confused with the *actors* described in Chapter 36. Listing 29.4 shows how to use process objects in the visualization pipeline.

Listing 29.4 A basic example showing how to create a sphere and render it.

```
#include "vtkSmartPointer.h"
#include "vtkSphereSource.h"
#include "vtkPolyDataMapper.h"
#include "vtkActor.h"
#include "vtkRenderer.h"
#include "vtkRenderWindow.h"
#include "vtkRenderWindowInteractor.h"

int main() {
  // Create a sphere source process object
```

```
    vtkSmartPointer<vtkSphereSource> sphereSource = vtkSmartPointer<vtkSphereSource>::New();
    sphereSource->SetRadius(1.0);

    // Create a mapper to convert the sphere source's output to graphics primitives
    vtkSmartPointer<vtkPolyDataMapper> mapper = vtkSmartPointer<vtkPolyDataMapper>::New();
    mapper->SetInputConnection(sphereSource->GetOutputPort());

    // Create an actor to represent the sphere in the scene
    vtkSmartPointer<vtkActor> actor = vtkSmartPointer<vtkActor>::New();
    actor->SetMapper(mapper);

    // Create a renderer to render the scene
    vtkSmartPointer<vtkRenderer> renderer = vtkSmartPointer<vtkRenderer>::New();
    renderer->AddActor(actor);

    // Create a render window to display the scene
    vtkSmartPointer<vtkRenderWindow> renderWindow = vtkSmartPointer<vtkRenderWindow>::New();
    renderWindow->AddRenderer(renderer);

    // Create an interactor to handle user input
    vtkSmartPointer<vtkRenderWindowInteractor> interactor = vtkSmartPointer<vtkRenderWindowInteractor>::New();
    interactor->SetRenderWindow(renderWindow);

    // Start the event loop
    interactor->Initialize();
    interactor->Start();

    return 0;
}
```

In Listing 29.4, a `vtkSphereSource` is created, and its output is connected to a `vtkPolyDataMapper`. The mapper is then connected to the `vtkActor`, which is used to represent the sphere on screen. The `vtkRenderer` is used to render the actor, and the `vtkRenderWindow` is the window that is created to display the rendered scene. Lastly, the `vtkRenderWindowInteractor` is used to handle user interaction with the scene.

29.4 VTK Reading and Writing

29.4.1 VTK File Formats

There are two types of file formats associated with the VTK library. The first is called the *legacy* format. This format is the simpler of the two formats with which to work; however, files of this format only work in serial and are less flexible overall than the more recent alternative, the *XML* format. The XML format supports random access, parallel I/O, and data compression. For this reason, despite it being more complicated, the XML format is preferred.

The format of VTK files can be determined from its file extension. Legacy VTK files use the ".vtk" file extension, regardless of the dataset type contained in the file. XML files use an extension beginning with ".vt" followed by an additional character that indicates the dataset contained in the file. For example, a ".vtu" file extension indicates that the file contains unstructured grid data. A filename with ".vtr" indicates a rectilinear grid. For a full list of file extensions, see the VTK User's Guide.

XML data file types are either serial or parallel. Serial files contain data from one process. Parallel files do not contain data; rather, they contain structural information about corresponding serial files that do contain data, allowing for the data to be read from multiple processes at the time of visualization. If a VTK XML file is a parallel file, the file extension is preceded by a "p". For example, ".pvtu" is the parallel file for unstructured grids, and it reads from ".vtu" files. Similarly, ".pvti" reads from serial ".vti" files, which contain image data.

29.4.2 Reader and Writer Objects

Several different readers and writers exist in the VTK library to connect data structures of different types to the wider visualization pipeline. Readers read in VTK files of a certain dataset type, and writers subsequently write out data structures of a certain type that can be processed by the rest of the pipeline. Readers and writers are `vtkAlgorithms`, which are process objects. Each data structure type has 1–2 VTK native reader classes. The older reader classes can be used with ".vtk" files, whereas the newer ones read XML-based files with their corresponding ".vt?" extension. For example, the `vtkStructuredGridReader` class reads in ".vtk" legacy files containing structured grid data, whereas the `vtkXMLStructuredGridReader` reads in ".vts" XML-based files. Naturally, there is also a `vtkXMLPStructuredGridReader` that reads in ".pvts" files, the parallel files that reference serial ".vts" files. Writers, on the other hand, take in data objects and output data objects in a specified format. There are also many of these, and they are similarly called and associated with legacy or XML-based VTK files.

29.4.3 Example from `deal.II`

Listing 29.5 is an example of how the finite element library `deal.II` interfaces with VTK to write out a solution as VTK or VTU files. The primary object is the `Patch`, a struct defined in `deal.II` that describes a "patch" of the solution. Contained in the patch vector are the corner vertices of the patch, the data attached to each vertex, and the indices of neighboring patches. The other function parameters assist with readability and organization of the resulting visualization.

Listing 29.5 The function declaration for writing out VTU files from the `deal.II` finite element library. For the full function definition, see `https://www.dealii.org/current/doxygen/deal.II/data__out__base_8cc_source.html`.

```
void DataOutBase::write_vtu(const std::vector< Patch< dim, spacedim
> > & patches, const std::vector< std::string > & data_names, const
std::vector< std::tuple< /*...*/ >>& nonscalar_data_ranges, const
VtkFlags & flags, std::ostream & out)
```

Many libraries that are intended to work on large problems have similar functionality for outputting results to VTK because of the existence of high-quality, open-source, programs (such as ParaView) for visualizing these files.

29.4.4 VTK Rendering Engines

The VTK library also includes a rendering engine, a group of classes involved in taking in the final output of the visualization pipeline and displaying them on a screen. `ParaView` is a common open-source program built on VTK that utilizes the rendering engine to display data. It features a graphical user interface that allows the user to access components of the dataset and apply certain post-processing filters. `VisIt` is another well-known visualization tool that is able to process VTK files, as well as many other file types. Many libraries that utilize VTK for data visualization take care of the majority of steps in the visualization pipeline; in many situations, the user's only job may be to ensure that the interfacing objects are defined properly according to the library's syntax. Therefore, the rendering stage is the point where the user re-engages with VTK. In these cases, the user simply opens the resulting ".vtk", ".vt?", or ".pvt?" files with `ParaView`, `VisIt`, or the renderer of their choice. Depending on the situation, they may apply post-processing filters as needed using the application's graphical user interface. More complex processing algorithms can be applied as well if desired at this stage, using `ParaView`'s or `VisIt`'s scripting algorithms.

29.5 Additional reference material

- VTK main page: `https://vtk.org/Wiki/VTK`
- VTK tutorials: `https://vtk.org/Wiki/VTK/Tutorials`
- VTK C++ examples: `https://kitware.github.io/vtk-examples/site/Cxx/`
- VTK user guide: `https://vtk.org/wp-content/uploads/2021/08/VTKUsersGuide.pdf`
- ParaView main page:`https://www.paraview.org/`
- VisIt main page: `https://visit-dav.github.io/visit-website/index.html`

Part VII
Debugging and profiling ESC applications

Consisting of three chapters, this part dives into concepts and tools for debugging and analyzing the performance of applications.

First, we consider debugging CPU code in Chapter 30. This process is explored through the GNU debugger (`gdb`) and the Valgrind toolset. Although the tools described are specific to code running on CPU, the concepts underlying their most useful components generalize to GPU debugging as well.

Next, we consider profiling CPU code in Chapter 31. Profiling is the act of determining and quantifying where a computer program is spending its time. For this, we consider the GNU profiler, the Valgrind toolset, and Intel VTune.

Finally, we consider debugging and profiling GPGPU applications in Chapter 32 using NVIDIA's tools (specifically Nsight and `nvprof`) and Kokkos Tools.

Chapter 30
Debugging with GDB

Finding and fixing bugs is the most time-consuming part of the software development process. No one writes perfect code, and even the most experienced programmers make mistakes. The process of finding and fixing bugs is called *debugging*. Debugging, like many other aspects of software development, is a skill that requires practice and experience to master. Before one becomes familiar with more sophisticated debugging software, programmers often use *print statement debugging* to find and fix bugs. Occasionally, print statement debugging is a useful technique. However, it is often more desirable to use dedicated debugging software, especially when working on complex software projects. In this chapter, we learn about a series of tools that help streamline the debugging process. The first tool is GDB, which is a powerful open-source tool used to interactively step through programs for the purposes of debugging. The second tool is `Valgrind`, which is an open-source *instrumentation framework* with various tools that can help find memory errors and threading bugs.

30.1 GDB

30.1.1 Configuring GDB

We begin by exploring the GNU debugger (GDB), a powerful tool that lets us inspect our program while it is running or after it has crashed. GDB is supported by many programming languages, including C, C++, and Fortran, and runs on both Linux and Windows-based systems. We concentrate on the traditional method of using GDB—through the command line. However, GDB has also been integrated into many code editors and IDEs. Exploring GDB through the command line first gives us a comprehensive overview of how to use GDB and, by extension, provides a solid foundation for learning the more visual-based integrations of GDB.

In order to use GDB effectively, we need to compile our program with *debugging symbols*, which give GDB additional information such as function names, file names,

© The Author(s), under exclusive license to Springer Nature Switzerland AG 2025

R. J. Spiteri and K. Klenk, *Extreme-Scale Computing*, CMS/CAIMS Books in Mathematics 15, https://doi.org/10.1007/978-3-031-89033-8_30

and line numbers. GDB can still run a program without debugging symbols, but it would then lack the ability to decipher things like specific locations where crashes occur, the values of variables, and the call stack. Compiling with debugging symbols is done by providing the `-g` flag to the compiler. Additionally, for the purposes of debugging, it is useful to compile with optimizations turned off via the `-O0` compiler flag. Although GDB can still be used effectively when optimizations are turned on, it may appear to jump around the source code in an unintuitive manner. This is because when optimizations are turned on, the compiler may re-sequence the source code of the program to provide better performance.

30.1.2 The GDB Environment

After compiling a program for use with GDB, we must ensure that the environment of the GDB session is configured correctly. The environment of a GDB session is the set of environment variables that are available to a program when it executes. GDB inherits the environment variables from the terminal in which it is launched. To launch a GDB session, we simply call the `gdb` executable and supply the program name as an argument, e.g., `gdb <our_program>`. Once GDB has launched, it will look similar to a terminal session while it waits for us to enter a command. Before running the program, if we forgot to set up our environment beforehand or need to adjust our environment mid-session, we can use the `set env` command to say add a directory to the `LD_LIBRARY_PATH` environment variable. The complete syntax for the above command is `set env LD_LIBRARY_PATH ~/.local/lib`. Alternatively, we can use `unset environment var_name` to unset an environment variable or `show environment var_name` to show the value of an environment variable.

30.1.3 Common Commands

With our program compiled and our environment configured, we can explore some common GDB commands. These commands control the flow of a GDB session. They help systematically navigate a program's execution to extract information about its state and behavior. Commands are generally a single word and, for the most part, intuitively named. Short forms of commands are also available, often the first letter of the command's name. Below, we provide a list of some of the most common commands that are used to interact with GDB.

- `run` or `r`: Execute the program from the beginning.
- `break` or `b`: Set a breakpoint.
- `continue` or `c`: Continue execution until the next breakpoint.
- `step` or `s`: Step to the next line of execution.
- `backtrace` or `bt`: Print the backtrace/call stack of the current position.
- `print` or `p`: Print the value of an expression.

There are many more commands available in GDB, and most of them also have multiple options. That being said, the above commands should be sufficient for most debugging needs. The majority of the time is typically spent setting breakpoints and stepping through a program line by line. We cover this workflow in more detail in Section 30.1.3.1 and Section 30.1.3.2. But first, we explore the basic way in which we can use GDB to find the cause of a *segmentation fault* (or segfault). Suppose a program `main` that takes two integer arguments experienced a segfault in an unknown location. We can use the following steps to find the location of the segfault with GDB:

1. Compile the program: `gcc -g -O0 main.cpp -o main`
2. Launch GDB and set the environment variables: `gdb main`
3. Run the program: `r [program-args]`
4. When the program segfaults, backtrace: `backtrace`

In the above example, when we run the program, GDB stops at the location of the segfault and provides the line of code that caused the segfault. We can then use the `backtrace` command to see the call stack at the time of the segfault. This allows us to see the function calls that led up to the segfault. With this information, we can then attempt to deduce the cause of the segfault. Finding the cause of a segfault is the difficult part of debugging because GDB can only tell us where the segfault occurred, not why. In Section 30.2, we look into a tool that can help us find the cause of a segfault.

30.1.3.1 Breakpoints

Breakpoints stop the execution of a program at specific lines of the code. The use of breakpoints allows us to inspect the current state of a program and step through the execution of a program line by line. We can also jump between breakpoints with the `continue` command. Breakpoints are set on specific line numbers, with `break filename:line_number`, or on functions, with `break filename:function_name`. At any time, we can use the `info break` command to see a list of all the breakpoints that are set. Breakpoints can be set at any point within a GDB session provided a program is not explicitly executing. If we need to pause a program before a breakpoint is reached, we can use `ctrl+c`. This is useful if, for example, we think the program is stuck, or we forgot to set a breakpoint before running GDB.

To delete breakpoints, we can use the `delete` (or `del`) command. The `delete` command takes a breakpoint number, which is found from the output of the `info break` command, as an argument. Alternatively, we can use the `delete` command without any arguments to delete all breakpoints. Deleting breakpoints removes any indication that they were ever set. To keep the breakpoint information but disable it, we can use the `disable` command. If we wish to use the breakpoint again, we can use the `enable` command. Disabling and enabling breakpoints can make it easier to remember the exact location of a breakpoint. Breakpoint management is essential to using GDB efficiently.

30.1.3.2 Stepping and Inspecting

Knowing how to set breakpoints and navigate between them is a great start to using GDB. Nonetheless, we often find ourselves wanting to step through a program line by line to better understand what is happening. There are a few commands that allow us to do this. The `step` and `next` commands both allow us to step through a program line by line. However, there is a subtle difference between how the two interact with function calls. The `step` command *steps into* function calls, whereas the `next` command *steps over* function calls. Another way to describe this behavior is that the `step` command executes the next line of code in a program, whereas the `next` command executes the next line of code in a program without leaving the current scope. If we need to further reduce the granularity of the stepping, we can use the `stepi` and `nexti` commands. These commands step to the next machine instruction and not the next line of code. The `nexti` command still steps over function calls in this case. Using these commands is not only useful for finding problems, but it also helps us understand how a program works by giving valuable insight into the exact execution path a program takes. One important note about the `step` command is that it steps into *every* function call. For example, if we use `step` at a `printf` statement, we will step into the source code of `printf`. This behavior is also important to remember when using libraries. If we are using libraries and they are not compiled with debugging symbols, GDB will step into them, but it will not provide us with any information about the source code. Therefore, it is important to be aware of the functions into which we are stepping.

While stepping through the code, we may want to inspect the state of the program. We have already introduced the `backtrace` command, which prints out the current call stack. However, we can add the `full` argument to the command to get a more detailed backtrace that includes the values of the local variables within the current function. Another useful command is the `print` or `p` command, which prints out the value of a specified variable. We can use the `print` command for both values and memory addresses. For example, if we have a variable `a`, we can print out the value of `a` by using the command `p a`. If we want to print out the address of `a`, we can use the command `p &a`. We can also dereference pointers by using `p *a` if `a` was initialized as a pointer.

30.1.4 Catchpoints and Watchpoints

Catchpoints are similar to breakpoints and are more precisely considered to be a special kind of breakpoint. Instead of specifying a line number or function at which to stop, we use catchpoints to stop at a specific *event*. Catchpoints for events of type `event` can be set using `catch event`. Examples of an event in C++ include `throw`, `catch`, and `rethrow`. A catchpoint that is enabled for a single stop can be set with the command `tcatch event`. Catchpoints enabled for a single stop are automatically deleted after they catch an event.

Watchpoints are another type of breakpoint that can be used to monitor the values of variables and expressions. Similar to catchpoints, watchpoints stop execution when certain conditions are met. Instead of events, watchpoints stop execution when the value of a variable or expression changes. Watchpoints can be set with the `watch <thing_to_watch>`. The `watch` command also works for expressions that use variables. For example, `watch var_1 + var_2` halts execution when the value of `var_1` or `var_2` changes. We can list all the watchpoints and catchpoints, in addition to the breakpoints, that we have set with the `info break` command. Deleting and disabling catchpoints and watchpoints is carried out in the same way as breakpoints.

30.2 Valgrind

The second tool we introduce, `Valgrind`, is more of a series of tools packaged together. `Valgrind` is often referred to as an *instrumentation framework*. We focus on two specific tools that debug memory management and multi-threading bugs. These tools are called `memcheck` and `drd`, respectively. `Valgrind` is not interactive like `GDB`. Instead, it runs the entirety of a program and produces a report on what it finds. `Valgrind` is widely available and can be installed on Linux through your distribution's package manager.

30.2.1 Setup and Usage

Similar to `GDB`, programs that aim to use `Valgrind` must be compiled with debugging symbols enabled (`-g`) and optimizations disabled (`-O0`) to allow the `Valgrind` tools to be maximally effective. In practice, some programs may take too long to run with optimizations disabled. Fortunately, `Valgrind` can still be an effective tool even with optimizations enabled.

To use `Valgrind`, we launch the executable by calling `valgrind` with the desired options, including the specific tool and the executable to be analyzed. For example, we can use `valgrind --tool=memcheck ./main` to debug memory management and `valgrind --tool=drd ./main` to debug multi-threading bugs. These tools are the present focus, but `Valgrind` also includes several other tools, one of which, Cachegrind, is covered in Chapter 31.

30.2.2 Memcheck

When using `Memcheck`, there are a few options that we can use to control the level of detail that `Memcheck` reports. The first option is `--leak-check`, which tells

`Memcheck` to do a detailed memory leak analysis. The downside of using this feature is that programs typically run 20–30 times slower and use a significant amount of memory. There are multiple options that can be used with the `--leak-check` argument that further control the level of detail. Some other useful options for `Memcheck` include `--show-leak-kinds`, which shows the type of memory leak found, and `--track-origins`, which shows where the memory was first allocated or freed that led to memory leak or memory bug. It is generally recommended to use, `--verbose` and `--log-file` to maximize the amount of reported information and to write the output to a file. Below, we show a complete example of how to run `Memcheck` with comprehensive options:

```
$ valgrind --tool=memcheck --leak-check=full  \
    --show-leak-kinds=all --track-origins=yes \
    --verbose --log-file=valgrind-out.txt     \
    ./my_program [program-args]
```

The above command provides the maximum information possible but at the cost of execution speed. The log file contains all `Memcheck` output and allows us to step away and return to the results later. The log file does not contain any output specific to the program being run, only the output from `Valgrind`. The log file (or the output of `Memcheck` in general if we do not use the `--log-file` option) contains a summary of the memory leaks with their various types along with other memory errors it finds. The types of memory leaks include *still reachable*, *definitely lost*, *indirectly lost*, and *possibly lost*. "Still reachable" memory is a memory leak that is still within the program's scope when it exits. "Definitely lost" memory leaks are those where memory was allocated but never freed, and the program has no way to reference the memory location on exit. In other words, the reference to the memory location was lost somewhere in the program. The `--track-origins=yes` option provides the origin of the memory location. "Indirectly lost" memory leaks are usually a result of "definitely lost" memory leaks in large pointer based data structures. For example, if a program allocates memory for a linked list and the program loses the pointer to the head of the list, then all the memory allocated for the linked list's nodes will be "indirectly lost". Finally, "possibly lost" memory leaks occur when a memory location was never freed, but the only reference to it references a location other than the beginning of the allocated memory.

The various other types of memory errors that `Memcheck` flags include invalid reads, writes, and frees. These errors occur when a program interacts with memory it should not. They are crucial to fix because they result in undefined behavior and can cause programs to crash in unexpected ways, such as on print statements. `Memcheck` provides the line number where the invalid read, write, or free occurs. It also provides the origin of the variable that is being read, written, or freed with the `--track-origins=yes` option. Uninitialized values are another type of memory error that `Memcheck` flags. These errors are also crucial to fix because otherwise the values are populated randomly and lead to undefined behavior.

Further details about `Valgrind` can be discovered using man pages or help output:

```
$ man valgrind
```

```
$ valgrind -h
```

30.2.3 Memcheck with MPI

Another useful feature of Memcheck is that it supports debugging MPI programs. To take advantage of this feature, Valgrind must first be compiled with MPI support; otherwise it will be unable to understand the memory state of a multi-process program and will have a lot of false positives—reports that are not actually problems.

Next, a specialized MPI wrapper library must be linked into the executable for proper functioning. The wrapper addresses standard MPI functions, allowing memory state changes to be observed for MPI calls. In addition, it also provides size checking on the data buffers being used in MPI communication routines.

If not linked into the executable, the wrapper library can be specified with a LD_PRELOAD to ensure that the wrapped MPI functions are used in place of the standard MPI functions.

The following command runs a program through Valgrind with multiple MPI processes:

```
$  LD_PRELOAD=/path/to/wrapper.so                \
    mpirun [mpi-args] valgrind [valgrind-args] \
    ./my_program [program-args]
```

We note that most standard Linux desktop repository installations of Valgrind are not compiled with this functionality by default. So it most likely needs to be built from source in order to use. Some useful references to this end are found on the Valgrind's project website:

- Obtaining Valgrind: https://valgrind.org/downloads/repository.html
- Memcheck MPI: https://valgrind.org/docs/manual/mc-manual.html#mc-manual.mpiwrap

30.2.4 DRD

The Data Race Detector (DRD) is used to find common multi-threading errors in programs. A quick review of some common multi-threading errors are as follows:

- Data Races: Two or more threads access the same memory location without proper locking; hence the data within memory change depending on the order of the thread execution. Bugs of this nature can be extremely difficult to find because they are often subtle and may not even lead to incorrect results for one or a small number of threads.

- Lock Contention: Two or more threads try to access the same resource, forcing ones that do not obtain the lock to wait. This can cause a program to run slower than expected.
- Deadlock: Two or more threads are stuck waiting for each other. This behavior can be described as the program *hanging*.
- False Sharing: Two or more threads are accessing memory locations that are close to each other (generally in the same cache line) causing the threads to continually (and unnecessarily) validate the cache line. This can cause a program to run more slowly than expected.

Like Memcheck, DRD provides detailed output about the errors that it finds. DRD provides the line number of where an error occurs, the type of error, and the threads that are involved in the error. A complete example of how to run DRD (with our recommended options) is shown below:

```
$ valgrind --tool=drd --read-var-info=yes                    \
    --exclusive-threshold=10 --log-file=valgrind-out.txt \
    ./my_program [program-args]
```

The above code runs the DRD tool and produces a detailed log file. It is a good idea to always log the output of Valgrind, especially because recreating the data can take a long time. The --read-var-info=yes option tells DRD to read debug information about global and stack variables. This information is not recorded by default, but it can help produce more informative error messages. The --exclusive-threshold=10 option tells DRD to report anytime a lock is held for more than 10 milliseconds. This information is useful for detecting lock-contention errors. Both Memcheck and DRD can be used in conjunction with each other, and sometimes it may be necessary to alternate between the two to identify as many errors as possible in a program.

30.3 MPI debugging

Debugging MPI programs is generally a much more involved process than serial or even OpenMP debugging. When a bug arises in MPI code, the first consideration is the problem size. If it is a small problem running on a few processes, you may be able to just recompile the executable with debugging flags and fire up GDB and Valgrind with MPI on your workstation or laptop. This approach is considered in Section 30.3.1.

If it is a large problem running on a cluster, it may be possible to figure out whether the problem can be reproduced within a smaller instance that can fit onto a local workstation. If not, you may have to lean on a parallel debugger. Some options for that are considered in Section 30.3.2.

30.3.1 Using GDB and Valgrind

GDB and Valgrind interact in a fairly straightforward way to allow debugging of multi-process applications while running Valgrind tools. This type of interaction is particularly useful when an MPI application was terminated by a segmentation fault. Valgrind's GDB command-line tool can be used with vgdb.

We consider the case of debugging a code with 2 MPI processes. To do so requires three terminals to i) launch the MPI Valgrind instance, ii) launch a debugging process for rank 0, and iii) launch a debugging process for rank 1.

The first terminal runs the command

```
$  mpirun -np 2 valgrind --vgdb=yes --vgdb-error=0 ./my_program
```

to tell Valgrind to prepare Valgrind GDB (vgdb). There is a bunch of output, but most important is the relevant information for attaching GDB to this process:

```
==122154== TO DEBUG THIS PROCESS USING GDB: start GDB like this
==122154==   /path/to/gdb ./my_program
==122154== and then give GDB the following command
==122154==   target remote | /usr/libexec/valgrind/../../bin/vgdb --pid=122154
==122154== --pid is optional if only one valgrind process is running
```

There should be two such statements with different pids specified. In the second and third terminals, we can then run the commands

```
$ gdb ./my_program
```

and when their gdb prompts become available, we can attach them to the running MPI job with

```
(gdb) target remote | /usr/libexec/valgrind/../../bin/vgdb --pid=122154
```

with the pid value changed for the second process. Now, the second and third terminals can be manipulated using the commands that were discussed in Chapter 30.

When using this approach, care must be taken to advance both of the separate debugging processes as the program progresses. In order to to debug an executable running with more processes, a debugging terminal is needed for each additional MPI process.

30.3.2 Debugging large problems

Sometimes, bugs cannot be reproduced with smaller problems or using smaller numbers of processes. In this case, one may need to turn towards a debugger that is built to operate in parallel directly on a cluster. Parallel debuggers are designed to simplify the above process, offering an environment where it is not necessary to manually attach separate debugging shells to running MPI processes. Another benefit is that parallel debuggers tend to provide a local graphical interface connected to the running MPI job on a cluster. There is a marked difference in quality, however, between open-source and commercial parallel debuggers, to the point that we do not currently recommend the use of any open-source parallel debugger in general.

For commercial software, two popular choices of parallel debugger are `Arm DDT` (formerly Allinea DDT) and `TotalView`.

Despite its name, `Arm DDT` can be used for debugging MPI jobs running on more than just Arm architectures. It can be used to debug applications that are multi-process CPU or GPU. This capability goes for `TotalView` as well. The choice of which tool to use may come down to what is installed on the cluster you are using; the cluster documentation should give details on how to initiate parallel debugging.

30.4 Additional resources

- GDB user manual: https://sourceware.org/gdb/current/onlinedocs/gdb
- GDB online (gdb browser-based gui): https://www.onlinegdb.com/
- `Valgrind` user manual https://valgrind.org/docs/manual/manual.html
  - Intro/summary Ch. 1–2
  - `Memcheck` Ch. 4
  - DRD Ch. 8
- Debugging MPI applications: https://www.open-mpi.org/faq/?category=debugging

Chapter 31
Profiling serial and shared-memory code

Once a program is working as expected, we can shift our focus to performance optimization. One common way to improve the performance of a program is to identify bottlenecks in its execution. The method of identifying how a code spends its time is called *profiling*. In this chapter, we introduce two tools to help profile programs. The first profiling tool is `gprof`, which is a GNU tool. The second tool is `Cachegrind`, which is a profiling tool included in the `Valgrind` instrumentation framework. `Cachegrind` is used to determine cache and branch predictor performance. Learning to use both of these tools in conjunction with each other increases our performance as programmers as well.

31.1 gprof

31.1.1 Usage

Unlike GDB, `gprof` is not an interactive tool. Instead, it runs programs to completion and generates detailed profiling information. To use `gprof`, we first compile a program with the `-pg` flag, i.e.,

```
$ g++ -pg -o my_program my_program.cpp
```

The `-p` flag tells the compiler to generate profiling information when the resulting executable is run. This is combined with the `-g` flag to enable debugging symbols so that the resulting profiling data are human readable.

When we run a program, the profiling information is generated in a file labeled `gmon.out`:

```
$ ls
  my_program my_program.cpp
$ ./my_program
$ ls
```

© The Author(s), under exclusive license to Springer Nature Switzerland AG 2025
R. J. Spiteri and K. Klenk, *Extreme-Scale Computing*, CMS/CAIMS Books in Mathematics 15, https://doi.org/10.1007/978-3-031-89033-8_31

```
gmon.out my_program my_program.cpp
```

Once we have obtained the profiling information, we can use the gprof executable to inspect it. This is done using the gprof tool:

```
$ gprof ./my_program gmon.out
```

The output from gprof is a summary of the collected profiling information and printed to stdout.

31.1.2 Profiling summaries

The gprof profiler has two primary outputs, a *flat profile* and a *call graph*. The flat profile shows the total execution time spent in each function. A flat profile is typically sorted from the highest total execution time spent (outside the main function) to the lowest. An example flat profile looks like[1]

```
Flat profile:

Each sample counts as 0.01 seconds.
  %   cumulative   self              self     total
 time   seconds   seconds    calls  ms/call  ms/call  name
 52.63      0.10     0.10    10001     0.01     0.01  sum_sequence(unsigned long, unsigned long)
 47.37      0.19     0.09    10001     0.01     0.01  sum_sequence_squared(unsigned long, unsigned long)
  0.00      0.19     0.00        1     0.00     0.00  __static_initialization_and_destruction_0(int, int)
  0.00      0.19     0.00        1     0.00     0.02  func1()
  0.00      0.19     0.00        1     0.00   189.98  func2(unsigned long)
```

This flat profile shows that 52.63% of the execution time is spent in one function, with the remaining 47.37% spent in a second function.

A call graph shows the execution time spent in each function and its children. The entries are again typically sorted from the highest total execution time spent to the lowest. Within the call graph, each entry is broken down into a function and its children. This format produces a clear view of *how* the execution time was spent. The number of times each function was called is also shown in both the flat profile and the call graph. In gprof call-graph output, each function is separated into an *entry*, with each entry separated by dashed lines. An example call graph looks like

```
          Call graph (explanation follows)


granularity: each sample hit covers 4 byte(s) for 5.26% of 0.19 seconds

index % time    self  children    called     name
                                                 <spontaneous>
[1]    100.0    0.00    0.19                 main [1]
                0.00    0.19       1/1           func2(unsigned long) [2]
                0.00    0.00       1/1           func1() [5]
-----------------------------------------------
                0.00    0.19       1/1           main [1]
[2]    100.0    0.00    0.19       1         func2(unsigned long) [2]
                0.10    0.00   10000/10001       sum_sequence(unsigned long, unsigned long) [3]
                0.09    0.00   10000/10001       sum_sequence_squared(unsigned long, unsigned long) [4]
-----------------------------------------------
                0.00    0.00       1/10001       func1() [5]
                0.10    0.00   10000/10001       func2(unsigned long) [2]
[3]     52.6    0.10    0.00   10001         sum_sequence(unsigned long, unsigned long) [3]
-----------------------------------------------
                0.00    0.00       1/10001       func1() [5]
                0.09    0.00   10000/10001       func2(unsigned long) [2]
[4]     47.4    0.09    0.00   10001         sum_sequence_squared(unsigned long, unsigned long) [4]
```

[1] The profiling information shown is from the code in Listing 31.2 at the end of this chapter.

```
-----------------------------------------------
                0.00    0.00       1/1           main [1]
[5]      0.0    0.00    0.00       1         func1() [5]
                0.00    0.00       1/10001       sum_sequence(unsigned long, unsigned long) [3]
                0.00    0.00       1/10001       sum_sequence_squared(unsigned long, unsigned long) [4]
-----------------------------------------------
                0.00    0.00       1/1           _GLOBAL__sub_I_main [13]
[12]     0.0    0.00    0.00       1         __static_initialization_and_destruction_0(int, int) [12]
-----------------------------------------------
```

From this call graph, we can see the relevant information about our code is that 52.6% of execution time is spent in the `sum_sequence` function, and 47.4% of execution time is spent in the `sum_sequence_squared` function. The call graph also shows us that both `sum_sequence` and `sum_sequence_squared` are called 10001 times each, with one call coming from the `func1` function and the remaining 10000 calls coming from the `func2` function.

Additionally, we can perform an *annotation* of the source code to determine which lines of code are run more frequently within the executables. With older versions of `gcc`, annotation is done by compiling a program with the `-a` compiler flag to enable block counting:

```
$ g++ -a -pg -o my_program my_program.cpp
```

We then run the executable,

```
$ ./my_program
```

and then run `gprof` with the `-A` command-line option:

```
$ gprof -A ./my_program gmon.out
```

Newer versions of gcc no longer have the `-a` option and instead rely on a separate tool, `gcov`, for annotating code execution. Compiling in preparation for `gcov` requires the following command-line options:

```
$ g++ -pg -fprofile-arcs -ftest-coverage -o my_program my_program.cpp
```

where `-fprofile-arcs` keeps track of how many times branches are taken throughout an execution and `-ftest-coverage` records the lines that were actually executed. Running a program that was compiled in this way results in two output files having `gcda` and `gcno` extensions:

```
$ ls
  my_program my_program.cpp
$ ./my_program
$ ls
  my_program my_program.cpp my_program.gcda my_program.gcno
```

The annotation of the source code is then performed with the `gcov` utility,

```
$ gcov my_program
```

which produces the annotaed output of the source code in `my_program.cpp.cov`. An example of a portion of this output looks like

```
    10001:   64:unsigned long int sum_sequence_squared(unsigned long int start_val,
        -:   65:            unsigned long int end_val)
        -:   66:{
    10001:   67:   int sum=0;
    10001:   68:   int i=start_val;
 50026502:   69:   for(;i<=end_val;i++){
 50016501:   70:     sum += i*i;
        -:   71:   }
        -:   72:   //  sum = end_val*(end_val+1)*(2*end_val+1)/6 - start_val*(start_val+1)*(2*start_val+1)/6;
    10001:   73:   return sum;
        -:   74:}
```

where the first column gives how many times the line of code was executed in the program, and the second column gives the line number of the source code.

To conclude, line-by-line annotation gives deeper insights into which lines of codes are executed the most. This information is helpful for determining which branching statements are executed more often than others within functions, as well as how often loops are executed, to name a few examples.

31.1.3 Run-time estimation

Although `gprof` reports the total execution time spent in each function, it does not provide the exact run time of the program. This is because the total execution time is an approximation. This approximation is achieved through *sampling* and thus introduces *statistical sampling error*. After running a profile, we can determine the statistical sampling error that arises in a function's run time from the number of samples that occur in that function and the sampling period. The formula for computing the statistical sampling error is

$$\mathrm{err}_{\mathrm{ss,func}} = \sqrt{n_{s,\mathrm{func}}}\, T_{\mathrm{period}} = \sqrt{T_{\mathrm{func}} T_{\mathrm{period}}}, \tag{31.1}$$

where $\mathrm{err}_{\mathrm{ss}}$ is the statistical sampling error, $n_{s,\mathrm{func}}$ is the number of samples that occur in the function, T_{func} is the time spent in the function, and T_{period} is the sampling period.

The flat profile gives the sampling period that was used in the profiling run, for example,

```
Each sample counts as 0.01 seconds.
```

Looking at the flat profile also allows us to determine how many samples were evaluated in each of the functions. For example, the lines

```
  %   cumulative   self              self     total
 time   seconds   seconds    calls  ms/call  ms/call  name
 52.63      0.10     0.10    10001     0.01     0.01  sum_sequence(unsigned long, unsigned long)
 47.37      0.19     0.09    10001     0.01     0.01  sum_sequence_squared(unsigned long, unsigned long)
```

indicate that there were 0.10/0.01 = 10 samples taken in `sum_sequence()` and 0.09/0.01 = 9 samples taken in `sum_sequence_squared()`. Using this information, we can determine the statistical sampling error for those two functions using equation (31.1):

$$\mathtt{err}_{\mathrm{ss,sum\_sequence}} = 0.0316\ \mathrm{s},$$

$$\mathtt{err}_{\mathrm{ss,sum\_sequence\_squared}} = 0.0300\ \mathrm{s}.$$

The times measured for both are only accurate to about 30%!

One conclusion drawn from this exercise is that many samples are needed to minimize the run-time error of the statistical profiling approach used in `gprof`. The longer the computations that are being profiled, the more accurate their timing results will be.

In contrast, the block counts, or the number of times that a function or line of code is executed, are exact because they are determined simply by counting.

31.2 Cachegrind

Program performance is not limited to simply the number of times a function is called. There are other factors that can play into the performance of a program. Two of these factors are *caching performance* and *branch prediction*. An accumulation of cache misses and poor branch prediction can have a significant impact on performance because in both cases, the CPU has to spend extra cycles to fetch the required instructions. To help us identify where cache misses and poor branch prediction occur, we can use the `Cachegrind` tool provided by the `Valgrind` instrumentation framework. `Cachegrind` can be used to simulate performance on a machine's cache hierarchy and branch predictor.

31.2.1 Cache Misses

Cache misses are a common source of program inefficiency. A cache miss occurs when a CPU core attempts to execute an instruction or operate on data not present in its cache. The inefficiency is a result of the CPU expending extra cycles to fetch the required instructions or data from the appropriate memory location. L1-cache misses are the least expensive at around 10 cycles per miss. However, higher-level cache misses are more expensive and can exceed 200 cycles per miss, depending on the CPU architecture. There are four different types of cache misses, and although `Cachegrind` does not distinguish between them, it is helpful to understand the different types. The first type is a *compulsory miss*, which occurs when the data requested have not yet been stored in cache. The second is a *capacity miss*, which occurs when the cache is full and the data needed are too large to fit in the cache. The third is a *conflict miss*, which occurs when the different data items map to the same cache location and cause the replacement of the data already present in the cache. The last type of cache miss is a *coherence miss*, which occurs when the data in the cache are not consistent because of multiple cores accessing the same data.

Knowing the different types of cache misses is helpful for better understanding why `Cachegrind` may be reporting a cache miss in a program's execution. To use `Cachegrind`, we again should first compile the program with the `-g` flag to enable human-readable (debug) symbols. Then we can run `Valgrind` with

```
$ valgrind --tool=cachegrind [cachegrind-args] \
    <executable> [executable-args]
```

Cachegrind then simulates the performance using a two-level hierarchical caching system with the capacities of the system on which it is running. Although many modern CPUs have multiple levels of cache, Cachegrind only simulates the first-level and last-level caches. The key statistics that Cachegrind measures are instruction or data reads and read misses, along with data writes and write misses. Because Cachegrind does not distinguish between the different types of cache misses, the onus is on the programmer to understand them. The approach to fix a program with many cache misses depends on the type of cache misses that occur. The output from Cachegrind is supplied to stdout, and additional information is written to the file cachegrind.out.<pid>. A more complete breakdown of the output with annotated source code is produced from running the supplementary tool, cg_annotate, on the output file, as in

```
$ cg_annotate cachegrind.out.<pid>
```

Each line of the annotated source code contains a count of the events that occurred on that line. Events are the number of instructions executed, the number of data reads and read misses, and the number of data writes and write misses. This should give a good idea of where reduced cache performance is occurring in a program and suggest where optimization efforts should focus.

Running Listing 31.2 produces the following output from Cachegrind

```
==64668==
==64668== I   refs:      11,041,312,504
==64668== I1  misses:             2,330
==64668== LLi misses:             2,258
==64668== I1  miss rate:           0.00%
==64668== LLi miss rate:           0.00%
==64668==
==64668== D   refs:       4,558,009,692  (1,655,090,543 rd   + 2,902,919,149 wr)
==64668== D1  misses:       277,390,149  (  227,301,633 rd   +    50,088,516 wr)
==64668== LLd misses:       257,449,850  (  207,447,909 rd   +    50,001,941 wr)
==64668== D1  miss rate:            6.1% (         13.7%     +           1.7%  )
==64668== LLd miss rate:            5.6% (         12.5%     +           1.7%  )
==64668==
==64668== LL refs:          277,392,479  (  227,303,963 rd   +    50,088,516 wr)
==64668== LL misses:        257,452,108  (  207,450,167 rd   +    50,001,941 wr)
==64668== LL miss rate:             1.7% (          1.6%     +           1.7%  )
```

The important pieces here are the data cache miss rates. The D1 (level-1 data cache) misses 13.7% of its attempted reads, and the LLd (last-level data cache) misses 12.5% of its attempted reads.

Although the percentage of data cache misses is important, it is also important to observe the absolute number of events occurring. The summary view does not necessarily do justice to particular regions where the number of cache misses is high.

Running the next step with cg_annotate results in line-by-line annotations. For example, annotated output of the running Listing 31.2 contains the following information:

```
Dr                      D1mr                    DLmr

          0                       0                       0          double sum = 0.0;
          0                       0                       0          auto tic = chrono::high_resolution_clock::now();
          0                       0                       0          for (auto const & ind : indices) {
100,000,000 ( 6.04%) 99,995,366 (43.99%) 97,364,713 (46.93%)            sum += data[ind];
          .                       .                       .          }
          0                       0                       0          auto toc = chrono::high_resolution_clock::now();
```

where the percentages are given with respect to the total number of that type of event; i.e., 100,000,000 represents 6.04% of the total number of data cache references (1,655,090,543) in the program execution. The level-1 data cache miss rate for that particular line of code accounts for 43.99% of all `D1` misses, and the last-level data cache miss rate accounts for 46.93% of all `DL` misses.

For comparison, running Listing 31.2 without the randomized access results in the following Cachgrind output

```
==66164==
==66164== I   refs:      2,853,285,228
==66164== I1  misses:            2,298
==66164== LLi misses:            2,219
==66164== I1  miss rate:          0.00%
==66164== LLi miss rate:          0.00%
==66164==
==66164== D   refs:      1,951,016,278  (250,765,695 rd   + 1,700,250,583 wr)
==66164== D1  misses:       75,020,736  ( 25,018,191 rd   +    50,002,545 wr)
==66164== LLd misses:       75,011,434  ( 25,009,645 rd   +    50,001,789 wr)
==66164== D1  miss rate:           3.8% (       10.0%     +           2.9%  )
==66164== LLd miss rate:           3.8% (       10.0%     +           2.9%  )
==66164==
==66164== LL refs:          75,023,034  ( 25,020,489 rd   +    50,002,545 wr)
==66164== LL misses:        75,013,653  ( 25,011,864 rd   +    50,001,789 wr)
==66164== LL miss rate:            1.6% (        0.8%     +           2.9%  )
```

with the annotated output

```
Dr                      D1mr                    DLmr
          0                       0                       0          double sum = 0.0;
          0                       0                       0          auto tic = chrono::high_resolution_clock::now();
          0                       0                       0          for (auto const & ind : indices) {
100,000,000 (39.88%) 12,500,001 (49.96%) 12,500,001 (49.98%)            sum += data[ind];
          .                       .                       .          }
          0                       0                       0          auto toc = chrono::high_resolution_clock::now();
```

It is important to note here that although the miss rates are still relatively high percentages of the total misses in the program execution, the absolute number of misses for that particular line of code has decreased radically. This shows up in the execution time of that for loop when the codes (randomized access and sequential access) are run outside of `Valgrind`:

```
$ ./vector_random
  Expected: 4999999950000000
  Sum:      5e+15
  Time:     1013 ms
$ ./vector_sequential
  Expected: 4999999950000000
  Sum:      5e+15
  Time:     185 ms
```

31.2.2 Branch Prediction

To increase performance, modern CPUs attempt to guess the outcome of a branching statement before it is executed. This guessing is done by using a special piece of hardware called a *branch predictor*. Successful branch predictions reduce stalls in the CPU pipeline by fetching the next instruction into the pipeline before the branching statement is executed. However, if the branch prediction is incorrect, the CPU will stall, then flush the pipeline and fetch the next instruction into the pipeline. The delay in waiting for the next instruction to be fetched can vary from 10–30 cycles.

To measure branch prediction performance, we specify the `--branch-sim=yes` option to `Cachegrind`. The branch prediction simulator in `Cachegrind` is not enabled by default because it significantly increases the run time of a program. The key statistics that `Cachegrind` collects are the number of conditional branches executed and mispredicted, along with the number of indirect branches executed and mispredicted. The output from `Cachegrind`'s branch prediction simulator is supplied to the file `cachegrind.out.`<pid>. A complete breakdown is contained in this file, where the lines are grouped into their respective files and functions.

Running Listing 31.3 produces the following output from `Cachegrind` with its branch simulator active:

```
==66817==
==66817== I   refs:      11,335,792,747
==66817== I1  misses:             2,206
==66817== LLi misses:             2,088
==66817== I1  miss rate:           0.00%
==66817== LLi miss rate:           0.00%
==66817==
==66817== D   refs:       2,055,490,393  (1,353,958,624 rd   + 701,531,769 wr)
==66817== D1  misses:            20,425  (       17,880 rd   +       2,545 wr)
==66817== LLd misses:            10,914  (        9,207 rd   +       1,707 wr)
==66817== D1  miss rate:            0.0% (          0.0%     +         0.0%  )
==66817== LLd miss rate:            0.0% (          0.0%     +         0.0%  )
==66817==
==66817== LL refs:               22,631  (       20,086 rd   +       2,545 wr)
==66817== LL misses:             13,002  (       11,295 rd   +       1,707 wr)
==66817== LL miss rate:             0.0% (          0.0%     +         0.0%  )
==66817==
==66817== Branches:       1,425,139,252  (1,425,132,735 cond +       6,517 ind)
==66817== Mispredicts:      186,030,216  (  186,028,945 cond +       1,271 ind)
==66817== Mispred rate:            13.1% (         13.1%     +        19.5%   )
```

There is now an additional section of information about branch prediction. Annotated output of the code looks like

```
Bc                    Bcm                     Bi   Bim
          0                     0             0    0          double sum{0.0};
          .                     .             .    .          // Choose randomly between 4 different branches
100,000,000 ( 7.02%)            9 ( 0.00%)    0    0          for (size_t i=0; i<N; ++i) {
          .                     .             .    .            auto num = dis(g);
100,000,000 ( 7.02%) 29,998,578 (16.13%)      0    0            if (num<0.25)
          0                     0             0    0              sum += 0.25;
 74,999,065 ( 5.26%) 30,007,972 (16.13%)      0    0            else if (num < 0.5)
          0                     0             0    0              sum -= 0.25;
 49,997,127 ( 3.51%) 24,997,377 (13.44%)      0    0            else if (num < 0.75)
          0                     0             0    0              sum += 0.5;
          .                     .             .    .            else
          0                     0             0    0              sum -= 0.5;
          .                     .             .    .          }
          0                     0             0    0          cout << "Sum:        " << sum << endl;
```

where we have trimmed the output to only show the relevant data. In this particular code, we see that the branch mispredictions amount to 25 million to 30 million for each of the branches evaluated.

31.2.3 Cachegrind Behavior

An important note about using Cachegrind is that it runs programs in a simulated environment. As a result, Cachgrind does not provide a true measure of cache behavior. For example, it does not account for either kernel or other process activity and will likely schedule threads differently than they would be natively. Cachegrind also does not record cache misses that occur outside the instruction level. Some examples of what Cachegrind does not record are misses that occur in the translation lookaside buffer (TLB) or misses as a result of speculative execution, which is a technique used by CPUs to attempt to predict the outcome of a branching statement. Speculative execution is slightly different from branch prediction because speculative execution occurs after the branch has been predicted.

31.2.4 Beyond Cachegrind

CPUs are generally equipped with hardware counters within them that keep track of the quantities covered in this chapter. The problem is that they are not standardized across manufacturers nor even across generations of a single manufacturer's chips. This makes the simulated environment of Valgrind a useful tool for first learning and discussing tools for looking at performance.

A good open-source tool for looking at cache and branching performance is perf, in particular the command

```
$ perf stat -d ./my_program
```

This type of invocation prints out a *detailed* view that includes information beyond cache misses. A problem with this command, however, is that it may display incorrect cache information, and you may have to dig in deeper to see how to specify the counter types/format required for assessment of the CPU. A comprehensive list of counters available on the CPU can be found with

```
$ perf list
```

which is organized by types of counters to help determine what is suitable.

31.3 Codes for profiling demonstrations

Listing 31.1 A simple C++ code that calls a few different functions many times. Example profiling output from this code is given in Section 31.1.2.

```
#include<iostream>
using namespace std;

/////////////////////////////////////////////////////////////////////////
// Function prototypes
void func1(void);
```

```
void func2(unsigned long int N);
unsigned long int sum_sequence(unsigned long int start_val,
                               unsigned long int end_val);
unsigned long int sum_sequence_squared(unsigned long int start_val,
                                       unsigned long int end_val);
unsigned long int sum_sequence_cubed(unsigned long int start_val,
                                     unsigned long int end_val);
///////////////////////////////////////////////////////////////////////////
// Main program
int main(void)
{
    cout << "\n Inside main()\n";

    func1();
    func2(10000);

    return 0;
}
///////////////////////////////////////////////////////////////////////////
void func1(void)
{
    cout << "\nInside func1\n";

    unsigned long int answer;
    answer = sum_sequence(1000,2500);
    cout << "  Sum of natural numbers from 1000 to 2500: " << answer << endl;
    answer = sum_sequence_squared(1000,2500);
    cout << "  Sum of squares of natural numbers from 1000 to 2500: "
         << answer << endl;

    return;
}
void func2(unsigned long int N)
{
    cout << "\nInside func2\n";
    unsigned long int i = 1;
    unsigned long int answer;
    for(;i<=N;i++){
      answer = sum_sequence(0,i);
      cout << "Sum of first " << N << " natural numbers: " << answer << endl;
      answer = sum_sequence_squared(0,i);
      cout << "Sum of first " << N << " squares of natural numbers: " << answer << endl;
    }

    return;
}
unsigned long int sum_sequence(unsigned long int start_val,
                               unsigned long int end_val)
{
  unsigned long int sum=0;
  unsigned long int i=start_val;
  for(;i<=end_val;i++){
    sum += i;
  }
  //  sum = end_val*(end_val+1)/2 - start_val*(start_val+1)/2;
  return sum;
}
unsigned long int sum_sequence_squared(unsigned long int start_val,
                                       unsigned long int end_val)
{
  int sum=0;
  int i=start_val;
  for(;i<=end_val;i++){
    sum += i*i;
  }
  //  sum = end_val*(end_val+1)*(2*end_val+1)/6 - start_val*(start_val+1)*(2*start_val+1)/6;
  return sum;
}
unsigned long int sum_sequence_cubed(unsigned long int start_val,
                                     unsigned long int end_val)
{
  unsigned long int sum=0;
  unsigned long int i=start_val;
  for(;i<=end_val;i++){
    sum += i*i*i;
  }
  return sum;
}
```

Listing 31.2 A C++ code that accesses a `std::vector`'s entries in a random order. This example is designed to intentionally show cache misses in Cachegrind/perf. Output from running this code is found in Section 31.2. Removing the `std::shuffle` call results in a program that accesses the vector data in sequential order and should perform much better.

```
#include <iostream>
```

```
#include <random>
#include <vector>
#include <algorithm>
#include <chrono>
using namespace std;

int main(int argc, char* argv[])
{
  const size_t N{100000000};
  vector<double> data(N);
  iota(data.begin(), data.end(), 0);

  vector<size_t> indices(N);
  iota(indices.begin(), indices.end(), 0);

  // Random permutation of indices
  random_device rd;
  mt19937 g(rd());
  shuffle(indices.begin(), indices.end(), g);

  double sum = 0.0;
  auto tic = chrono::high_resolution_clock::now();
  for (auto const & ind : indices) {
    sum += data[ind];
  }
  auto toc = chrono::high_resolution_clock::now();
  double ms = chrono::duration_cast<std::chrono::milliseconds>(toc-tic).count();

  cout << "Expected: " << N*(N-1)/2 << endl;
  cout << "Sum:      " << sum << endl;
  cout << "Time:     " << ms << " ms" << endl;

  return 0;
}
```

Listing 31.3 A C++ code that randomly selects between two branches. The intention is to show up in the branch mispredictions in, for example, `Cachegrind`.

```
#include <iostream>
#include <random>
#include <vector>
#include <algorithm>
#include <chrono>
using namespace std;

int main(int argc, char* argv[])
{
  const size_t N{100000000};

  // Random number genreation
  int seed{1907};
  mt19937 g(seed);
  uniform_real_distribution<double> dis(0.0, 1.0);

  double sum{0.0};
  // Choose randomly between 4 different branches
  for (size_t i=0; i<N; ++i) {
    auto num = dis(g);
    if (num<0.25)
      sum += 0.25;
    else if (num < 0.5)
      sum -= 0.25;
    else if (num < 0.75)
      sum += 0.5;
    else
      sum -= 0.5;
  }
  cout << "Sum:      " << sum << endl;

  return 0;
}
```

31.4 Additional resources

- gprof manual: `https://sourceware.org/binutils/docs-2.42/gprof.pdf`.

- Hacking `gprof` to record data from all threads: `http://sam.zoy.org/writings/programming/gprof.html`.
- Cachegrind: `https://valgrind.org/docs/manual/cg-manual.html`.
- perf: `https://perf.wiki.kernel.org/index.php/Main_Page`

Chapter 32
Debugging and profiling compute kernels

As GPU programming becomes increasingly popular, there is a greater need for tools to help debug and profile GPU code. Elegant tools for profiling and debugging GPU code are crucial for ensuring developer productivity and code quality. In this chapter, we look at two such tools, `cuda-gdb` and `NVIDIA Nsight Compute`. We begin by introducing `cuda-gdb`, which is a GDB extension developed by NVIDIA for interactively debugging CUDA code. The `cuda-gdb` tool may feel familiar to those who have used GDB, introduced in Chapter 30. The second tool is `NVIDIA Nsight Compute`, which is the replacement for the deprecated `nvprof` tool. The NVIDA Volta architecture is the final one supported for `nvprof`. Future GPU architectures are to be centered around `NVIDIA Nsight Compute` for profiling CUDA code.

32.1 CUDA-GDB

NVIDIA developed the CUDA-GDB extension to GDB to facilitate interactive debugging of both device (GPU) kernels and host (CPU) code within the same debugging session. The standard GDB interface is used to debug the host code, while the CUDA extension provides additional commands (all prefixed with `cuda`) for working with device kernels. The main advantage of using CUDA-GDB is that its interface is highly similar to the standard GDB interface. Only a few additional commands and concepts are required to get started debugging CUDA code. CUDA-GDB supports all CUDA applications built with C, C++, and Fortran.

32.1.1 Compiling CUDA Code for Debugging

To prepare a CUDA application for debugging, it must first be compiled with the appropriate flags. Similarly, to compile host code for debugging, CUDA code must

© The Author(s), under exclusive license to Springer Nature Switzerland AG 2025

R. J. Spiteri and K. Klenk, *Extreme-Scale Computing*, CMS/CAIMS Books in Mathematics 15, https://doi.org/10.1007/978-3-031-89033-8_32

be compiled with debugging symbols enabled and optimizations disabled. This is achieved through specifying the `-g -G` flags to the `nvcc` compiler. The `-g` flag instructs the compiler to include debugging symbols in the executable. The `-G` flag turns off all optimizations, similar to specifying the `-O0` flag to the `gcc/g++` compiler. After successfully compiling the executable, a CUDA-GDB session can be launched with the command `cuda-gdb <executable>`.

32.1.2 Breakpoints

The essential task for efficiently debugging applications is breakpoint management. Breakpoints halt the execution of device threads when they are reached, allowing for quick maneuvering throughout the source code. However, when a device thread reaches a breakpoint, it is not guaranteed that all device threads have reached the same breakpoint. This situation results in the breakpoint being hit multiple times by different threads, stopping the execution of the program each time. It is important to be aware of such situations in order to correctly identify the desired thread to inspect. Once the desired thread is identified, the `disable <breakpoint_number>` command can be used to disable the breakpoint from being triggered by subsequent threads. The `<breakpoint_number>` can be found using the `info breakpoints` command. Some examples of setting breakpoints include

```
break desired_function
break desired_class::desired_method
break int desired_template_function<int>(int)
break desired_file.cu:65
```

Manually identifying the desired thread can be a tedious task. Happily, breakpoints can be set using conditionals. Any variable can be used for the conditional, including built-in variables such as `threadIdx` and `blockIdx`. This allows for breaking on specific threads or blocks, reducing the need to find them manually. The only restriction is that function calls cannot be used in the breakpoint conditional.

32.1.3 Focus

With GPUs containing many threads, it can be difficult to precisely control the execution of CUDA-GDB. To help simplify the debugging process, execution is controlled from a chosen point of *focus*. The focus is the current context of the debugger that specifies the current GPU device, kernel, block, and thread. If a breakpoint has been set, the focus is automatically set to the context of the thread that triggered the breakpoint. Inspecting the current point of focus is done by specifying the desired *coordinates*. For example, `cuda thread block` returns the thread and block coordinates of the current focus.

There are two types of coordinates that can be used to specify the point of focus. The first type is the set of software coordinates, which consist of a kernel, block, and thread. The second type is the set of hardware coordinates, which consist of a device, streaming multiprocessor (SM), warp, and lane. Issuing the `cuda` command followed by the desired coordinate or combination of coordinates returns its value. Changing the point of focus is done by specifying the numeric ID of the desired coordinates. When specifying the focus, both software and hardware coordinates can be used interchangeably. If a coordinate is omitted, the debugger assumes the value is the same as the current focus. An example of changing the focus to block 5, thread 3, is done with the command `cuda block 5 thread 3`. Lastly, a *grid* is an additional software coordinate that is sometimes used in favor of a kernel. A grid differs from a kernel in its scope; a grid has an ID that is unique to a GPU, whereas the kernel ID is unique across all GPUs within the host system.

32.1.4 Program Execution

Once the executable has been compiled successfully and breakpoints have been set, running the program is performed with the `run` (or `r`) command followed by the required arguments for the executable. If breakpoints are not set, the program runs as far as it can, normally until it reaches the end or crashes. In the event of a crash, the debugger generally provides the exact line on which the crash occurred. In this situation, the `backtrace` (or `bt`) command can be used to inspect the call stack. At any time during execution, `ctrl+c` can be used to interrupt the program and pause its execution. This is useful when the debugger appears to be frozen, or something has gone wrong. Once paused, the current state of the program can be inspected, the program can be stepped through, or execution can be resumed.

Stepping through the program can be performed using the `step` and `next` commands. The `step` command steps into functions it encounters, whereas the `next` command steps over functions. One consideration to keep in mind is that stepping is done at the warp level. When a single step is taken, all threads within the warp of the current focus advance. Any divergent threads in the warp, i.e., those threads that are not executing the same instruction, do not advance. To advance multiple warps, breakpoints must be used along with the `continue` command to resume execution until the next breakpoint is reached. However, if the code contains a *thread barrier call*, such as `__syncthreads()`, a single step advances all threads/warps. This is because a temporary breakpoint is set after the barrier when it is encountered. One aspect to keep in mind is that *inline functions* cannot be stepped into, over, or out of. In order to step into an inline function, the function must use the `__noinline__` keyword in its function declaration.

32.1.5 Inspecting State

We now give more detail behind inspecting the current state of a program. The `print` command can be used to print the value or address of a variable in the host or device code. The methods introduced in Section 30.1.3.2, cover how to inspect the state of the host code specifically. In device code, variables can be stored in multiple locations, such as registers, local memory, shared memory, or global memory. The command `print &variable` outputs the location of where `variable` is stored. Additionally, the `info` command can be used to print information about the GPU or application state. This command has many potential arguments for gathering information. Some example arguments include `info devices`, which prints information about all GPU devices, `info threads`, which prints information about all threads in the current kernel, and `info kernels`, which prints information about all the active kernels.

32.2 Profiling

Profiling is the process of collecting performance data from an application to determine its performance characteristics. In particular, these data can provide significant insight into where an application is spending the most time and where developers can focus their efforts to improve performance. Profiling CUDA code was done with the `nvprof` tool; however, the NVIDIA Volta architecture has brought with it increased performance and the last leg of support for `nvprof`. NVIDIA has released a new set of "next-generation" profiling and debugging tools. These tools are the NVIDIA Nsight tools, which include NVIDIA Nsight Systems and NVIDIA Nsight Compute. Nsight Systems is aimed at providing an all-encompassing performance analysis tool for visually analyzing the performance of an application across both the CPU and GPU. Nsight Compute is aimed at providing lower-level performance analysis for the GPU at the kernel level. We focus on Nsight Compute in this section because it is the tool most similar to profiling with `nvprof`.

32.3 NVIDIA Nsight Compute

Once installed, Nsight Compute is used from the command line with `ncu`. The command-line interface can be used to print information directly to `stdout` and to a file. Nsight Compute launches the target application, instruments the target API, and collects profiling results for the specified kernels. To quickly get started, `ncu -o profile_name target_application` can be used to collect the default set of *metrics* from a program. Metrics are the data collected during profiling. After running Nsight Compute, the collected metrics are written to `profile_name.nsight-cuprof`. The file separates each kernel launch with `==PROF==`. Within each `==PROF==` section, the name of the kernel function is recorded, followed by the proportion of metrics collected for that kernel in terms of a percentage. After running Nsight Compute once, the percentages may not add up to 100%. This is because the collection of some metrics require a *replay* of the kernel to collect all the information the metric requires. Options are available to specify which kernel metrics should be collected. Some examples of these options include `-c`, which limits the number of kernel launches collected, and `-s`, which skips the given number of kernels before metric collection starts. Additionally, the `-k` option allows the filtering of kernels by a `regex` match of their names, and `--kernel-id` allows for filtering kernels by context, stream, name, and invocation, similar to `nvprof`.

32.3.1 Metrics

Before diving into replays, we first introduce metrics more formally because they are the reason replays are needed in the first place. Metrics are the fundamental performance measurements that are collected during profiling. The collection of metrics is the key feature offered by Nsight Compute. However, it is not recommended to collect all metrics at once because it requires a significant amount of time and resources. To help streamline the collection of metrics, Nsight Compute uses *section sets* to provide a pre-configured group of metrics to collect. Section sets are a concept provided by Nsight Compute to help group specific metrics together into a hierarchical structure, where a *set* is a collection of one or more *sections*, and a *section* is a collection of one or more *metrics*. Sections are intended to collect a group of metrics that are useful in answering a specific performance question. For example, the pre-defined section `SchedulerStats` provides a set of metrics that are useful in understanding the activities performed by the scheduler. Sets, on the other hand, group sections together to help users quickly answer multiple performance questions. Sets do not need to be collected explicitly, meaning that the user can choose to collect specific sections, or even individual metrics. The command-line options for collecting metrics, sections, and sets are `-metric`, `-section`, and `-set`, respectively. Section sets are just a convenient way to group metrics together; they can even be customized by the user to collect specific metrics that are useful to them.

32.3.2 Replays

It is normally not possible to collect all metrics in one pass. This is because the GPU is limited in how much data it can physically collect in a single pass. It may take multiple runs of the kernel to collect all desired metrics. There are three types of replays: *Kernel Replay*, *Application Replay*, and *Range Replay*. In a Kernel Replay, specific kernels are replayed multiple times to collect all desired metrics. On the first pass of the kernel, The kernel saves all accessible memory and determines the subset that was interacted with. On subsequent passes, the memory that was interacted with is stored in the original location so that the kernel accesses the same memory locations on each pass for more accurate profiling information. The more a kernel interacts with memory, especially writes, the longer it takes to collect all the desired metrics. On the other hand, application replays rerun the entire application multiple times. Finally, range replays are replays that occur within a defined range of kernel launches and CUDA API calls. For this replay option, metrics are generally applied to the entire range rather than to specific kernels. To specify a replay, the `--replay-mode` option can be used.

32.3.3 Supplementary Information

In this chapter, we have introduced the command-line interface for Nsight Compute. However, NVIDIA has made great strides in providing a GUI experience for Nsight Compute. It is difficult to provide a comprehensive overview of the GUI in text form. NVIDIA provides helpful tutorials on their website at `https://developer.nvidia.com/nsight-compute`. We highly recommend them and the use of the GUI, which is a powerful and intuitive tool for profiling.

32.4 Additional resources

- CUDA-GDB: `https://developer.nvidia.com/cuda-gdb`
- `nvvp`, `nvprof`: `https://docs.nvidia.com/cuda/profiler-users-guide/index.html`
- Nsight: `https://developer.nvidia.com/nsight-compute`

Part VIII
Numerical libraries for ESC

This section consists of four chapters that introduce libraries commonly used in ESC applications.

Chapter 33 looks at the Basic Linear Algebra Subroutines (BLAS), the Linear Algebra Package (LAPACK), and the Fastest Fourier Transform in the West (FFTW).

Chapter 34 looks at the portable, extensible toolkit for scientific computation (PETSc), which contains a suite of tools for parallel (distributed) linear algebra, nonlinear solvers, and timestepping methods.

Chapter 35 looks at parts of the Trilinos software stack. Trilinos contains a much wider suite of tools/libraries than PETSc, with their base foundation on the MPI+X parallelism paradigm. We consider only the general tools, Teuchos and the linear algebra components in Tpetra, sub-libraries on which many of the higher-level sub-libraries depend.

Finally, Chapter 36 looks at the Actor model of concurrent programming, and in particular, its usage through the C++ Actor Framework (CAF).

Chapter 33
Linear algebra and FFTW

Linear algebra is fundamental to many scientific computing applications. It is therefore crucial to have high-quality linear algebra libraries that are optimized for specific hardware architectures. In this chapter, we discuss the de facto standard for implementing basic vector and matrix operations, BLAS, and the high-quality implementation of linear algebra routines, LAPACK. LAPACK leverages the BLAS to provide optimized routines for solving linear algebra problems. We also discuss the FFTW library, which provides optimized routines for computing discrete Fourier transforms (DFT). Discrete Fourier transforms are another common operation in scientific computing applications that are often expensive to compute.

33.1 BLAS

BLAS is short for *Basic Linear Algebra Subprograms* and provides the go-to standard specification for basic vector and matrix operations. The BLAS standard is used by many high-quality linear algebra libraries such as ATLAS, OpenBLAS, Intel MKL, and LAPACK. With the focus on performance, BLAS is often the foundation of these libraries because it provides the specification for their underlying routines. Through the foundation provided by BLAS, the library implementations aim to optimize their routines for specific hardware architectures. It was important to have a standard for basic vector and matrix operations to avoid constantly rewriting the implementations of these operations within specific programming applications. This focus on curating a standard allowed for more performant implementations of these operations and enabled developers to focus on their application specifics rather than underlying vector and matrix operations.

BLAS is divided into three levels of functionality. Each level categorizes routines based on their time complexity, which is apparent through the data structures outlined in the operations at each of the levels. For data of size n, level-1 BLAS routines run in linear time (i.e., $O(n)$) and provide scalar, vector, and vector-vector operations. Level-2 BLAS routines run in quadratic time (i.e., $O(n^2)$) and provide matrix-vector

© The Author(s), under exclusive license to Springer Nature Switzerland AG 2025
R. J. Spiteri and K. Klenk, *Extreme-Scale Computing*, CMS/CAIMS Books in Mathematics 15, https://doi.org/10.1007/978-3-031-89033-8_33

operations. Level-3 BLAS routines run in cubic time (i.e., $O(n^3)$) and provide matrix-matrix operations. Each level can run with time complexity better than the stated value if a special matrix type can be used. For example, if working with band matrices (see Listing 33.1 for instance), the level-2 routines scale with $O(kn)$ complexity for matrices with a bandwidth of k.

Each routine in each level of BLAS is designed to work with four different data types: single precision, double precision, complex single precision, and complex double precision. The precision of the data type is indicated by the first letter of the routine name. The letter `s` denotes single-precision routines, the letter `d` denotes double precision, the letter `c` denotes complex single precision, and the letter `z` denotes complex double precision.

Some examples of BLAS routines are

- Level-1 BLAS
  - `xSCAL` (vector scaling)
  - `xAXPY` (constant times a vector plus a vector)
  - `xDOT` (computing the dot product of two vectors)
- Level-2 BLAS
  - `xGEMV` (matrix-vector multiplication)
  - `xSYMV` (symmetric matrix-vector multiplication)
  - `xTRMV` (matrix-vector multiplication with a triangular matrix)
- Level-3 BLAS
  - `xGEMM` (matrix-matrix multiplication)
  - `xSYMM` (symmetric matrix-matrix multiplication)
  - `xTRMM` (matrix-matrix multiplication with a triangular matrix)

where `x` can be `s`, `d`, `c`, or `z`.

BLAS originally defined its routines for Fortran but now provides an API that is more easily called from C (and hence C++) through the `cblas` interface. This avoids the legacy issues of having to deal with the column-major ordering of Fortran while still providing fast in-place memory assignments with additional C enums for legible simplicity of the library's usage.

33.1.1 BLAS example codes

An example of using the BLAS level-2 routine `cblas_ssbmv` is found in Listing 33.1. When compiling a code that uses `cblas`, linking with `cblas` can be done by adding the `-lcblas` flag to the command line, i.e.,

```
$ g++ -o blas_example blas_example.cpp -lcblas
```

Listing 33.1 An example of how to use the BLAS level-2 routine `cblas_ssbmv`. This routine works with scalars α and β, a symmetric band matrix $\mathbf{A}$, and vectors $\mathbf{x}$ and $\mathbf{y}$ to compute $y := \alpha\mathbf{A}\mathbf{x} + \beta\mathbf{y}$. We note the more efficient storage of the *symmetric band* matrix type that ignores the zero values off of the banded diagonals.

```
#include <iostream>
#include "cblas.h"

// Example program that computes y = alpha*A*x + beta*C using
// SSBMV (i.e., A is a symmetric band matrix)

template<typename T>
void output_mat(T* A, const int M, const int N, const char* leftpad);

int main(int argc, char* argv[]) {
  const int N = 4; // number of rows and columns of A
  const int K = 1; // number of super-diagonals in A

  const float alpha = 0.1;
  const float beta  = 0.2;

  // Banded form of A
  float Asb[N][K+1] = {
    {-2, 1},
    {-2, 1},
    {-2, 1},
    {-2, 5000}}; // Last entry does not matter because it is never used

  // General form of A
  float Age[N][N] = {
    {Asb[0][0], Asb[0][1],         0,         0},
    {Asb[0][1], Asb[1][0], Asb[1][1],         0},
    {        0, Asb[1][1], Asb[2][0], Asb[2][1]},
    {        0,         0, Asb[2][1], Asb[3][0]}};

  float y[N] = {1, 2, 3, 4};
  float x[N] = {1, 1, 1, 1};

  // expected result of the sbmv
  float expected[N] = {
    -0.1 + 0.2,
       0 + 0.4,
       0 + 0.6,
    -0.1 + 0.8 };

  // Computation: y := alpha*A*x + beta*y
  cblas_ssbmv(CblasRowMajor, // Layout of matrices
              CblasUpper,    // Upper or lower part?
              N,             // Number of rows/columns of A
              K,             // Number of super- (and sub-) diagonals
              alpha,         // alpha value in sgbmv
              &Asb[0][0],    // data for matrix A
              K + 1,         // size of leading dimension of Asb
              x,             // data for vector x
              1,             // increment for vector x
              beta,          // beta value in sgbmv
              y,             // data for vector y     !! IN/OUT PARAM !!
              1);            // increment for vector y

  // Output
  std::cout << "For the matrix A:\n";
  output_mat(&Age[0][0], N, N, " ");
  std::cout << "\nExpected result of y := alpha*A*x + beta*y:\n";
  for(int i=0;i<N;++i) std::cout << " " << expected[i] << "\n";
  std::cout << "\ncblas_sgbmv result:\n";
  for(int i=0;i<N;++i) std::cout << " " << y[i] << "\n";

  return 0;
}

template <typename T>
void output_mat(T* A, const int M, const int N, const char* leftpad){
  std::cout << leftpad;
  for(int i=0; i<N; ++i){
    for(int j=0; j< M; j++){
      std::cout << A[j*N+i] << " ";
    }
    std::cout << "\n" << leftpad;
  }
}
```

The output from Listing 33.1 is

```
For the matrix A:
 -2 1 0 0
```

```
 1 -2 1 0
 0 1 -2 1
 0 0 1 -2

Expected result of y := alpha*A*x + beta*y:
 0.1
 0.4
 0.6
 0.7

cblas_sgbmv result:
 0.1
 0.4
 0.6
 0.7
```

An example of performing general matrix-matrix multiplication using the BLAS level-3 `cblas_dgemm` routine is found in Listing 33.2.

Listing 33.2 An example of how to use the BLAS level-3 routine `cblas_dgemm`. This routine works with scalars α and β, and general matrices **A**, **B**, and **C** to compute $\mathbf{C} := \alpha\mathbf{AB} + \beta\mathbf{C}$.

```
#include <iostream>
#include "cblas.h"

// Example program that computes C = alpha*A*B + beta*C using
// DGEMM in BLAS

template<typename T>
void output_mat(T* A, const int M, const int N, const char* leftpad);

int main(int argc, char* argv[]) {
  const int M = 2; // number of rows of A
  const int N = 3; // number of columns of A

  const double alpha = 0.1;
  const double beta  = 0.2;

  double A[M][N] = {
    {1, 2, 3},
    {4, 5, 6}};
  double B[N][M] = {
    {1, 4},
    {2, 5},
    {3, 6}};
  double C[M][M] = {
    {1, 1},
    {1, 1}};

  // expected result of the GEMM
  double expected[M][M] = {
    {1.4+0.2, 3.2+0.2},
    {3.2+0.2, 7.7+0.2}};

  // Computation: C := alpha*A*B + beta*C
  cblas_dgemm(CblasRowMajor, // Layout of matrices
              CblasNoTrans,  // Is A transposed?
              CblasNoTrans,  // Is B transposed?
              M,             // Number of rows of A (and C)
              M,             // Number of columns of B (and C)
              N,             // Number of columns of A (and rows of B)
              alpha,         // alpha value in dgemm
              &A[0][0],      // data for matrix A
              N,             // size of leading dimension of A
              &B[0][0],      // data for matrix B
              M,             // size of leading dimension of B
              beta,          // beta value in dgemm
              &C[0][0],      // data for matrix C   !! IN/OUT PARAM !!
              M);            // size of leading dimension of C


  // Output
```

```
  std::cout << "Expected result of C := alpha*A*B + beta*C:\n";
  output_mat(&expected[0][0],M,M," ");
  std::cout << "\ncblas_gemm result:\n";
  output_mat(&C[0][0],M,M," ");
  std::cout << std::endl;

  return 0;
}

template <typename T>
void output_mat(T* A, const int M, const int N, const char* leftpad){
  std::cout << leftpad;
  for(int i=0; i<N; ++i){
    for(int j=0; j< M; j++){
      std::cout << A[j*N+i] << " ";
    }
    std::cout << "\n" << leftpad;
  }
}
```

The output from Listing 33.2 is

```
$ ./blas_example
Expected result of C := alpha*A*B + beta*C:
 1.6 3.4
 3.4 7.9

cblas_gemm result:
 1.6 3.4
 3.4 7.9
```

33.2 Multi-precision GPU computation

Artificial intelligence (AI) and machine learning (ML) have become an increasingly impactful part of daily life. Because of the large amounts of computation involved in creating AI/ML models, AI and ML have also had an outsized impact on ESC, most notably in the use of multi-precision computations and the role of GPUs. In the exascale era, the use of GPUs for extreme-scale software is an imperative.

Early GPUs were optimized with graphics and gaming applications in mind. Because single-precision computations were typically sufficient for these applications, GPUs often delivered significantly higher performance for single-precision computations compared to double-precision. The spread in the use of GPUs for general-purpose computing, however, has led to the performance of single- and double-precision computations to now be comparable.

AI/ML-inspired computations are also able to utilize lower-precision formats to achieve significant speed and memory improvements while maintaining sufficient accuracy for training and inference. In addition to the floating-point types described in Chapter 5, AI/ML computations often mix in *half-precision* (added to the IEEE 754 standard in 2008 to generally reduce memory usage and computation time) and `bfloat16` (introduced by Google in 2014 specifically for neural networks). The trend of using lower-precision formats also comes with the advantage of requiring less energy, ameliorating some concerns around the general energy-intensive nature of AI/ML.

We now describe *iterative refinement* as a classic example of a multi-precision calculation performed on a GPU. The general idea of iterative refinement is to solve a linear system with a low-precision solver but then refine the solution via corrections computed in a higher precision. Although the example described uses single-precision as the low-precision and double as the higher precision, the algorithm can be extended to other precisions. Iterative refinement was particularly effective on early GPUs, where lower-precision arithmetic was significantly faster, but it is also well-suited for modern hardware architectures such as NVIDIA Tensor Cores, which are specialized hardware units that natively support mixed-precision computations.

The algorithmic description of iterative refinement to solve a linear system $\mathbf{Ax} = \mathbf{b}$ is given in Algorithm 33.1.

Algorithm 33.1 Iterative refinement algorithm for solving $\mathbf{Ax} = \mathbf{b}$

Compute **LU** factorization of **A** in single precision
Compute initial solution **x** using **LU** factorization in single precision
repeat
    Compute residual $\mathbf{r} = \mathbf{b} - \mathbf{Ax}$ in double precision
    Compute correction **c** from $\mathbf{Ac} = \mathbf{r}$ by back substitution in single precision
    Update $\mathbf{x} \leftarrow \mathbf{x} + \mathbf{c}$ in double precision
until **x** satisfies a given tolerance

Code to run iterative refinement on a GPU using CUDA is given in Listing 33.3. The code solves the `mult_dcop_01` linear system from the SuiteSparse Matrix Collection, which represents a linear system of dimension 25 187 with a coefficient matrix having 193 276 nonzero entries.

Listing 33.3 Iterative refinement applied to a linear system $\mathbf{Ax} = \mathbf{b}$.

```
#include <cuda_runtime.h>
#include <cusolverDn.h>
#include <cublas_v2.h>
#define THREADS 128

__global__
static void double_to_float(double * A, float* sA, int N) {
  unsigned int idx = blockIdx.x * blockDim.x + threadIdx.x;

  if (idx < N) {
    sA[idx] = (float) A[idx];
  }
}

__global__
static void float_to_double(float * A, double* dA, int N) {
  unsigned int idx = blockIdx.x * blockDim.x + threadIdx.x;

  if (idx < N) {
    dA[idx] = (double) A[idx];
  }

}

void solve_mixed(double * dA, double * dB, double * dX, int N, double tolerance) {
  float * dS;
  int * d_info;
  float * d_work;
  float * dSB;
  double * dR;
  int* dPiv;

  double residual;
```

```
  double alph, beta;

  int lwork;

  cublasHandle_t blas_handle;
  cusolverDnHandle_t cusolverH;

  cudaError_t cuerr;
  cusolverStatus_t solvererr;

  cuerr = cudaMalloc((void**)&dS, N*N * sizeof(float));

  dim3 grid((N*N+THREADS)/THREADS);
  dim3 block(THREADS);
  double_to_float<<<grid, block>>>(dA, dS, N*N);

  // compute LU factorization in single precision
  solvererr = cusolverDnCreate(&cusolverH);

  cudaMalloc((void **)&d_info, sizeof(int));

  solvererr = cusolverDnSgetrf_bufferSize(cusolverH, N, N, dS, N, &lwork);

  cudaMalloc((void **) &d_work, sizeof(float) * lwork);
  cudaMalloc((void **) &dPiv, sizeof(int) * N);

  solvererr = cusolverDnSgetrf(cusolverH, N, N, dS, N, d_work, dPiv, d_info);

  cudaMalloc((void **) &dSB, sizeof(float) * N);

  dim3 grid2((N+THREADS)/THREADS);
  dim3 block2(THREADS);
  double_to_float<<<grid2, block2>>>(dB, dSB, N);

  // A x0 = b using LU factorization
  cusolverDnSgetrs(cusolverH, CUBLAS_OP_N, N, 1, dS, N, dPiv, dSB, N, d_info);

  alph = -1;
  beta = 1;

  cudaMalloc((void **) &dR, N * sizeof(double));
  float_to_double<<<grid2, block2>>>(dSB, dX, N);

  cublasCreate(&blas_handle);

  do {
    // compute residual r = b - A xi in double precision
    cudaMemcpy(dR, dB, N * sizeof(double), cudaMemcpyDeviceToDevice);
    cublasDgemv(blas_handle, CUBLAS_OP_N, N, N, &alph, dA, N, dX, 1, &beta, dR, 1);
    cublasDasum(blas_handle, N, dR, 1, &residual);

    // Solve Aci = ri using LU factorization (computed in single and cast to double)

    double_to_float<<<grid2, block2>>>(dR, dSB, N);
    solvererr = cusolverDnSgetrs(cusolverH, CUBLAS_OP_N, N, 1, dS, N, dPiv, dSB, N, d_info);
    float_to_double<<<grid2, block2>>>(dSB, dR, N);

    // Update x(i+1) = xi + ci (in double)
    cublasDaxpy(blas_handle, N, &beta, dR, 1, dX, 1);

  } while (residual > tolerance);


}

void solve(double * dA, double * dB, double * dX, int N) {
  cudaError_t cuerr;
  cusolverStatus_t solvererr;
  int * d_info;
  int lwork;
  double * d_work;
  int* dPiv;


  // compute LU factorization
  cusolverDnHandle_t cusolverH = NULL;

  solvererr = cusolverDnCreate(&cusolverH);
  cudaMalloc((void **)&d_info, sizeof(int));

  solvererr = cusolverDnDgetrf_bufferSize(cusolverH, N, N, dA, N, &lwork);

  cudaMalloc((void **) &d_work, sizeof(double) * lwork);

  solvererr = cusolverDnDgetrf(cusolverH, N, N, dA, N, d_work, dPiv, d_info);


  cusolverDnDgetrs(cusolverH, CUBLAS_OP_N, N, 1, dA, N, dPiv, dB, N, d_info);
```

```
}

int main() {
  int N;
  double * A, *B, *X;
  double * dA, *dB, *dX;
  cudaError_t cuerr;
  N = 25187;
  A = (double *) malloc(N*N*sizeof(double));
  B = (double *) malloc(N*sizeof(double));
  for (int i=0; i< N; i++) {
    for (int j=0; j<N; j++) {
      A[i*N+j] = 0;
    }
    B[i] = 0;
  }
  #include "mult_dcop.h"

  X = (double *) malloc(N*sizeof(double));

  cuerr = cudaMalloc((void**)&dA, N*N * sizeof(double));
  cuerr = cudaMalloc((void **)&dB, N * sizeof(double));
  cuerr = cudaMalloc((void **)&dX, N * sizeof(double));

  clock_t begin = clock();

  cuerr = cudaMemcpy(dA, A, N*N*sizeof(double), cudaMemcpyHostToDevice);
  cuerr = cudaMemcpy(dB, B, N*sizeof(double), cudaMemcpyHostToDevice);
  cuerr = cudaMemcpy(dX, X, N*sizeof(double), cudaMemcpyHostToDevice);

  solve_mixed(dA, dB, dX, N, 1e-10);

  cuerr = cudaMemcpy(X, dX, N*sizeof(double), cudaMemcpyDeviceToHost);

  clock_t end = clock();
  printf("iterative refinement with mixed precision: %f\n",
        (double)(end - begin) / CLOCKS_PER_SEC);

  // Reset solution for direct solve
  for (int i=0; i<N; i++) {
      X[i] = 0;
  }

  begin = clock();

  cuerr = cudaMemcpy(dA, A, N*N*sizeof(double), cudaMemcpyHostToDevice);
  cuerr = cudaMemcpy(dB, B, N*sizeof(double), cudaMemcpyHostToDevice);
  cuerr = cudaMemcpy(dX, X, N*sizeof(double), cudaMemcpyHostToDevice);
  solve(dA, dB, dX, N);

  cuerr = cudaMemcpy(X, dX, N*sizeof(double), cudaMemcpyDeviceToHost);

  end = clock();

  printf("direct solve in double precision: %f\n",
        (double)(end - begin) / CLOCKS_PER_SEC);

}
```

The main function begins by allocating memory for the coefficient matrix **A** and right-hand side vector **b** and initializing them to zero. It then reads in **A** and **b** (whose dimension N has been hardcoded to 25 187 — the size of the `mult_dcop_01` system) via the `#include` statement and allocates memory for the solution vector **x**.

Memory is then allocated on the GPU for **A**, **b**, and **x**. The timer is started, and **A**, **b**, and **x** are copied over to the GPU. The `solve_mixed` function is then called which, after some initializations that include casting **A** and **b** to `float`, uses the cuSOLVER library functions `cusolverDnSgetrf` to compute the **LU** factorization and `cusolverDnSgetrs` to obtain the initial solution, both in single precision. The `solve_mixed` function is called with a tolerance of 1e−10.

Having obtained the initial solution, we enter the iterative refinement loop to improve it to within the specified tolerance. As described in Algorithm 33.1, the first step is to compute the residual $\mathbf{r} = \mathbf{b} - \mathbf{Ax}$ in double precision. This computation is performed via the `cublasDgemv` function, which performs general matrix-vector operations in double precision. The 1-norm of $\mathbf{r}$ is then computed using `cublasDasum` in double precision. This value is compared to the tolerance to determine whether the loop should be exited. The correction $\mathbf{c}$ is then computed by back substitution using the previously computed $\mathbf{U}$ factor of $\mathbf{A}$ in single precision and then cast to double precision. Finally, the solution is updated using `cublasDaxpy` in double precision. We note the use of constants `alph` and `beta` passed to the cuBLAS functions `cublasDgemv` and `cublasDaxpy`, to execute the desired level-2 and level-1 BLAS operations $\alpha\mathbf{A}\mathbf{x} + \beta\mathbf{y} = -\mathbf{A}\mathbf{x} + \mathbf{b}$ and $\alpha\mathbf{A}\mathbf{x} + \mathbf{y} = \mathbf{x} + \mathbf{c}$. The loop continues until the tolerance is met, after which the timer is stopped to record the time taken to solve the linear system in this way.

For the purposes of comparison, the same problem is then solved on GPU using a standard direct **LU** decomposition via `cusolverDnSgetrf` and `cusolverDnSgetrs` as called from the `solve` function. The time it takes to produce a solution in this way is also recorded.

As a final note, the listing contains two kernel functions `double_to_float` and `float_to_double` to cast data between double- and single-precision.

Using an NVIDIA RTX A6000 GPU, we find the mixed-precision solver took 2.05 seconds to run, whereas the direct double-precision solver took 21.72 seconds.

33.3 LAPACK

Originally written in Fortran 77, the Linear Algebra Pack (LAPACK) is a library of optimized subroutines for solving linear algebra problems. It has since been updated to the Fortran 90 standard and also includes a C interface that was formerly its own library. The design of LAPACK is based on the BLAS standard, using it for many of the implemented computations. LAPACK can solve systems of linear equations, linear least-squares problems, and eigenvalue problems and compute singular-value decompositions. These problems are a higher-level abstraction than the BLAS standard, which is why LAPACK uses BLAS for its foundations.

LAPACK separates its routines into three categories: driver routines, computational routines, and auxiliary routines. Each category of routine provides a different level of abstraction, with the higher-level routines generally making calls to the lower-level ones. The driver routines are the highest level of abstraction and are designed to solve specific types of problems such as linear least-squares problems and computing eigenvalues. The computational routines are the middle level of abstraction and are designed to perform a specific computational task. Multiple computational routines are often called by single driver routine, but they can also be called directly by the programmer. Some examples of computational routines include factoring general rectangular matrices and computing singular-value decompositions. The auxiliary routines form the lowest level of abstraction. These routines are designed to compute specific subtasks or common low-level computations. Some examples of auxiliary routines include performing a subtask of a block algorithm, scaling a matrix, and computing a matrix norm. These routines are often called by the programmer directly to achieve a desired task. Computational tasks can also be used individually to solve more problems than the ones provided by the driver routines.

Routines in LAPACK are also implemented for working on different data types. The supported data types are single-precision and double-precision for both real and complex numbers. Almost all the routines supported by LAPACK are implemented for all four data types, except for a few routines that only work with complex data types. LAPACK also supports both dense and banded matrices, but there is no support for sparse matrices.

The routines in LAPACK follow a strict naming convention that outline the data type, the type of matrix, and the computation that is to be performed. Each name is 6 characters long and follows the format `XYYZZZ` where `X` is the data type, `YY` is the type of matrix, and `ZZZ` is the computation that is to be performed. The supported data types are the same as the supported data types in BLAS, i.e., real single precision, which is denoted by the letter `s`, real double precision, which is denoted by the letter `d`, complex single precision, which is denoted by the letter `c`, and double complex precision, which is denoted by the letter `z`. There are many supported types of matrices in LAPACK, examples of which include but are not limited to

- GE: General matrix
- SY: Symmetric matrix
- TR: Triangular matrix
- PO: symmetric or Hermitian positive definite
- HE: Hermitian matrix
- HP: Hermitian, packed storage
- GB: General band matrix
- TB: Triangular band matrix

Some examples of the computations that can be performed are

- GESV: Solves a system of linear equations with a general matrix
- GELS: Solves overdetermined or underdetermined real linear systems involving an $m \times n$ matrix $\mathbf{A}$, or its transpose, using a $\mathbf{QR}$ or $\mathbf{LQ}$ factorization of $\mathbf{A}$
- GESVD: Computes the singular-value decomposition (SVD) of a real $m \times n$ matrix $\mathbf{A}$, optionally computing the left or right singular vectors
- TRF: Factorize to triangular factors.
- TRS: Solve (forward/back substitution of triangular factors).
- QRF: Compute $\mathbf{QR}$ factorization.
- SYEV: Computes all eigenvalues and, optionally, eigenvectors of a real symmetric matrix

Similar to BLAS, there is a modern implementation of LAPACK that simplifies the C usage of the library, LAPACKE. Example usage of using LAPACKE to factor and solve general linear systems (with pivoting) is given in Listing 33.4 and to compute eigenvalues and eigenvectors of a symmetric matrix is given in Listing 33.5.

33.3.1 LAPACKE examples

An example code that factors and solves a square linear system using LAPACKE is given in Listing 33.4. Compiling code that uses LAPACKE should link with the command-line flag `-llapacke`, i.e.,

```
$ g++ -o lapack_example lapack_example.cpp -llapacke
```

Listing 33.4 An example of how to use LAPACKE to factor and solve a square linear system (with pivoting) using `LAPACKE_dgetrf` and `LAPACKE_dgetrs`.

```
#include <iostream>
#include "lapacke.h"

// Example program that factors and solves Ax = b
// using DGETRF(S) in LAPACKE

template<typename T>
void output_mat(T* A, const int M, const int N, const char* leftpad);

int main(int argc, char* argv[]) {
  const int N = 3; // number of columns of A

  double A[N][N] = {
    { 1, -2,  1},
    {-2,  1,  0},
    { 0,  1, -2}};
  double x[N] = {0, -1, -1};
  int ipiv[N] = {0, 0, 0};
  // expected result of the GEMM
  double expected[N] = {1, 1, 1};
  char transpose[] = "N";

  //Display original matrix
  std::cout << "For the matrix A:\n";
  output_mat(&A[0][0], N, N, " ");

  // Computation: Factor and solve Ax = b
  // Factorization -- O(N^3)
  LAPACKE_dgetrf(LAPACK_ROW_MAJOR, // Layout of matrices
                 N,                // Number of rows of A
                 N,                // Number of colums of A
                 &A[0][0],         // data for matrix A
                 N,                // size of leading dimension of A
                 ipiv);            // Pivot vector for stability

  // Storage after factorization
  std::cout << "\nFactorized A (L and U overwrite A):\n";
  output_mat(&A[0][0], N, N, " ");
  std::cout << "With permutation:\n";
  for(int i=0;i<N;++i) std::cout << " " << ipiv[i] << "\n";

  // Solve -- O(N^2)
  LAPACKE_dgetrs(LAPACK_ROW_MAJOR, // Layout of matrices
                 transpose[0],             // Transpose? ("N"=no, "T"=transpose, "C"=conjugate transpose)
                 N,                // Number of rows of A
                 1,                // Number of right-hand sides to solve for
                 &A[0][0],         // data for matrix A
                 N,                // size of leading dimension of A
                 ipiv,             // Pivot vector for stability
                 x,
                 1);// RHS (and solution on return!)    !!! IN/OUT vector


  // Output
  std::cout << "\nExpected result of solving Ax = b\n";
  for(int i=0;i<N;++i) std::cout << " " << expected[i] << "\n";
  std::cout << "\nLAPACKE_dgetrf(s) result:\n";
  for(int i=0;i<N;++i) std::cout << " " << x[i] << "\n";
  std::cout << std::endl;

  return 0;
}

template <typename T>
void output_mat(T* A, const int M, const int N, const char* leftpad){
  std::cout << leftpad;
  for(int i=0; i<N; ++i){
    for(int j=0; j< M; j++){
      std::cout << A[j*N+i] << " ";
    }
```

```
    std::cout << "\n" << leftpad;
  }
}
```

The output of Listing 33.4 is

```
For the matrix A:
 1 -2 0
 -2 1 1
 1 0 -2

Factorized A (L and U overwrite A):
 -2 -0.5 -0
 1 -1.5 -0.666667
 0 1 -1.33333
 With permutation:
 2
 2
 3

Expected result of solving Ax = b
 1
 1
 1

LAPACKE_dgetrf(s) result:
 1
 1
 1
```

Similarly, an example code that computes eigenvalues and eigenvectors of a symmetric matrix using LAPACKE is given in Listing 33.5. Compiling code that uses LAPACKE should link with the command-line flag `-llapacke`, i.e.,

```
$ g++ -o lapack_example2 lapack_example2.cpp -llapacke
```

Listing 33.5 An example of how to use LAPACKE to solve for eigenvalues and eigenvectors of a symetric matrix using the LAPACKE_dsyev routine.

```
#include <iostream>
#include "lapacke.h"

// Example program that computes eigenvalues and eigenvectors
// A*v = lambda*v using DSYEV in LAPACKE

template<typename T>
void output_mat(T* A, const int M, const int N, const char* leftpad);

int main(int argc, char* argv[]) {
  const int N = 3; // number of columns of A

  double A[N][N] = {
    {-2,  1,  0},
    { 0, -2,  1},
    { 0,  0, -2}};
  double w[N];

  char evec[] = "V";
  char uplo[] = "U";
```

```
  //Display original matrix
  std::cout << "For the matrix A:\n";
  output_mat(&A[0][0], N, N, " ");

  // Computation: Factor and solve Ax = b
  // Factorization -- O(N^3)
  LAPACKE_dsyev(LAPACK_ROW_MAJOR, // Layout of matrices
                'V',              // 'V'  =compute eigenvectors too
                'U',              // 'U' = upper storage, 'L' = lower storage
                N,                 // Number of rows of A
                &A[0][0],          // Number of colums of A
                N,                 // size of leading dimension of A
                w);                // Pivot vector for stability

  // Output
  std::cout << "\nEigenvectors of A (columns):\n";
  output_mat(&A[0][0], N, N, " ");
  std::cout << "\nWith eigenvalues:\n";
  for(int i=0;i<N;++i) std::cout << " " << w[i] << " ";
  std::cout << "\n";

  return 0;
}

template <typename T>
void output_mat(T* A, const int M, const int N, const char* leftpad){
  std::cout << leftpad;
  for(int i=0; i<N; ++i){
    for(int j=0; j< M; j++){
      std::cout << A[j*N+i] << " ";
    }
    std::cout << "\n" << leftpad;
  }
}
```

The expected output of Listing 33.5 is

```
For the matrix A:
 -2 0 0
 1 -2 0
 0 1 -2

Eigenvectors of A (columns):
 -0.5 0.707107 -0.5
 0.707107 1.31839e-16 -0.707107
 -0.5 -0.707107 -0.5

With eigenvalues:
 -3.41421  -2  -0.585786
```

33.4 FFTW

The *Fastest Fourier Transform in the West* (FFTW) is a library that is optimized for computing discrete Fourier transforms (DFTs) using $O(m \log m)$ algorithms for vectors of size m. To use FFTW, the user is recommended to learn one basic concept about the internal structure of FFTW, namely, that FFTW does not use a fixed algorithm for its computations. Instead, it adapts the DFT algorithm to the underlying hardware in order to maximize performance. FFTW does this by using a two-phase approach to the computation.

The first phase is a planning phase, where FFTW employs a planner to learn the fastest way to compute the transform on the target machine. The result is a data structure called a *plan* that can then be used in the second phase, which is the execution of the plan. The same plan can be reused multiple times. The two-phase structure of FFTW is used because when multiple transforms are required to be computed, the transforms are often the same size. Therefore, the planning phase can endure the expensive one-time cost of learning the fastest way to compute the transform. If the number of transforms that are to be computed is small, then FFTW has fast planners that utilize heuristics or previously computed plans for this situation.

FFTW also aims to be adaptable to support user needs. The FFTW interface is separated into three levels. Each level is intended to be more general than its predecessor. The first and most basic level is aptly named the *basic interface*. This interface is designed to perform transforms on contiguous data. The second and more general level is the *advanced interface*. This interface is designed to perform multiple transforms on strided arrays. The third and most general level is the *guru interface*. This interface is designed to support the most general data layouts, multiplicities, and strides. Most users have their needs met by the basic interface, but the advanced and guru interfaces are available for users who need more control over the data layout. The more advanced interfaces do, however, require the user to have a deeper understanding of FFTW to use them effectively.

33.4.1 Using FFTW

To use FFTW, the user must first include the header file `fftw3.h`. Transforms can be performed on arrays of any dimension. The first step is to allocate the correct data structures. FFTW provides its own `malloc` function, `fftw_malloc`, which attempts to align the data structures to leverage SIMD instructions whenever possible. After allocating the data types, the planning stage can be done with calls to `fftw_plan_dft_1d`, `fftw_plan_dft_2d`, or `fftw_plan_dft_3d`, depending on the dimensionality of the data. If a user needs more than three dimensions, then the `fftw_plan_dft` function, which takes an array of integers that specify the dimensions of the data, can be used. The plan must be executed before any data are initialized within the data structures. After the plan has been run, the data can be initialized, and the `fftw_execute` function can be called using the plan as many times as needed. When finished, the plan can be destroyed with the `fftw_destroy_plan`, and the data structures can be freed with the `fftw_free` function. An example of using FFTW to compute the 2D DFT of a 10×10 matrix is shown in Listing 33.6.

Listing 33.6 An example of how to use fftw

```
#include <fftw3.h>

int main(int argc, char **argv) {
  fftw_complex *in, *out;
  fftw_plan p;
  int dims[2] = {10, 10};
  int num_dims = 2;
```

```
  p = fftw_plan_dft(num_dims, dims, in, out, FFTW_FORWARD, FFTW_ESTIMATE);

  // Data Setup goes here
  fftw_execute(p);
  fftw_destroy_plan(p);
  fftw_free(in); fftw_free(out);

  return 0;
}
```

If a user needs to use something other than a complex data type as input or output, then the `fftw_plan_dft_r2c`, `fftw_plan_dft_c2r`, and `fftw_plan_dft_r2r` functions can be used, respectively, for real input, real output, or both. Each of these plan functions can be used with an array of any dimensionality. When using a C compiler that adheres to more recent standards, including `<complex.h>` before the `<fftw3.h>` header file allows the use of the native complex type instead of the one implemented by FFTW.

33.4.2 Parallel FFTW

FFTW supports both shared- and distributed-memory parallelism. For shared-memory parallelism, FFTW supports both the native thread libraries for Linux and Windows, as well as the OpenMP standard. The FFTW library just needs to be compiled with the correct flags to enable multithreading. The `--enable-threads` option can be used to enable the OS's native thread library, and `--enable-openmp` can be used to enable OpenMP. If desired, the user can compile with both because the names to the functions are different when using native threads vs. OpenMP. When using threads, the user must call `fftw_init_threads()` before any other FFTW functions. Before calling the function to create a plan, the user must call `fftw_plan_with_nthreads(int nthreads)` to set the proper number of threads. Creating a plan and executing it are done as usual. The plan utilizes all the threads that are set. When finished, the user must call `fftw_cleanup_threads()` before calling `destroy_plan`.

For distributed-memory parallelism, FFTW supports MPI. The interface for distributed-memory parallelism is different than the interface for shared-memory parallelism because each MPI process stores only a portion of the data that are being transformed. However, the benefit with using MPI is that larger data structures can be split across multiple processes in the event the data are too numerous to fit on a single process. To compile FFTW with MPI support, the user must compile with the `--enable-mpi` flag. When using FFTW with MPI, the user must call `MPI_Init()` first and then `fftw_mpi_init()` before calling any other FFTW functions. At the end of the program, before calling `MPI_Finalize()`, the user must call `fftw_mpi_cleanup()`. Listing 33.7 shows how to use FFTW with MPI.

Listing 33.7 An example of how to use fftw with MPI

```
#include <complex.h>
#include <fftw3-mpi.h>

int main(int argc, char **argv) {
```

```
  const ptrdiff_t N0 = 10, N1 = 10;
  fftw_plan plan;
  fftw_complex *data;
  ptrdiff_t local_proc_size, local_n0, local_0_start, i, j;

  MPI_Init(&argc, &argv);
  fftw_mpi_init();

  // Get the local data size
  local_proc_size = fftw_mpi_local_size_2d(N0, N1, MPI_COMM_WORLD,
                                           &local_n0, &local_0_start);

  data = fftw_alloc_complex(local_proc_size);

  // Create the plan
  plan = fftw_mpi_plan_dft_2d(N0, N1, data, data, MPI_COMM_WORLD,
                              FFTW_FORWARD, FFTW_ESTIMATE);

  /* initialize data to some_function(x,y) */
  for (i = 0; i < local_n0; ++i) {
    for (j = 0; j < N1; ++j) {
      data[i*N1 + j] = some_function(local_0_start + i, j);
    }
  }
  /* compute transforms, in-place, as many times as desired */
  fftw_execute(plan);

  fftw_destroy_plan(plan);

  fftw_mpi_cleanup();

  MPI_Finalize();
  return 0;
}
```

In Listing 33.7, it can be seen that the interface for MPI is different than the shared-memory/serial interface. Before creating the plan, the user must call `fftw_mpi_local_size_2d` to get the local size of the data. Then this size must be allocated for the processes. After this is set up, the plan can be created and executed as normal. A special note about the `ptrdiff_t` type is that it is used in place of `int` to allow for its width to match the number of bits in the machine on which it is running. For a 32-bit machine, `ptrdiff_t` is 32 bits, and for a 64-bit machine, `ptrdiff_t` is 64 bits. The execution and cleanup of the plan is done similarly to the serial/shared-memory interface. One final point about using FFTW in parallel is that both MPI and multi-threading can be used together.

33.5 Additional Resources

- BLAS: https://www.netlib.org/blas/
- LAPACK: https://www.netlib.org/lapack/
  - LAPACKE (C interface) https://www.netlib.org/lapack/lapacke.html
- FFTW: http://www.fftw.org/fftw3_doc/

Chapter 34
PETSc

Computing the numerical solution of partial differential equations (PDEs) is a hallmark of scientific computing. The *Portable Extensible Toolkit For Scientific Computing* (PETSc) is a robust and scalable library that provides a suite of classes for developing large-scale scientific applications centered around the solution to PDEs and other numerical problems. The class structure of PETSc is designed in a hierarchical fashion, with classes building on each other in increasing levels of abstraction. At the lowest level, PETSc provides a set of classes for creating data structures, specifically vectors and matrices. At the highest level, the classes are designed to solve time-dependent nonlinear problems. Each class implemented in PETSc is designed around an abstract interface that defines how the class should be interacted with but not how it is actually implemented. This design keeps the interaction with the classes simple and consistent while allowing for a variety of underlying implementations, including user-defined ones. Through the use of an abstract interface, PETSc relies on run-time polymorphism to allow easier modification of both data storage formats and algorithms without having to re-compile or re-design the application. This chapter introduces PETSc and demonstrates how to leverage its capabilities in scientific applications.

34.1 PETSc Setup

PETSc provides great documentation for installation at `https://petsc.org/release/install/install_tutorial/`. After installing PETSc, the environment variables `$PETSC_DIR` and `$PETSC_ARCH` must be set in order to use it. The `$PETSC_DIR` variable should be set to the location of the PETSc installation. The `$PETSC_ARCH` variable should be set to the configuration of PETSc that is being used. The `$PETSC_ARCH` can be used to switch between different installations of PETSc, for example, a debug and release version.

It is important to note that PETSc relies on MPI. Thus, the same procedure for using MPI must be used within a PETSc application. However, PETSc does provide

© The Author(s), under exclusive license to Springer Nature Switzerland AG 2025

R. J. Spiteri and K. Klenk, *Extreme-Scale Computing*, CMS/CAIMS Books in Mathematics 15, https://doi.org/10.1007/978-3-031-89033-8_34

some convenient functions for initializing and finalizing MPI along with PETSc itself. The `PetscInitialize` function initializes PETSc and makes the call to `MPI_Init`, while the `PetscFinalize` function is used to finalize PETSc and MPI. These functions conveniently contain all the initialization and finalization calls for PETSc and MPI in one function. If a user wishes to use MPI before PETSc, the `MPI_Init` function can be called before the `PetscInitialize` function. However, if this is done, the user must also call `MPI_Finalize` because the `PetscFinalize` function will not call it in this situation.

Additionally, PETSc provides accessible functionality to print out help information and set program options through the command line. By specifying the `-help` option to the command line of a PETSc program, all the options available to the program are printed out. Command-line options can also be used to control certain aspects of a PETSc application through the use of the `PetscOptionsGetValue` function. This function allows users to retrieve the value of a command-line option to be stored in a variable or control the behavior of a PETSc object. Finally, PETSc also uses its own error-handling type, `PetscErrorCode`, which is an integer type used to indicate whether an error has occurred. The error code is set to 0 upon success, and any other value is considered an error. In general, the error code should always be checked after a function call. Happily, PETSc provides a function, `PetscCall`, that can check the error code automatically and print out a message if an error has occurred.

34.2 PETSc Vectors and Matrices

PETSc provides `Vec` classes for holding one-dimensional data in a distributed container. Vectors are used to store things like PDE solutions and the right-hand sides of linear systems. PETSc also provides `Mat` classes, which are distributed containers for holding dense or sparse matrices. Parallel sparse matrices are stored such that each rank locally stores a continuous set of rows.

Creating vectors and matrices can be done through both a low-level interface, where the user directly interacts with the `Vec` and `Mat` classes to instantiate their objects, and a high-level interface, where the user interacts with the Data Management (`DM`) object, which is the abstraction PETSc uses to organize and distribute data in distributed-memory systems. With the low-level interface, vectors are created with the `VecCreate` constructor, and matrices are created with the `MatCreate` constructor. Both vectors and matrices have a `VecSetSizes` and `MatSetSizes` method respectively, that are used to set the local and global sizes of the vector or matrix. The local size is the number of elements that are stored within an MPI rank, and the global size is the total number of elements that belong to the respective data structure.

Values are set using the `VecSetValues` and `MatSetValues` functions. Any rank can set the values of any location within the vector or matrix. The data are distributed throughout a structure; however, they are not completely isolated. The values are sorted between the MPI ranks by the calls to `VecAssemblyBegin` and `VecAssemblyEnd` for vectors and `MatAssemblyBegin` and `MatAssemblyEnd` for matrices. Below is an example of how to create a vector and a matrix and populate them with values.

Listing 34.1 An example of how create to vectors and matrices in PETSc.

```
#include petscvec.h
#include petscmat.h

int main(int argc, char argv[]) {
  // Declare PETSc objects
  Vec first_vector;
  Mat first_matrix;
  PetscInt i; col[3];

  // Size of Vector, and Matrix (square)
  PetscInt size = 10;
  PetscScalar value[3];

  // Initialize PETSc, MPI, and check for errors
  PetscCall(PetscInitialize(&argc, &argv, NULL, NULL));

  // Create a vector
  PetscCall(VecCreate(PETSC_COMM_WORLD, &first_vector));
  // Set the size of the vector
  PetscCall(VecSetSizes(first_vector, PETSC_DECIDE, size));

  // Populate the vector with values 0...size - 1
  for (i = 0; i < size; i++) {
    value = (PetscScalar)(i);
    VecSetValues(first_vector, 1, &i, &i, INSERT_VALUES);
  }

  // Allows ranks to pass data to each other
  PetscCall(VecAssemblyBegin(first_vector));
  PetscCall(VecAssemblyEnd(first_vector));

  // Create a matrix
  PetscCall(MatCreate(PETSC_COMM_WORLD, &first_matrix));
  // Set the size of the matrix
  PetscCall(MatSetSizes(first_matrix, PETSC_DECIDE, PETSC_DECIDE, size, size));


  // Assemble matrix
  value[0] = -1.0;
  value[1] = 2.0;
  value[2] = -1.0;
  for (i = 1; i < n - 1; i++) {
    col[0] = i - 1;
    col[1] = i;
    col[2] = i + 1;
    PetscCall(MatSetValues(our_first_matrix, 1, &i, 3, col, value, INSERT_VALUES));
  }
  i      = n - 1;
  col[0] = n - 2;
  col[1] = n - 1;
  PetscCall(MatSetValues(our_first_matrix, 1, &i, 2, col, value, INSERT_VALUES));
  i        = 0;
  col[0]   = 0;
  col[1]   = 1;
  value[0] = 2.0;
  value[1] = -1.0;

  // Allows ranks to pass data to each other
  PetscCall(MatAssemblyBegin(our_first_matrix, MAT_FINAL_ASSEMBLY));
  PetscCall(MatAssemblyEnd(our_first_matrix, MAT_FINAL_ASSEMBLY));

  // Destroy the objects
  PetscCall(VecDestroy(&our_first_vector));
  PetscCall(MatDestroy(&our_first_matrix));

  // Finalize PETSc
  PetscCall(PetscFinalize());

  return 0;
}
```

In Listing 34.1, a vector `first_vector` and a matrix `first_matrix` are created and populated with values. The first function called is `PetscInitialize`, which initializes PETSc and MPI. Then `first_vector` is created, and its global size is set to 10. For the local size, `PETSC_DECIDE` is used when creating the vector as well as the matrix. This is a special value used in PETSc that lets PETSc decide how to distribute the data structures between the MPI ranks. If the user knows or has a preference of how the data structure should be distributed, they can specify the distribution themselves. In addition to the `PETSC_DECIDE` flag, the population of data within the vector and matrix uses the `INSERT_VALUES` flag. This flag indicates that the values at each specified index should be overwritten, if they exist, by the new values. An alternative flag is `ADD_VALUES`, which indicates that the new values should be added by performing an addition operation to the existing values at the specified indices. Some other functions that were not used in the example include `VecGetArray`, which returns the underlying array allowing it to be modified, and `VecRestoreArray`, which returns a read-only array. Conversely, for matrices, the `MatGetRow` and `MatGetLocalSubMatrix` functions are used to obtain portions of the matrix. There are many operations that can be performed on vectors and matrices in PETSc. The details for vectors can be found at `https://petsc.org/release/docs/manual/vec/#basic-vector-operations`, and the details for matrices can be found at `https://petsc.org/release/docs/manual/mat/#basic-matrix-operations`.

To help manage the distribution of vectors and matrices, PETSc provides the `DM` class. There are multiple implementations of `DM` classes, each associated with a different family of problems. The first family of problems involves structured grids. Here, a `DM` object is instantiated to manage the communication of data between MPI ranks before local computations can occur. The `DM` object for this family of problems is referred to specifically as a `DMDA` object. A data structure consisting of 1, 2, or 3 dimensions is created with the `DMDACreate1d`, `DMDACreate2d`, and `DMDACreate3d` functions, respectively. Some other families of problems that apply the `DM` object are staggered grids (`DMStag`) and unstructured grids (`DMPlex`). Using the `DM` object is intended for more advanced users and is not covered in detail for this reason. However, PETSc provides well-written documentation for those interested in using DM objects.

34.3 PETSc Solvers

34.3.1 Linear Solvers

The linear solver interface is provided through a `KSP` object. It provides the general interface for both direct and iterative linear system solvers. The `KSP` object is intended to solve systems of the form $\mathbf{Ax} = \mathbf{b}$, where $\mathbf{A}$ is a matrix, $\mathbf{b}$ is the right-hand side vector, and $\mathbf{x}$ is the solution vector. The calling sequence is the same whether using direct or iterative solvers. After creating a linear system with vectors and matrices, the user can use the `KSP` object to solve the linear system. It is best to jump right into an example of how to use the `KSP` object to properly introduce it.

Listing 34.2 An example of how to use the KSP object to solve a linear system.

```
#include "petscksp.h"
#include "petscmat.h"

int main(int argc, char argv[]) {
  PetscInt num_iterations;

  KSP ksp; // Linear solver context
  Vec x, b; // solution vector, RHS
  Mat A; // Linear system coefficient matrix

  // Initialize PETSc and MPI and check for errors
  PetscCall(PetscInitialize(&argc, &argv, NULL, NULL));

  /*
  *  Matrix and vector setup and population omitted for brevity
  */

  // Create the Context for the Linear Solver
  PetscCall(KSPCreate(PETSC_COMM_WORLD, &ksp));

  // Set operators
  // A serves as the coefficient matrix and preconditioner
  PetscCall(KSPSetOperators(ksp, A, A));

  // Set runtime options
  PetscCall(KSPSetFromOptions(ksp));

  // Set the tolerances using the default values for most parameters
  PetscCall(KSPSetTolerances(ksp, 1.e-5, PETSC_DEFAULT, PETSC_DEFAULT, PETSC_DEFAULT));

  // Solve the linear system
  PetscCall(KSPSolve(ksp, b, x));

  // Get the number of iterations it took to solve the linear system
  KSPGetIterationNumber(ksp, &num_iterations);

  // Clean up
  PetscCall(KSPDestroy(&ksp));
  PetscCall(PetscFinalize());

  return 0;
}
```

In Listing 34.2, the KSP object is used to solve a linear system $\mathbf{Ax} = \mathbf{b}$. The KSP context is created with the `KSPCreate` function. The operators are then set with the `KSPSetOperators` function. In Listing 34.2, the operators are the same: $\mathbf{A}$ is the linear system matrix and also the preconditioner matrix. The PETSc documentation states that, for most cases, the preconditioner matrix will be the same as the linear system matrix. However, there are cases where the preconditioner matrix will be different. For example, when a linear system matrix is obtained from a higher-order method, a simpler preconditioner matrix may be used. Once the context in which the linear system is being solved is set up, the `KSPSolve` function is called. This function takes in the right-hand side vector $\mathbf{b}$ and the solution vector $\mathbf{x}$ as arguments. The solution vector $\mathbf{x}$ is then populated with the solution to the linear system. The KSP context can then be destroyed with the `KSPDestroy` function. If the preconditioner matrix need not change, `KSPSolve` can be solved multiple times if needed before destroying the KSP context. However, if the preconditioner matrix changes, then the `KSPSetOperators` function must be called again before calling `KSPSolve`.

Preconditioners can be applied to KSP solves through a PC object. This component is crucial for ensuring that iterative methods converge in a reasonable number of iterations. Preconditioners can be applied with the `PCSetType` function. The `KSPSolve` function defaults to left preconditioning. Calling the `KSPSetPCSide` function with the argument `PC_RIGHT` changes the `KSPSolve` function to use right preconditioning. However, not all methods work with right preconditioning, and PETSc issues

an appropriate error message in such a case, e.g.,
KSPSetUp_Richardson:No right preconditioning for KSPRICHARDSON.

The final part of the KSP object that needs to be discussed is the convergence of the linear solve. Two parameters are used to control the convergence of linear iterator solvers, atol and rtol. The absolute tolerance, atol, is the maximum allowable error in the solution. The relative tolerance, rtol, is the maximum allowable error in the solution relative to the norm of the right-hand side vector. These values, along with the maximum number of iterations, can be set with the KSPSetTolerances function. The default values are rtol = 1e−5 and atol = 1e−50[1] and can be set with the PETSC_DEFAULT flag. The default maximum number of iterations is 10 000.

34.3.2 Nonlinear Solvers

The SNES class is used to create objects that solve nonlinear systems of algebraic equations of the form $\mathbf{F}(\mathbf{x}) = \mathbf{0}$ where $\mathbf{F}(\cdot)$ is a function that takes a vector of dimension m and returns a vector of the same size. This class provides numerous iterative solver algorithms such as line-search Newton and nonlinear Richardson. Both nonlinear Krylov methods and methods based on problem decomposition are supported. The implementation and use of the nonlinear solvers should feel similar to what has been shown with the linear solvers.

Newton's method is the default nonlinear solver. The algorithm for Newton's method is as follows

```
Choose initial guess x_0
for k = 0, 1, 2, ...
  x_{k+1} = x_k - J(x_k)^{-1} F(x_k)
  if abs(F(x_{k+1})) < tol
    break
   end
 end
```

where $\mathbf{F}(\mathbf{x})$ is a system of nonlinear equations and $\mathbf{J}(\mathbf{x})$ is the Jacobian of $\mathbf{F}(\mathbf{x})$. The above algorithm loops for a maximum number of iterations or until it has converged to a solution with a specified tolerance (if abs(F(x_{k+1})) < tol from the above algorithm). When using the nonlinear solver, the user must provide a function that computes the nonlinear function. The Jacobian can either be evaluated analytically by a user-defined function, or it can be approximated by using finite differences. The SNES context is created with the SNESCreate function. Choosing a nonlinear solution method can be done with the SNESSetType function. The problem can then be solved with a call to the SNESSolve function. The SNESSolve function takes in the right-hand side vector and an initial guess for the solution vector. Again, we can solve multiple non-linear systems before calling the SNESDestroy function.

[1] Scientific notation is being used here, where 1e−5 ≡ 1×10^{-5} and 1e−50 ≡ 1×10^{-50}.

Listing 34.3 An example of how to use the SNES class to solve a nonlinear system.

```
#include "petscsnes.h"

int main(int argc, char argv[]) {
  KSP ksp; // Linear solver context
  SNES snes; // nonlinear solver

  Vec r;
  Mat J;
  PC pc;   // preconditioner context
  PetscCall(PetscInitialize(&argc, &argv, (char *)0, help));

  // Create a non-linear solver context
  PetscCall(SNESCreate(PETSC_COMM_WORLD, &snes));
  PetscCall(SNESSetType(snes, SNESNEWTONLS));

  // Set the function to be solved
  PetscCall(SNESSetFunction(snes, r, function_to_solve, NULL));
  // Set the Jacobian matrix structure and evaluation routine
  PetscCall(SNESSetJacobian(snes, J, J, function_jacobian, NULL));

  // Get the default linear solver context
  PetscCall(SNESGetKSP(snes, &ksp));
  PetscCall(KSPGetPC(ksp, &pc));
  PetscCall(PCSetType(pc, PCNONE));
  PetscCall(KSPSetTolerances(ksp, 1.e-4, PETSC_DEFAULT, PETSC_DEFAULT, 20));

  // Create an initial guess for the solution
  // ...
  // ...
  // ...
  // ...

  // Solve the system
  PetscCall(SNESSolve(snes, NULL, x));

  PetscCall(SNESDestroy(&snes));
  PetscCall(PetscFinalize());
  return 0;
}
```

The example shown in Listing 34.3 demonstrates how to set up the PETSc calls to solve a nonlinear system. Most of the implementation is left out for brevity, with only the most important parts being shown. The SNES context is created with the `SNESCreate` function, and the `SNESSetType` function sets the nonlinear solver to use Newton-based line search with the SNESNEWTONLS flag. Calls to set the nonlinear function and the function that computes the Jacobian are also made with `SNESSetFunction` and `SNESSetJacobian`, respectively. The default linear solver context is then extracted from the nonlinear solver context. This allows for the setting of specific options to the KSP and PC objects. Then the initial guess for the solution vector is obtained and followed by a call to the `SNESSolve` function to solve the nonlinear system. Finally, the example is cleaned up with calls to `SNESDestroy` and `PetscFinalize` functions.

There is a default convergence test for most nonlinear solvers; however, users are free to define their own convergence test. The default convergence test uses default parameters that are aimed at being general enough for a wide range of problems. There is an additional convergence parameter that is used in the nonlinear solvers, `stol`, which is the solution tolerance. The step tolerance is used as way to check if a solution has converged by comparing the difference between the solution at the current and previous iterations. When this difference is less than the specified tolerance, the solution is deemed to have converged. To set the tolerances for the nonlinear solver, the `SNESSetTolerances` function is used. The maximum number of iterations and function evaluations can also be set within this function. If trust regions are being used, the user must set a minimum allowable trust region radius, which can be done using the `SNESSetTrustRegionTolerances` function.

34.3.3 Time-Steppers

The TS class provides the interface for creating objects for time-stepping algorithms in PETSc. These objects operate on systems of ordinary differential equations (ODEs) and differential-algebraic equations (DAEs) that typically arise from method-of-lines discretizations of PDEs.[2] The ODEs/DAEs are written as initial-value problems (IVPs) in the form $\mathbf{F}(t, \mathbf{u}, \dot{\mathbf{u}}) = \mathbf{G}(t, \mathbf{u})$, $\mathbf{u}(t_0) = \mathbf{u}_0$. As for algorithms, the time-steppers range from fully explicit methods, which only require direct function evaluations, to fully implicit methods, which generally require the solution of a nonlinear system of equations at each time step. Most of the time-stepping functions use adaptive timesteps to control the error.

Listing 34.4 Example of using the TS class to solve a system of ODEs.

```
#include <petscdm.h>
#include <petscdmda.h>
#include <petscts.h>

int main(int argc, char argv[]) {
  PetscInitialize(&argc, &argv, (char *)0, help);

  // Create time-stepping context
  TSCreate(PETSC_COMM_WORLD, &ts);
  TSSetDM(ts, da);
  TSSetProblemType(ts, TS_NONLINEAR);
  TSSetRHSFunction(ts, NULL, FormFunction, da);

  // Customize nonlinear solver
  TSSetType(ts, TSBEULER);
  TSGetSNES(ts, &ts_snes);

  // Set Initial Conditions
  FormInitialSolution(da, x);
  TSSetTimeStep(ts, .0001);
  TSSetExactFinalTime(ts, TS_EXACTFINALTIME_STEPOVER);
  TSSetSolution(ts, x);

  // Solve linear system
  TSSolve(ts, x);
  TSGetSolveTime(ts, &ftime);
  TSGetStepNumber(ts, &steps);
  VecViewFromOptions(x, NULL, "-final_sol");

  VecDestroy(&x);
  VecDestroy(&r);
  TSDestroy(&ts);
  DMDestroy(&da);

  PetscFinalize();
  return 0;

}
```

34.4 Parallelism

PETSc aims to support a wide range of parallelism models. The model with the strongest support is distributed memory parallelism through MPI. Users can implement MPI directives to control the behavior of their program. However, by default, PETSc hides the message-passing details from the user inside its parallel objects,

[2] Although not necessary, the method of lines is assumed to be performed in the "spatial" dimensions, leaving time as the dimension to be integrated.

vectors, matrices, and solvers. When creating these objects, an MPI communicator is passed in to indicate which set of processors the objects are to be distributed over. In addition to MPI, OpenMP can also be used within PETSc. This is especially helpful when additional libraries within PETSc (such as BLAS/LAPACK) are used that can benefit from OpenMP. One must be cautious when using OpenMP in combination with MPI so as to not oversubscribe the number of processors available. GPU support is also available and supported for CUDA, HIP, and Kokkos. PETSc provides specific classes of `Vec` and `Mat` written for efficient execution on GPUs.

34.5 PDE Example

As a final example, we demonstrate how to use PETSc to solve a PDE. This example is taken from the PETSc source repository at `https://gitlab.com/petsc/petsc/-/blob/main/src/ksp/ksp/tutorials/ex50.c`, which is distrubuted under the BSD 2-Clause License. In this example, we solve a Poisson equation in 2D. The code for this example appears in Listing 34.5.

Listing 34.5 Example of solving a PDE with KSP and DMDA.

```
/*   Example code from: https://gitlab.com/petsc/petsc/-/blob/main/src/ksp/ksp/tutorials/ex50.c
     Contributed by Michael Boghosian <boghmic@iit.edu>, 2008,
         based on petsc/src/ksp/ksp/tutorials/ex29.c and ex32.c
*/

static char help[] = "Solves 2D Poisson equation using multigrid.\n\n";

#include <petscdm.h>
#include <petscdmda.h>
#include <petscksp.h>
#include <petscsys.h>
#include <petscvec.h>

extern PetscErrorCode ComputeJacobian(KSP, Mat, Mat, void *);
extern PetscErrorCode ComputeRHS(KSP, Vec, void *);

typedef struct {
  PetscScalar uu, tt;
} UserContext;

int main(int argc, char **argv) {
  KSP         ksp;
  DM          da;
  UserContext user;

  PetscFunctionBeginUser;
  PetscCall(PetscInitialize(&argc, &argv, NULL, help));
  PetscCall(KSPCreate(PETSC_COMM_WORLD, &ksp));
  PetscCall(DMDACreate2d(PETSC_COMM_WORLD, DM_BOUNDARY_NONE, DM_BOUNDARY_NONE, DMDA_STENCIL_STAR, 11, 11,
         PETSC_DECIDE, PETSC_DECIDE, 1, 1, NULL, NULL, &da));
  PetscCall(DMSetFromOptions(da));
  PetscCall(DMSetUp(da));
  PetscCall(KSPSetDM(ksp, (DM)da));
  PetscCall(DMSetApplicationContext(da, &user));

  user.uu = 1.0;
  user.tt = 1.0;

  PetscCall(KSPSetComputeRHS(ksp, ComputeRHS, &user));
  PetscCall(KSPSetComputeOperators(ksp, ComputeJacobian, &user));
  PetscCall(KSPSetFromOptions(ksp));
  PetscCall(KSPSolve(ksp, NULL, NULL));

  PetscCall(DMDestroy(&da));
  PetscCall(KSPDestroy(&ksp));
  PetscCall(PetscFinalize());
  return 0;
}

PetscErrorCode ComputeRHS(KSP ksp, Vec b, void *ctx) {
  UserContext  *user = (UserContext *)ctx;
```

```
  PetscInt      i, j, M, N, xm, ym, xs, ys;
  PetscScalar   Hx, Hy, pi, uu, tt;
  PetscScalar **array;
  DM            da;
  MatNullSpace  nullspace;

  PetscFunctionBeginUser;
  PetscCall(KSPGetDM(ksp, &da));
  PetscCall(DMDAGetInfo(da, 0, &M, &N, 0, 0, 0, 0, 0, 0, 0, 0, 0, 0));
  uu = user->uu;
  tt = user->tt;
  pi = 4 * atan(1.0);
  Hx = 1.0 / (PetscReal)M;
  Hy = 1.0 / (PetscReal)N;

  PetscCall(DMDAGetCorners(da, &xs, &ys, 0, &xm, &ym, 0)); /* Fine grid */
  PetscCall(DMDAVecGetArray(da, b, &array));
  for (j = ys; j < ys + ym; j++) {
    for (i = xs; i < xs + xm; i++) array[j][i] = -PetscCosScalar(uu * pi * ((PetscReal)i + 0.5) * Hx) *
           PetscCosScalar(tt * pi * ((PetscReal)j + 0.5) * Hy) * Hx * Hy;
  }
  PetscCall(DMDAVecRestoreArray(da, b, &array));
  PetscCall(VecAssemblyBegin(b));
  PetscCall(VecAssemblyEnd(b));

  /* force right-hand side to be consistent for singular matrix */
  /* note this is really a hack, normally the model would provide you with a consistent right handside */
  PetscCall(MatNullSpaceCreate(PETSC_COMM_WORLD, PETSC_TRUE, 0, 0, &nullspace));
  PetscCall(MatNullSpaceRemove(nullspace, b));
  PetscCall(MatNullSpaceDestroy(&nullspace));
  PetscFunctionReturn(PETSC_SUCCESS);
}

PetscErrorCode ComputeJacobian(KSP ksp, Mat J, Mat jac, void *ctx) {
  PetscInt     i, j, M, N, xm, ym, xs, ys, num, numi, numj;
  PetscScalar  v[5], Hx, Hy, HydHx, HxdHy;
  MatStencil   row, col[5];
  DM           da;
  MatNullSpace nullspace;

  PetscFunctionBeginUser;
  PetscCall(KSPGetDM(ksp, &da));
  PetscCall(DMDAGetInfo(da, 0, &M, &N, 0, 0, 0, 0, 0, 0, 0, 0, 0, 0));
  Hx    = 1.0 / (PetscReal)M;
  Hy    = 1.0 / (PetscReal)N;
  HxdHy = Hx / Hy;
  HydHx = Hy / Hx;
  PetscCall(DMDAGetCorners(da, &xs, &ys, 0, &xm, &ym, 0));
  for (j = ys; j < ys + ym; j++) {
    for (i = xs; i < xs + xm; i++) {
      row.i = i;
      row.j = j;

      if (i == 0 || j == 0 || i == M - 1 || j == N - 1) {
        num  = 0;
        numi = 0;
        numj = 0;
        if (j != 0) {
          v[num]      = -HxdHy;
          col[num].i = i;
          col[num].j = j - 1;
          num++;
          numj++;
        }
        if (i != 0) {
          v[num]      = -HydHx;
          col[num].i = i - 1;
          col[num].j = j;
          num++;
          numi++;
        }
        if (i != M - 1) {
          v[num]      = -HydHx;
          col[num].i = i + 1;
          col[num].j = j;
          num++;
          numi++;
        }
        if (j != N - 1) {
          v[num]      = -HxdHy;
          col[num].i = i;
          col[num].j = j + 1;
          num++;
          numj++;
        }
        v[num]      = (PetscReal)numj * HxdHy + (PetscReal)numi * HydHx;
        col[num].i = i;
        col[num].j = j;
        num++;
```

```
        PetscCall(MatSetValuesStencil(jac, 1, &row, num, col, v, INSERT_VALUES));
      } else {
        v[0]     = -HxdHy;
        col[0].i = i;
        col[0].j = j - 1;
        v[1]     = -HydHx;
        col[1].i = i - 1;
        col[1].j = j;
        v[2]     = 2.0 * (HxdHy + HydHx);
        col[2].i = i;
        col[2].j = j;
        v[3]     = -HydHx;
        col[3].i = i + 1;
        col[3].j = j;
        v[4]     = -HxdHy;
        col[4].i = i;
        col[4].j = j + 1;
        PetscCall(MatSetValuesStencil(jac, 1, &row, 5, col, v, INSERT_VALUES));
      }
    }
  }
  PetscCall(MatAssemblyBegin(jac, MAT_FINAL_ASSEMBLY));
  PetscCall(MatAssemblyEnd(jac, MAT_FINAL_ASSEMBLY));

  PetscCall(MatNullSpaceCreate(PETSC_COMM_WORLD, PETSC_TRUE, 0, 0, &nullspace));
  PetscCall(MatSetNullSpace(J, nullspace));
  PetscCall(MatNullSpaceDestroy(&nullspace));
  PetscFunctionReturn(PETSC_SUCCESS);
}
```

We begin by initalizing PETSc and MPI. Then, we create PETSc-specific objects, such as a KSP object `ksp` for the linear solver and a DM object da for managing the MPI computation on a 2D grid. Next, we allow `da` to be set from the command line with `DMSetFromOptions` followed by a call to `DMSetUp` to create the underlying data structures. We then set `ksp` to use the `da` object and associate a user-defined context with it to help facilitate any user-defined data with the da object.

In the next step, we set some of the user-defined data for the problem in `user.uu` and `user.tt`. Then, we set the function for the RHS of the linear system and set the function for the Jacobian of the linear system. We also use `KSPSetFromOptions` to allow the user to configure the solver settings from the command line.

Finally, we solve the linear system with `KSPSolve` and clean up the objects with `KSPDestroy` and `DMDestroy`. We can run the example with

```
mpiexec -n 1 ./petsc_pde -da_grid_x 256 -da_grid_y 256.
```

We can change the size of the grid by changing the values of `da_grid_x` and `da_grid_y`.

34.5.1 Direct vs. Iterative Linear System Solvers

The solution to a Poisson equation boils down to the solution to a linear system of equations, where each variable in the linear system represents the solution to the Poisson equation at a point in the domain.

There are two flavors of solvers for such systems: *direct* and *iterative*. In principle, direct solvers apply a finite number of deterministic steps to produce a solution to a linear system that has a residual on the order of roundoff error. This description suggests the advantages of direct methods. Unfortunately, there is no mention of how large this "finite number of deterministic steps" is (or how it scales) or how much memory is required. Poor scalability and memory considerations are in fact the factors that often limit the use of direct methods in ESC.

Iterative methods, on the other hand, are generally more efficient in time and memory than direct methods for large problems. In general, however, there is no guarantee that a given iterative method will even converge (for given settings) for a given problem. Furthermore, it is common in practice for iterative methods to require a *preconditioner*, i.e., a transformation to an equivalent system with more favorable numerical properties, in order to realize convergence within a reasonable number of iterations. This practical requirement can add to the uncertainty and patience involved in achieving satisfactory performance from iterative solvers.

The PETSc code Listing 34.5 by default uses the Generalized Minimal Residual (GMRES) method, which is an iterative method for solving linear systems of equations. The default preconditioner applied in this case is the *incomplete LU factorization with zero fill-in* (ILU(0)) preconditioner, which approximates the LU factorization of the original matrix but enforces the same sparsity pattern (i.e., allows no fill-in).

By making some changes to the driver code, we can use a direct solver like MUMPS instead. We only need to modify the main function. Here is what the modified main function looks like:

Listing 34.6 Example of solving a PDE with KSP and DMDA.

```
int main(int argc, char **argv) {
  KSP         ksp;
  DM          da;
  UserContext user;
  PC          pc;

  PetscFunctionBeginUser;
  PetscCall(PetscInitialize(&argc, &argv, NULL, help));
  PetscCall(KSPCreate(PETSC_COMM_WORLD, &ksp));
  PetscCall(DMDACreate2d(PETSC_COMM_WORLD, DM_BOUNDARY_NONE, DM_BOUNDARY_NONE, DMDA_STENCIL_STAR, 11, 11,
         PETSC_DECIDE, PETSC_DECIDE, 1, 1, NULL, NULL, &da));
  PetscCall(DMSetFromOptions(da));
  PetscCall(DMSetUp(da));
  PetscCall(KSPSetDM(ksp, (DM)da));
  PetscCall(DMSetApplicationContext(da, &user));

  user.uu = 1.0;
  user.tt = 1.0;

  PetscCall(KSPSetComputeRHS(ksp, ComputeRHS, &user));
  PetscCall(KSPSetComputeOperators(ksp, ComputeJacobian, &user));

  // Get the preconditioner context
  PetscCall(KSPGetPC(ksp, &pc));

  // Set the preconditioner type to LU
  PetscCall(PCSetType(pc, PCLU));

  // Set the factorization solver type to MUMPS
  PetscCall(PCFactorSetMatSolverType(pc, MATSOLVERMUMPS));

  PetscCall(KSPSetFromOptions(ksp));
  PetscCall(KSPSolve(ksp, NULL, NULL));

  PetscCall(DMDestroy(&da));
  PetscCall(KSPDestroy(&ksp));
  PetscCall(PetscFinalize());
  return 0;
}
```

In Listing 34.6, we modified the main function to use MUMPS instead of GMRES. To achieve this, we use a preconditioner object, pc, to configure the direct sovler. Although this approach may seem counterintuitive, it demonstrates PETSc's unified interface, which allows a direct solver to be implemented as a "preconditioner" within an iterative solver. Alternatively, a direct solver can also be used on its own. We get the pc object with KSPGetPC, and then set the type of the preconditioner to use LU factorization. Lastly, we set the matrix solver type to MATSOLVERMUMPS use MUMPS.

Now running the two codes in Listings 34.5 and 34.6, we can see some differences in the time it takes each solver to solve the same linear system.

```
$ time mpiexec -n 1 ./petsc_pde -da_grid_x 512 -da_grid_y 512
real    0m25.677s
user    0m25.607s
sys     0m0.070s
```

```
$ time mpiexec -n 1 ./petsc_pde_mumps -da_grid_x 512 -da_grid_y 512

real    0m6.632s
user    0m6.088s
sys     0m0.544s
```

From the comparison, we can see that the direct solver MUMPS is about four times faster than the iterative solver GMRES for this problem. On the other hand, we observed that the MUMPS solver uses approximately 298 MB of memory, which is almost double the approximately 160 MB of memory used by the GMRES solver.

In light of the problem size, neither of these observations is surprising. The problem being solved is small, and the direct solver ends up requiring fewer floating-point operations because it does not need to iterate. On the other hand, the direct solver stores many more intermediate quantities—and hence requires more memory—than its iterative counterpart.

34.6 Additional Resources

- https://petsc.org/release/docs/manual/
- https://petsc.org/release/overview/nutshell/

34.7 Addendum

For completeness, we also mention *randomized* numerical linear algebra (RNLA), where techniques such as random projections, sampling, and sketching, are used to accelerate or make tractable linear algebra problems in ESC (even beyond what is considered to be ESC at the time of writing). A comprehensive treatment of RNLA is beyond the scope of this work. However, RNLA is expected to gain increasing attention and prominence in ESC over time.

Chapter 35
The Trilinos ecosystem

Trilinos, which is originally a Greek term that translates to "a string of pearls", is a collection of libraries for ESC. The translation is meant to signify that each package is a pearl in its own right (in the sense of a beautiful building block) but that the packages are also connected by a common infrastructure (the string). The Trilinos project contains over 50 independent packages, each aimed at solving a primary class of scientific computing problem. At the core of Trilinos packages are parallel solver algorithms designed to solve large-scale, complex multi-physics applications from science and engineering.

Trilinos provides a collection of mathematical software libraries into a unified object-oriented framework. Each package is integrated into the Trilinos framework using a common set of interfaces, allowing each package to be used consistently. A package within the Trilinos framework is defined as a numerical library that uses state-of-the-art algorithms within its problem domain, is developed by domain experts, and is self-contained, configurable, and individually well documented. Trilinos ties together multiple core packages, of which we focus on three: Teuchos, Tpetra, and Kokkos. Teuchos provides memory management and MPI communicators. Tpetra provides distributed data structures for linear algebra. Finally, Kokkos provides performance portability on CPU and GPU nodes. Because Kokkos is covered in Chapter 24, the focus of this chapter is on the Teuchos and Tpetra packages.

35.1 Obtaining Trilinos

The Trilinos source code can be obtained from the Trilinos GitHub repository at `https://github.com/trilinos/trilinos`. Within the repository, there is a file called `INSTALL` that contains instructions for how to build Trilinos and enable many of its packages. The basic instructions for building Trilinos with MPI enabled are

```
$ cmake \
  -DTPL_ENABLE_MPI=ON \
  -DMPI_BASE_DIR=<path to mpi install> \
```

© The Author(s), under exclusive license to Springer Nature Switzerland AG 2025
R. J. Spiteri and K. Klenk, *Extreme-Scale Computing*, CMS/CAIMS Books in Mathematics 15, https://doi.org/10.1007/978-3-031-89033-8_35

```
    -DTrilinos_ENABLE_ALL_PACKAGES=ON \
    -DCMAKE_INSTALL_PREFIX=<path to install Trilinos into> \
    <path to Trilinos source>
$ make -j<n_proc> install
```

35.2 Teuchos

Teuchos (which is a Greek word meaning "tool" or "instrument" and pronounced `te-fos`) is a package that provides many useful tools within various sub-packages. These sub-packages are provided through a collection of classes and service software that are useful to almost all Trilinos packages. Some of the included sub-packages are the Core, ParameterList, and Communication[Numerics, Parser] sub-packages. In addition to these sub-packages, the Teuchos package also provides wrappers for the BLAS/LAPACK libraries and includes a parser for parsing various text-based file formats.

35.2.1 Teuchos Core

The Teuchos Core sub-package is an important sub-package because it provides classes that help with memory management. These classes include `Teuchos::Prt`, `Teuchos::RCP`, `Teuchos::Array`, and `Teuchos::ArrayRPC`. `Teuchos::Prt` is a simple smart pointer that is similar to `std::unique_ptr`. `Teuchos::RCP` is a reference-counted smart pointer like `std::shared_ptr`. `Teuchos::Array` is a replacement for `std::vector`, and `Teuchos::ArrayRPC` is the same as `Teuchos::Array`, except that it is wrapped within `Teuchos::RCP` to provide automatic deallocation of `Teuchos::ArrayRPC` objects when they go out of scope. To enable thread safety for any of these objects, the `-DTrilinos_ENABLE_THREAD_SAFE=ON` option must be enabled when configuring and building Trilinos. Listing 35.1 is an example of how to use the Teuchos Core sub-package to create the objects introduced above.

Listing 35.1 Teuchos Core classes

```cpp
#include "Teuchos_RCP.hpp"
#include "Teuchos_Version.hpp"
#include "Teuchos_CommandLineProcessor.hpp"
#include <Teuchos_ArrayRCPDecl.hpp>
#include <string>

using namespace Teuchos;

int main(int argc, char *argv[]) {
  // Examples of how to use Teuchos::RCP
  RCP<int> rpc_int = rcp(new int(42));
  RCP<std::string> rpc_str = rcp(new std::string("Hello, World!"));

  // Examples of how to use Teuchos::ArrayRCP with 10 doubles
  ArrayRPC<double> new_array = arcp<double>(10);

  return 0;
}
```

In Listing 35.1, the Teuchos::RCP class is used to wrap an integer pointer around the value 42 into rpc_int, along with a pointer to a string "Hello, World!" into rpc_str. The Teuchos::ArrayRPC class is used to create an array object, new_array, and the arcp constructor is used to specify that it should hold 10 double values. The classes within the Teuchos Core sub-package are usually used to create objects that wrap around other objects that are created with other Trilinos packages.

35.2.2 Teuchos ParameterList class

The ParameterList class creates objects that store a database set up within a hierarchy to enable serialization. A ParameterList is a collection of key-value pairs, where the key is a string that is mapped to a value. The value can be of any type, and the ParameterList itself is then passed to an object that can use the key-value pairs to configure itself. The hierarchy is set up by nesting ParameterList objects within each other. Each ParameterList object can be queried for a key-value pair including other nested ParameterList objects. The ParameterList sub-package also includes the Teuchos::CommandLineProcessor, which helps obtain command-line arguments from the argc and argv variables. Using the Teuchos::CommandLineProcessor class is recommended because it adds the ability to use the --help command-line option to print out helpful information about the options that can be set.

Listing 35.2 Teuchos ParameterList class

```
#include "Teuchos_ParameterList.hpp"
using namespace Teuchos;

int main(int argc, char *argv[]) {
  // example of how to use Teuchos::CommandLineProcessor
  CommandLineProcessor clp;
  clp.setDocString("This is an example of how to use Teuchos::CommandLineProcessor");

  int some_option = 0;
  clp.setOption("some_option", &some_option, "An option");

  std::string some_other_option = "default";
  clp.setOption("some_other_option", &some_other_option, "Another option");

  // Examples of how to use Teuchos::ParameterList
  ParameterList parm_list;
  // Set some options
  parm_list.set("some_option", 42);
  parm_list.set("some_other_option", "Hello World!");

  // Set some nested options
  ParameterList &nested_pl = parm_list.sublist("nested");
  nested_pl.set("nested_option", 3);

  // Query some options
  bool has_some_option = parm_list.isParameter("some_option");

  // Query some nested options
  bool has_nested_option = parm_list.isSublist("nested");

  // Get some options
  int some_option = parm_list.get("some_option");

  // Get some nested options (with default value if not set)
  double nested_option = nested_pl.get("nested", 3.14);

  return 0;
}
```

In Listing 35.2, a Teuchos::CommandLineProcessor object is used to first set a DocString that is printed out when the --help command-line option is used. Then the setOption method is called to obtain the values to set an int variable and a string variable. After the command-line arguments are parsed, a Teuchos::ParameterList object, param_list, is created. The set method is used to set an integer option named some_option and a string option named some_other_option. Following the setting of these first two options, a nested Teuchos::ParameterList object, nested_pl, is created. The nested_pl variable is used to set a parameter, nested_option. All parameters and nested ParameterLists within a ParameterList object can be queried with methods that begin with is. Such methods are used to check whether parameters or nested ParameterLists exist within the ParameterList object. Finally, the get method can be used to retrieve the values in the ParameterList. When called with a single argument—the name of a parameter—the function returns the value of that parameter, provided it exists. A second argument can also be given, which sets the value of the returned parameter to the second argument if the parameter does not exist within the ParameterList.

35.2.3 Teuchos Communication

Within the Teuchos communication sub-package are two utilities that each contain multiple classes. The first utility is centered around creating parallel programs that support single program multiple data (SPMD) operations. These classes are designed as wrappers to handle MPI communication. The wrappers within the Teuchos::Comm class are designed to be used by other Trilinos packages instead of the MPI communicator objects themselves. The second utility is centered around performance monitoring and includes the classes Teuchos::Time, Teuchos::TimeMonitor, and Teuchos::Flops. The Teuchos::Time class creates objects for measuring the wall-clock time of a process. The Teuchos::TimeMonitor class creates objects of increased functionality to allow the user to measure specific regions of code. Lastly, the Teuchos::Flops class creates objects that count the number of floating-point operations performed within objects based on the Teuchos::CompObject class. An example of how to use MPI within Teuchos is shown in Listing 35.3. The Teuchos documentation provides a more detailed example of how to use the time-performance monitoring classes at https://docs.trilinos.org/dev/packages/teuchos/doc/html/TimeMonitor_2cxx_main_8cpp-example.html#_a3

Listing 35.3 Teuchos Communication classes

```
#include "Teuchos_GlobalMPISession.hpp"
#include "Teuchos_RCP.hpp"

int main (int argc, char* argv[]) {
  using Teuchos::Comm;
  using Teuchos::DefaultComm;
  using Teuchos::RCP;

  // This replaces the call to MPI_Init.  If you didn't
```

```
  // build with MPI, this doesn't call MPI functions.
  Teuchos::GlobalMPISesssion session(&argc, &argv, NULL);
  // comm is the equivalent of MPI_COMM_WORLD.
  RCP<const Comm<int> > comm = DefaultComm<int>::getComm ();

  // Get the rank of this process in the default communicator.
  const int procRank = Teuchos::GlobalMPISession::getRank();

  const int count = 10; // Send 10 doubles

  if (procRank == destinationRank) {
    double values[10];
    const int sourceRank = 0; // Receive from Process 0

    Teuchos::receive<int, double> (*comm, sourceRank, 10, values);
  } else {
    double values[10] = ...;

    const int destinationRank = 1; // Send to Process 1
    Teuchos::send<int, double> (*comm, 10, values, destinationRank);
  }

  // No need to call MPI_Finalize, this is called in
  // the destructor of GlobalMPISesssion.
  return 0;
}
```

Instead of calling `MPI_Init` to initialize MPI, a `Teuchos::GlobalMPISession` object is created to handle the initialization and finalization of MPI. If MPI is not enabled, then the object can still be created; it will simply not call any MPI functions and act as if the program was running with one MPI process. After initializing MPI, the `comm` object is obtained by calling `Teuchos::DefaultComm<int>::getComm`. Next, obtaining the rank and sending and receiving an array are demonstrated. Finally, the call to `MPI_Finalize` is not needed because the `Teuchos::GlobalMPISession` object calls `MPI_Finalize` when it goes out of scope as part of its destructor.

35.2.4 Teuchos Numerics

Within the Teuchos Numerics sub-package are wrappers for BLAS and LAPACK routines. There is also a `Teuchos::SerialDenseMatrix` class that provides basic operations for dense matrices. The `Teuchos::SerialDenseMatrix` class is not a full-fledged matrix class and is commonly used as an interface for BLAS and LAPACK routines. However, it is does provide enough functionality to be used on its own to create and manipulate small dense matrices in serial. `Teuchos::BLAS` wraps the BLAS Fortran library to provide a C++ interface. The same is true for the `Teuchos::LAPACK` class that wraps the LAPACK Fortran library. Both of these classes are not guaranteed to behave properly because the interoperability between C++/Fortran is not standard across all platforms. To use these classes, the `-DTrilinos_ENABLE_TeuchosNumerics=ON` may need to be set during configuring and building of Trilinos to ensure their availability.

35.3 Tpetra

Tpetra is the replacement for the legacy Epetra core package aimed to provide a more flexible linear algebra package that can be used with more than just the double-precision floating-point type. Tpetra offers classes that create linear algebra objects that can be used in serial or in a distributed environment. The capabilities of Tpetra are similar to those offered in the Epetra package but with more flexibility from the use of templates. Tpetra's distributed data structures are implemented as MPI+X, meaning that MPI handles the distributed programming (memory and messaging) side, and X (which in this case refers to Kokkos) handles the shared programming (memory and threading) side. Tpetra is designed to be used for problems that can have billions of unknowns. It also attempts to exploit hybrid parallelism with the use of both the GPU and CPU.

35.3.1 Using Tpetra

Tpetra provides templated classes that can be used to create linear algebra objects. These classes are centered around creating container objects in the form of matrices and vectors that can be used to solve linear algebra problems. The classes are generally templated around three types: `Scalar`, `LocalOrdinal`, and `GlobalOrdinal`, which are used to define the type of data stored in the Tpetra objects. The `Scalar` type is used to define the type of the values stored in the matrix and vector objects. The `LocalOrdinal` type is used to define the integer type used for storing local IDs. The `GlobalOrdinal` type is used to define the integer type used for storing global IDs. The `GlobalOrdinal` and `LocalOrdinal` types do not need their underlying base types to be identical. In fact, it occasionally makes more sense to have them be different types. For example, it may be more efficient to use a smaller integer size for the `LocalOrdinal` type because the number of local IDs is usually much smaller than the number of global IDs. The `Scalar` type can be set to virtually any type on which basic arithmetic operations can be performed.

The most commonly used classes to create linear algebra objects are `Tpetra::MultiVector`, `Tpetra::Vector`, and `Tpetra::CrsMatrix`. The `Tpetra::MultiVector` class creates a collection of distributed dense vectors. The `Tpetra::Vector` class is a special case of the `Tpetra::MultiVector` class that creates a single distributed dense vector. The `Tpetra::CrsMatrix` class creates a distributed sparse matrix. Unlike PETSc, the local rows of distributed matrices in Tpetra do not have to have a continuous set of indices. Furthermore, the user is not limited to these classes, but they are the most commonly used because they work in most situations.

The most important class for constructing the linear algebra objects is `Tpetra::Map`. The `Tpetra::Map` class is used to manage the relationship between local and global IDs. This functionality is useful for determining communication patterns in distributed objects. The distribution of the data across MPI processes is crucial for performance. `Tpetra::Map` objects are passed to the constructors of vectors and matrices to define their distribution. These objects help hide certain details about the distribution of vectors and matrices and allow the user to only worry about the global indices of the data if they so choose. The map handles how global indices correspond to local indices. The user can choose to use global or local indices. It is more common to use global indices, but local indices are sometimes more efficient at reducing communication overhead.

When creating a vector, the length of the vector must be specified. The length of the vector is the number of global indices that the vector has. In most cases, the vector is distributed as evenly as possible across all available MPI processes. Matrices generally distribute their rows over the available MPI processes. However, both columns and rows can be specified. This is again to help with performance and reduce communication overhead. When creating a matrix with both rows and columns specified, the user must provide two `Tpetra::Map` objects, one for the rows and one for the columns, to the `Tpetra::CrsMatrix` constructor. If the user only specifies the rows, then only the row map needs to be passed to the constructor. Listing 35.4 shows how to create a map and use it to create a vector and a matrix.

Listing 35.4 Tpetra classes

```
#include <Tpetra_Core.hpp>
#include <Tpetra_Vector.hpp>
#include <Tpetra_Version.hpp>
#include <Teuchos_Comm.hpp>
#include <Teuchos_OrdinalTraits.hpp>

using std::endl;
using Teuchos::Array;
using Teuchos::ArrayView;
using Teuchos::RCP;
using Teuchos::rcp;

int main(int argc, char *argv[]) {
  // Initialize MPI
  Tpetra::ScopeGuard tpetraScope(&argc, &argv);
  {
    // Get the default communicator
    auto comm = Tpetra::getDefaultComm ();

    // Select the number of global entries
    const Tpetra::global_size_t numGlobalEntries = 50;

    // Select the starting index for the map
    // 0 for C style, 1 for Fortran style
    const Tpetra::Vector<>::global_ordinal_type indexBase = 0;

    // Create the map
    RCP<const typedef Tpetra::Map<>> first_map = rcp(new Tpetra::Map<>(numGlobalEntries, indexBase, comm));

    // Create a vector with the contigious map
    RCP<Tpetra::Vector<>> first_vector (new first_map);

    // Create a sparse matrix with at most 3 non-zeros per row
    RCP<Tpetra::CrsMatrix<>> first_matrix (new crs_matrix_type (map, 3));

    // Mpi Finalize is called in the destructor of tpetraScope
  }

  return 0;
}
```

Listing 35.4 shows how to initialize a `Tpetra::Vector`, `Tpetra::CrsMatrix`, and its corresponding `Tpetra::Map` object. The first step is to wrap the code inside a `Tpetra::ScopeGuard` object. This handles the initialization and finalization of the MPI or Kokkos environment, if applicable. Even if MPI or Kokkos are not enabled, `Tpetra::ScopeGuard` can still be used because it is aware to not set up these environments in this situation. Instead, `Tpetra::ScopeGuard` uses a serial communicator with one process. The destructor of the `Tpetra::ScopeGuard` object finalizes the MPI or Kokkos environment, so there is no need to do this explicitly. `Tpetra::ScopeGuard` is not required, and `MPI_Init` and `MPI_Finalize` can be called explicitly in its place. The MPI communicator is passed to the `Tpetra::Map` constructor via `Tpetra::getDefaultComm()`. The `Tpetra::getDefaultComm()` is aware if MPI is enabled or not and returns a communicator with rank 0 and size 1 if MPI is not enabled. The next step is to specify how many elements are going to be in the vector and the matrix. The number of elements in the vector is set to 50, and the number of rows and columns in the matrix is also set to 50. A variable, `indexBase`, is then set to specify the starting index of the global indices. The starting index can be set to either 0, for C/C++-style indexing, or 1, for Fortran-style indexing. The last step before creating the vector and matrix is to create the map. When no type is specified within the angle brackets <>, the default type is used. Finally, the vector and matrix are both initialized within a `Tuechos::RCP` object. For the vector, all we need to do is pass the map, and for the matrix, we pass the map and the number of non-zeros per row.

35.4 Additional Resources

- Getting started: `https://trilinos.github.io/getting_started.html`
- Packages: `https://trilinos.github.io/packages.html`
- Teuchos: `https://docs.trilinos.org/dev/packages/teuchos/doc/html/index.html`
- Tpetra: `https://docs.trilinos.org/dev/packages/tpetra/doc/html/index.html`

Chapter 36
The actor model of concurrent computing

We now discuss the actor model of concurrent computing. The actor model is used extensively in many real-world applications such as Facebook or X (Twitter), but it is not widely used in scientific computing. We start this chapter by providing the motivation for the use of the actor model in scientific computing. Then we introduce the basic concepts surrounding the actor model with examples using the C++ Actor Framework (CAF). We build upon the basics to demonstrate how the actor model is fault-tolerant and introduce how it can be leveraged in distributed systems. Finally, we finish this chapter by exposing some of the "hidden details" and other advanced features common within actor model implementations that are taken care of automatically.

36.1 Motivation

The expectations and demands of computational scientists and their cutting-edge models require that high-performance software run in distributed environments, scale well, use resources efficiently, and be fault-tolerant. Previous chapters introduce "classical" tools and techniques to meet these expectations. We now establish an alternative approach centered around leveraging concurrency to exploit parallelism. Concurrent software is software that is broken down into multiple tasks or executable units that can be *perceived* to be running at the same time, even though in reality they are not. In contrast, parallelism is when multiple tasks *are* running at the same time. By developing concurrent software, we attempt to compose a program such that as much of it as possible can be running in parallel. The actor model makes this process relatively painless compared to classical tools such as multithreading or MPI. More importantly, however, actors facilitate a task-based approach to programming, which is perhaps a more natural and potentially more flexible and powerful programming model for extreme-scale software than classical programming models. This is the main motivation for using the actor model and for its discussion in this book.

© The Author(s), under exclusive license to Springer Nature Switzerland AG 2025
R. J. Spiteri and K. Klenk, *Extreme-Scale Computing*, CMS/CAIMS Books in Mathematics 15, https://doi.org/10.1007/978-3-031-89033-8_36

36.2 The Actor Model

The actor model is a mathematical model of concurrency in which the *actor* is the universal primitive for concurrent computation. Carl Hewitt introduced the actor model in 1973 as a theoretical foundation for how concurrent systems behave. This theoretical foundation now serves as the backbone for modern implementations of the actor model. The actor model is similar to task-based programming; however, actors are built upon three fundamental characteristics that differentiate them from tasks. These three characteristics are:

1. *Send*: Actors communicate asynchronously by sending each other messages.
2. *Create*: Actors can create other actors.
3. *Become*: Actors can change their behavior to alter how they respond to messages.

Actors execute independently of each other and carry out any combination of the three characteristics listed above. Generally, actors are associated with additional characteristics such as schedulers, mailboxes, streams/flows, and so on. However, some of these additional characteristics, in particular schedulers and mailboxes, are intentionally hidden and handled by the *actor system* for the convenience of the user. Other characteristics such as streams/flows are considered to be advanced features of specific actor model frameworks. We touch on some of these other characteristics in Section 36.6.

The *actor system* is the context that manages the operations of the actors behind the user-exposed API. All behind-the-scenes items are handled by the actor system (sometimes called the *runtime* system), and actors cannot be used until this system is established. Hiding details apart from *Create*, *Send*, and *Become* from the user is a key feature of the actor model. When using the actor model, the software development process is simplified because users can focus on creating actors and defining the interactions between them. Thus, it is easy to create extremely scalable programs because we can easily create hundreds of thousands of actors that interact with each other. The actor system then handles the details of mapping and executing these actors on the underlying hardware. Hidden are the difficulties of manually managing CPU cores, threads, and other low-level primitives. To help illustrate these concepts further, see Figure 36.1 for a visual representation of the actor model. To begin using the actor model with CAF, we begin with an introductory example of how to create an actor system in Listing 36.1.

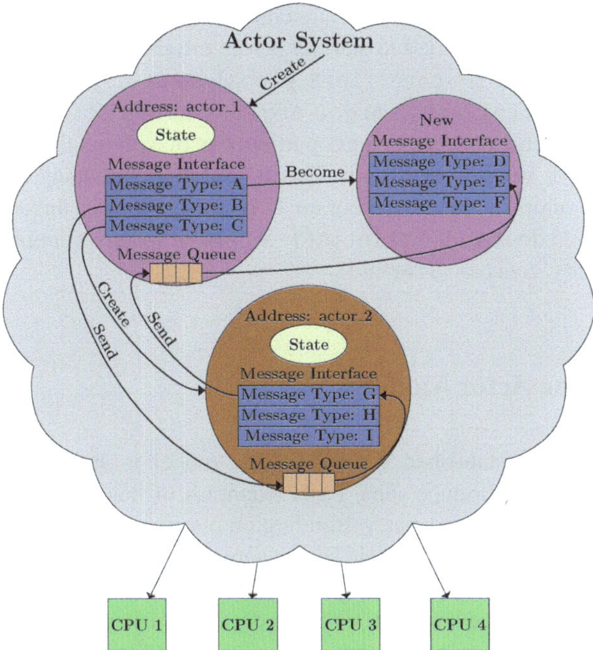

Fig. 36.1 A visual representation of the actor model. Actors exist within an actor system and are identifiable by an address. Actors contain a state, a message interface, and a message queue (mailbox). The actor system decides how to map actors to CPUs. Shown in the figure is the flow of the *Create*, *Send*, and *Become* operations, where actors can create other actors, send messages to other actors, and invoke functionality (change their behavior) based on the receivers' message interface.

Listing 36.1 A simple example to demonstrate how to create an actor system using CAF. CAF provides convenient features such as CAF_MAIN() that hides the boilerplate code required to start an actors program. In addition, CAF provides a `scoped_actor` that allows us to use the `actor_system` as if it were its own actor. This is useful for spawning and interacting with the initial actors to begin the program. CAF also provides their own method for printing to `stdout` called `println()`.

```
#include "caf/all.hpp"

using namespace caf;

void caf_main(actor_system& sys) {
  scoped_actor self{sys};
  self->println("Actor System Started");
}

CAF_MAIN()
```

In Listing 36.1, we create an `actor_system` within `caf_main` and create a `scoped_actor` to interact with the `actor_system`. The `scoped_actor` allows us to use the `actor_system` as if it were its own actor with the same `send` and `spawn` syntax. CAF provides a convenient macro, `CAF_MAIN`, that hides the boilerplate code required to start an actors program. This macro hides the calls to initialize the actor_system as well as parses any command line arguments that may have been provided. It is not needed to start an actors program. Instead, the user could explicitly call the function core::init_global_meta_objects() and then return exec_main<>(caf_main, argc, argv) to start the actor_system. Inside the caf_main function, we create a scoped_actor that allows the actor_system to be used as if it were its own actor. This is useful for spawning and interacting with the initial actors to begin the program. CAF also provides a thread-safe method for printing to stdout called println(). We use it in this example to print the message, "Actor System Started".

36.3 Using the Actor Model

Now that we have established what the actor model is and how to create an actor_system, we introduce some basic examples of how to use the three fundamental characteristics of an actor. Building on our previous example, we use our actor_system to spawn an actor and send it a message. Furthermore, we show how to change our new actor's response to a subsequent message.

A behavior is defined as what an actor does when it receives a message. In Listing 36.2, we define two behaviors, hello and goodbye. We spawn one actor from the actor system. When an actor is spawned, it returns a reference of type CAF::actor, which we store in the variable our_first_actor in order to interact with the newly spawned actor. This reference is similar to the actor's address, which is used to uniquely identify the actor. However, the reference is simply the variable that we use to interact with the actor. We use our_first_actor as an argument to the send method to specify which actor to communicate with. Before the send portion of the code, we need to define the payload or contents of the message by specifying arguments to the mail method. The arguments in the mail method must match the message handlers defined within the target actor's behavior to get a useful response. Sending a message that does not match any of the message handlers for an actor produces an *unexpected message* error by the receiving actor.

Listing 36.2 An example of an actors program with two behaviors. CAF provides two implementations of actors, *class-based actors* and *function-based actors*. In this example, we are using function-based actors. Our actors are defined as functions that return a behavior, i.e., a set of message handlers that define the messages an actor can receive and how it responds to them. Message handlers are defined as lambda functions and can only catch arguments by value.

```
#include "caf/all.hpp"
#include "caf/io/all.hpp"

using namespace caf;

behavior goodbye(event_based_actor *self) {
  // Here is where initialization code for the actor is run upon creation
  return {
    // We return lambda functions that define the messages and actor can receive
    [=](std::string name) {
      self->println("Goodbye, {}", name);
      self->quit(); // A graceful exit
    }
  };
}
```

```
behavior hello(event_based_actor *self){
  return {
    [=](std::string name){
      self->println("Hello, {}", name);
      self->become(goodbye(self)); // An actor can change its behavior
    }
  };
}

void caf_main(actor_system& sys){
  scoped_actor self{sys};
  self->println("Actor System Started");
  auto our_first_actor = self->spawn(hello);
  self->mail("Kyle").send(our_first_actor);
  self->mail("Kyle").send(our_first_actor);
}

CAF_MAIN()
```

In Listing 36.2, we send the same message to the same actor twice. However, we expect the actor to respond differently to the second message because we expect the actor to change its behavior after receiving the first message. The output from Listing 36.2 is

```
Actor System Started
Hello Kyle
Goodbye Kyle
```

In this example, we see that the actor did indeed change its behavior after receiving the first message. We also note the syntax for receiving messages in CAF. When using function-based actors, messages are received via lambda functions. We can define an arbitrary number of lambda functions within the return part of the actor's behavior to have the actors receive different kinds of messages. When sending a message to an actor, the message is pattern matched to the appropriate lambda function, or an *unexpected message* error is returned.

CAF introduces a convenient method for distinguishing between different messages that consist of arguments of the same type, called `atoms`. An `atom` is a non-numeric constant used as an argument to provide additional information about the contents of the message. `Atoms` are useful for situations like in the example shown in Listing 36.3. Here we want the actor to respond differently to two messages; however, both messages consist of a single string type argument. Using `atoms` allows us to avoid having to define two different behaviors for the same actor.

Listing 36.3 Example of how to use message atoms to distinguish between different messages in CAF.

```
#include "caf/all.hpp"
#include "caf/io/all.hpp"

using namespace caf;

CAF_BEGIN_TYPE_ID_BLOCK(example_project, caf::first_custom_type_id)
  CAF_ADD_ATOM(example_project, hello_atom)
  CAF_ADD_ATOM(example_project, goodbye_atom)
CAF_END_TYPE_ID_BLOCK(example_project)

behavior hello_goodbye(event_based_actor *self) {
  // Here is where the actor's inital behavior is defined
  return {
  // We return lambda functions that define the messages and actor can receive
    [=](hello_atom, std::string name) {
      self->println("Hello, {}", name);
    },
    [=](goodbye_atom, std::string name) {
      self->println("Goodbye, {}", name);
    }
```

```
  };
}

void caf_main(actor_system& sys) {
  scoped_actor self{sys};
  self->println("Actor System Started");
  auto our_first_actor = self->spawn(hello_goodbye);
  self->mail(hello_atom_v, "Kyle").send(our_first_actor);
  self->mail(goodbye_atom_v, "Kyle").send(our_first_actor);
}

CAF_MAIN(id_block::example_project)
```

The output from Listing 36.3 is:

```
Actor System Started
Hello Kyle
Goodbye Kyle
```

As we can see, the output produced by Listing 36.3 is identical to the output produced by Listing 36.2. Whether to change the behavior of an actor or use atoms is mostly a matter of preference and may depend on the application.

36.4 Fault Tolerance

With the fundamentals of the actor model introduced, we can now discuss how to use the actor model to create fault-tolerant programs. The actor model was designed to be highly fault-tolerant and is often described as having a *let-it-fail* mentality. Carl Hewitt defines the fault-tolerant behavior of the actor model more formally as a model that demonstrates *inconsistency robustness*. A system that demonstrates inconsistency robustness is one that continues to operate in the event that inconsistencies or errors occur during execution. The opposite of models that demonstrate inconsistency robustness are models that demonstrate *inconsistency denial*. A system that demonstrates inconsistency denial assumes that inconsistencies or errors never occur during execution. Typically, new programmers are taught to program defensively, which is to assume that errors do occur during execution and to design their programs around potential errors. Therefore, there are most likely no programs or models that demonstrate inconsistency denial in its purest form. Nonetheless, the actor model makes designing fault-tolerant programs relatively painless.

First and foremost, the execution of an actor is isolated from other actors. What this achieves for fault tolerance is that when actors fail or encounter errors, they have no *immediate* effect on the other actors within a program's execution. In spite of this, errors and failures do propagate through the program over time if other actors need to interact with the failed actor. Luckily, the actor model is equipped to handle these situations using *supervision*. When actors spawn other actors, they naturally create parent-child relationships between them. When we create many parent-child relationships, a *supervision hierarchy* starts to emerge. We can leverage these supervision hierarchies to create fault-tolerant programs by following a simple rule, which is that an error or failure in a child actor is to be handled by the child actor's immediate parent. This rule can be applied even to the situation where a parent cannot handle the error of the child actor. In this case, the parent's parent is responsible for the handling of the error. We can follow this rule all the way up our supervision hierarchy until we reach the root, which is typically the `actor_system`. It is not desirable, however, to have the `actor_system` be the last line of defense against a program crashing outright. Instead, we can use the *error kernel* pattern. A program's error kernel is responsible for ensuring the state of our program and handling its most important behaviors. Behaviors with a higher potential for failure are outsourced to child actors where errors can easily be handled, decreasing the likelihood of the program crashing. We provide an example of the error kernel pattern within a supervision hierarchy in Figure 36.2.

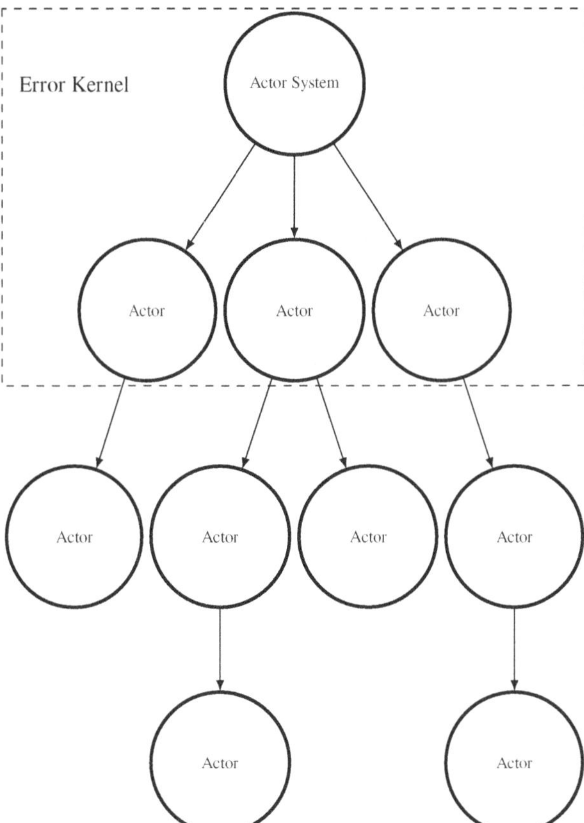

Fig. 36.2 A supervision hierarchy with an error kernel. Parent-child relationships are denoted by the arrows connecting the actors. An arrow that points to an actor signifies the parent spawning the child actor and creates the parent-child relationship. We also show an example of an error kernel. The error kernel is the set of actors that are responsible for ensuring consistency within the system. They handle the most important functions of the program as well as maintain the necessary state of the program for continued operation. More risky operations with a high potential for failure are outsourced to the child actors outside the error kernel.

The point of Figure 36.2 is to show that as we spawn actors, they naturally create a tree like structure of parent-child relationships. We can leverage this structure to create fault-tolerant programs by following the rule previously mentioned. In addition, we can designate an error kernel within our supervision hierarchy. The number of actors that make up an error kernel is dependent on the application. Our example depicts three actors plus the actor system as our error kernel. Other applications may have a larger or smaller error kernel. In any event, the error kernel should be made as robust as possible to prevent applications from crashing.

When handling errors, there are a few different options that a parent actor can take. We introduce these options along with an example of when we might want to use each option. The first option is to ignore the error. This option may be useful when multiple actors are exploring the same solution space to a problem. In this scenario, we may safely ignore these failures because other actors may still return a solution. A second option is to restart the child actor without changing its behavior or passing it any new parameters. This may be a useful option when using the actor model in distributed systems. For example, we may have an actor that is responsible for managing a node in a cluster. If the node fails, we may want to restart the actor as is on a different node. A third option is to change the actor's behavior or pass it new parameters. This option may be used when it is not known what parameters lead to success. In this situation, we can restart the actor with a different set of parameters.

36.5 Distributed Computing

Although formally a model of concurrent computation, the actor model is inherently a distributed model of computation. At its inception, the actor model's creation was motivated by the need to cluster machines together to solve large computational problems due CPU performance limitations at the time. We recall that every actor executes its actions in isolation and communicates with other actors through message passing. From this perspective, it is easy to conclude that actors on a single machine are in fact already behaving in a distributed manner. Therefore, it is natural to extend the actor model to distributed systems where we have actors interacting from different physical machines. To achieve this, an actor's reference is *network transparent* and is uniquely identifiable across machines connected in a network. This feature means an actor's reference is the only thing that is needed in order to send another actor a message, regardless of whether the actor is local or remote. The actor is a powerful abstraction for leveraging distributed systems because it again hides complexity from the user and lets them focus on implementing how actors interact regardless of their physical location. There is, however, some additional setup that is required to get actors to communicate across a network and some practical considerations that must be taken into account.

When using actors across distributed systems using CAF, we have to first connect the actor systems together. We cannot simply spawn an actor on a remote system without explicitly starting actor systems on the remote set of machines first. This limitation is because (as we have described) the actor system is responsible for hiding complexity from the programmer. When using actors across a network, this includes all the network I/O that is required for machines to communicate with each other. To set up the actor system, CAF provides a `middleman` component that is responsible for managing the network connections. To use the `middleman`, we publish it to a port on our local machine that then allows other actor systems to connect. In doing so, we can create a cluster of actor systems that behave as if they were just one system. This property allows us to get all the benefits of the actor model without having to worry about the network I/O. Specific examples of how to do this can be found in the CAF documentation at `https://actor-framework.readthedocs.io/en/stable/index.html`. The most pressing consideration when using actors across a network is network latency. Latency determines how much actors can and should communicate with actors on other machines. This characteristic is an implementation detail that depends on the application, the network on which it is running, and the physical distance between the machines.

36.6 Advanced Actor Model Features

As alluded to earlier, the actor model intentionally hides the details of how actors are implemented. However, it is still important to understand a bit about how actors work under the hood. Message passing is the central theme of the actor model. Messages are how actor programs are separated into logical units of execution. Sending a message to an actor is similar to calling a function. The main difference is that a message is sent asynchronously (in a non-blocking mode) to the receiving actor. In other words, the sender does not cede execution to the receiving actor when the message is sent. In most actor model implementations, the message that is sent is placed in the receiving actor's *mailbox*. Mailboxes and the details of how actors receive messages in general are implementation-specific, with some implementations allowing users to define their own mailbox data structures. Mailboxes in CAF are single-reader, multiple-writer queues. When an actor sends a message, the sender writes to the mailbox of the receiver. The receiver is then notified of this write and processes the message accordingly. The processing is generally done in FIFO (first-in-first-out) order; however, many frameworks allow developers to specify priority or special rules on when certain messages are processed.

In order for actors to respond to messages in their mailbox, a *scheduler* is used to coordinate the execution of actors to the system hardware. The job of the scheduler is to map actors to the available hardware threads or CPU cores on which the actor system is running. In CAF, worker threads are spawned at the start of a program that match the number of threads on the system. Each worker thread has a queue where work items are placed, such as spawning actors or executing messages. To balance the load on the system, the worker threads use a work-stealing algorithm. This means that once a worker threads queue is empty, it attempts to steal work items from another worker. This is a popular algorithm for load balancing in task-based programming models in general. For the most part, the implementation of *mailboxes* and *schedulers* are nothing to be concerned about as a user. However, it is important to understand a little about these elements because they can potentially affect the performance of an application depending on its needs.

In addition, modern implementations of the actor model have added new features, which although beyond the scope of the original actor model, are still useful to mention to provide awareness. One such feature is that actor implementations do not explicitly limit actors to only sending messages asynchronously. In fact, many implementations allow explicit receive statements or even actors that block execution. The use of blocking actors is generally discouraged because this practice re-introduces the problem of deadlocks, but it can be useful in some situations. Another feature that has gained popularity recently is actor streams or what is sometimes referred to as *flows*. Actor streams allow users to coordinate actors that send many messages between each other in a manner that reduces the amount of overhead needed for typical message passing. In actor streams, there is also the concept of backpressure, where the rate of sending is slowed if the receiver cannot process messages at the same rate.

36.7 Addtional Resources

- `https://actor-framework.readthedocs.io/en/stable/`

Index

A
Absolute tolerance (atol), 356
Abstractions, 45, 46, 189, 214, 216, 219, 243, 244, 273, 343, 351, 352, 380
Accelerator, 107, 121
Actor model of concurrent computation, 333, 373
Actors, 5, 295, 296, 333, 373–382
Actor system, 374–381
Adapters (STL containers), 88
ADD_VALUES flag, 354
Algorithms, 47, 79, 85, 87, 89–92, 95, 96, 98–101, 149, 153–155, 157, 158, 160, 170–172, 174, 178, 179, 189, 199, 200, 202–204, 214, 244, 247, 259, 293, 295, 297, 298, 323, 340, 342, 343, 347, 351, 356, 358, 365, 381
Amdahl's law, 178–180
API, 49, 136, 181, 217–219, 223, 231, 237, 275, 278, 279, 328, 330, 336, 374
Array, 59–61, 63, 64, 86, 87, 97, 114, 119, 158, 161, 166, 167, 170–174, 182, 184, 210, 223, 225, 226, 228, 229, 238, 240, 242, 244–246, 248, 249, 251, 252, 254, 255, 257, 259, 265, 266, 271, 274, 275, 278, 281, 282, 286, 294, 295, 348, 349, 354, 360, 366, 367, 369, 371
Associative containers (STL containers), 85, 87, 88, 99
Asynchronous, 149, 231, 237, 238, 239, 374, 381, 382
Atomic clause, 205, 206
Atoms, 377, 378
ATPESC, vi
auto (C++ keyword), 7, 71, 73, 75, 80, 81, 87–90, 100, 154, 157, 158, 160, 172, 194, 196, 197, 201, 209, 211, 218, 258, 261–263, 265, 266, 282, 283, 285–287, 319, 320, 323, 371, 377, 378

B
Barriers, 154, 157, 158, 160, 161, 167, 203, 204, 210, 212, 217, 262, 327
Behavior, 10, 12, 24, 33, 55, 56, 58, 62, 64, 78, 98, 121, 150, 151, 165, 197, 201, 202, 217, 233, 236, 279, 292, 304, 306, 308, 310, 321, 352, 358, 374–380
BLAS, 333, 335–344, 350, 359, 366, 369
Blocking, 143, 145–152, 181, 224, 231, 240, 260, 381, 382
Branch, 19, 21, 23–26, 28, 45, 46, 48, 54–56, 69, 147, 216, 233, 234, 313, 315–317, 320, 321, 323
Branch prediction, 317, 320, 321
Breakpoint, 304–307, 326, 327
Broadcasting, 153–155, 169, 171, 172
Build target, 29, 32, 35, 36

C
C++, 16, 27, 28, 31–33, 39, 43, 45–49, 51–61, 64–66, 69–71, 73–75, 78, 79, 81, 85, 89, 90, 91, 93, 95, 99, 103, 104, 135–137, 139, 140, 142, 149, 154, 169, 173, 175, 189, 191, 199, 210, 214, 215, 233, 243, 251, 271, 275, 278, 279, 281–283, 287–289,

© The Editor(s) (if applicable) and The Author(s), under exclusive license to Springer Nature Switzerland AG 2025
R. J. Spiteri and K. Klenk, *Extreme-Scale Computing*, CMS/CAIMS Books in Mathematics 15, https://doi.org/10.1007/978-3-031-89033-8

291, 292, 298, 303, 306, 321–323, 325, 333, 336, 369, 372, 373
Cache, 34, 59, 61, 102, 110–112, 114–117, 123, 124, 214, 217, 253, 260, 264, 310, 313, 317–319, 321, 322
Cachegrind, 307, 313, 317, 318, 320–324
Cache misses, 59, 111, 317–319, 321, 322
C++ Actor Framework (CAF), 333, 373–381
Carl Hewitt, 374, 378
Catchpoints, 306, 307
Central Processing Unit (CPU), 59, 61, 97–99, 102, 107, 109–118, 122–125, 133, 140, 143, 178, 183, 189, 192, 195, 202, 206, 213–216, 233, 237, 242–244, 246–248, 253, 259, 301, 312, 317, 318–321, 325, 328, 365, 370, 374, 375, 380, 381
Classes, 28, 58, 64, 65–69, 71–73, 77, 78, 80–82, 85, 88, 90–93, 173–175, 225, 251, 252, 265, 283, 286, 287, 291–293, 295, 297, 298, 326, 351, 352, 354, 356–359, 365–371, 376
Clauses, 195–197, 200, 201, 204–207, 210, 359
Clusters, 107, 118, 121, 122, 125, 126, 128, 129, 149, 150, 178, 192, 213, 288, 310–312, 380, 381
CMake, 31–38, 40, 192, 244, 291, 292, 365, 366
Collective communication, 136, 143, 153, 156, 161, 181, 182
Command line, 3, 7–10, 12, 13, 15–17, 24, 26, 28–30, 32–34, 37, 137, 139–141, 191, 192, 220, 279, 288, 289, 303, 311, 315, 328–330, 336, 345, 346, 352, 361, 367, 368, 376
Commit, 19–26, 47, 49, 142, 169–172, 174
Communicator, 139, 141–143, 145–148, 150, 153–156, 157, 159–161, 163–168, 181, 182, 279, 359, 365, 368, 369, 371, 372
Compiler, 27, 28, 32, 33, 43, 47–49, 54, 67, 69–71, 81, 95, 137, 178, 191, 192, 215, 223, 248, 254, 262, 263, 279, 304, 313, 315, 326, 349
Compiling, 27, 28, 33, 61, 135, 137, 140, 191, 193, 207, 211, 247, 279, 304, 315, 325, 326, 336, 345, 346
Computation, 5, 55, 96–99, 101, 107, 121, 123, 135, 138, 143, 149, 152, 175, 178–180, 199, 205, 206, 213, 214, 237, 239, 243–247, 256, 257, 265, 266, 317, 333, 337, 338, 339, 340, 342–345, 347, 354, 361, 373, 374, 380
Concurrency, 373, 374
Concurrent computation, 121, 374, 380
Containers, 47, 85–93, 99, 136, 173, 175, 199, 209, 244, 273, 352, 370
Contiguous datatypes, 169, 170
Critical clause, 206, 207
CUDA, 189, 213–219, 221–226, 228–230, 231, 233–235, 237–243, 246–248, 255, 256, 260, 263, 325–328, 330, 340, 359

D

Data structure, 59, 65, 75, 85, 86, 101, 141, 253, 256, 297, 308, 335, 348, 349, 351, 352, 354, 361, 365, 370, 381
Datatypes, 136, 139, 142, 145, 146, 150, 154–157, 159–161, 163, 168–175, 181, 182, 185, 251, 274–276, 278
Debugging, 30, 140, 243, 245, 301, 303–307, 309–313, 325, 326, 328
Deep copy, 175, 255, 256
Dense matrix, 369
Dependencies, 27, 29, 30, 32, 36, 38–40, 95–98, 125, 204, 216, 261, 265
Derived datatypes, 163, 168, 169, 174, 275, 276
Desktops, 10, 11, 107, 109, 113, 114, 116, 118, 122, 140, 309
Direct methods, 361, 362
Directory, 5, 7–22, 29, 32–34, 36–38, 40, 128, 273, 304
Direct solver, 361–363
Distributed computing, 380
Distributed file-system, 128, 129
Distributed machine, 380
DMDACreate1d, 354
DMDACreate2d, 354, 359, 362
DMDACreate3d, 354
DMDA object, 354
DMDestroy, 358, 359, 361, 362
DM objects, 352, 354, 361
DM option, 354
DMSetFromOptions, 359, 361, 362
DMSetUp, 359, 361, 362
DMStag, 354

E

Error handling, 221, 224, 230, 352
Error kernel, 379, 380

Exascale computing, 123, 213, 243
Exascale computing project, 243
Execution spaces, 214, 215, 229, 247–249, 261, 265
Extreme-Scale Computing (ESC), 3, 5, 27, 28, 43, 48, 49, 102, 107, 113, 122, 123, 125, 133, 177, 271, 333, 339, 362, 363, 365

F
Fastest Fourier Transform in the West (FFTW), 333, 335, 347–350
Fault-tolerance, 378
Fetch, 25, 61, 86, 111, 112, 114, 224, 317, 320
File permissions, 10, 12
Filesystem, 271, 273
First-In-First-Out (FIFO), 381
Floating-point, 53, 98, 123, 339, 363, 368, 370
Flow control, 18, 34, 54, 122
Flows, 18, 21, 34, 51, 54, 78, 95, 97, 98, 111, 112, 122, 123, 143, 149, 185, 304, 305, 374, 375, 382
Framework, 168, 251, 303, 307, 313, 317, 333, 365, 373, 374, 381, 382
Function objects, 80, 93
Function pointer, 73, 79, 80
Functions, 3–5, 8, 26, 28, 31, 33–37, 38, 40, 47, 49, 51, 53, 54, 57–59, 66, 67, 69–71, 73–81, 88, 91, 93, 97, 99, 100, 103, 110, 111, 135, 136, 138–141, 145, 146, 150, 153–155, 164, 175, 177–179, 193, 194, 214–217, 219, 222–225, 231, 238–241, 245–249, 252, 256, 257, 263–267, 277, 297, 303, 305, 306, 309, 314–317, 320, 321, 326, 327, 328, 335, 342, 343, 348–350, 352–362, 368–370, 375–377, 379, 381

G
Gathering, 153, 156, 159, 161, 328
Generalized Minimal Residual (GMRES), 362, 363
Git, 6, 19–26, 40
GlobalOrdinal, 370
GNU Autotools, 31
GNU Debugger (GDB), 301, 303–307, 310–313, 325, 326, 330
Graphics Processing Units (GPUs), 114, 123, 124, 189, 213–221, 223–226, 229, 233, 237–239, 241–243, 246–248, 253, 301, 312, 325–329, 339, 340, 342, 343, 359, 365, 370
Group communicator, 163

H
Hardware, 8, 9, 45, 46, 48, 61, 90, 95, 107, 109, 111, 114, 121–123, 136, 138, 153, 177, 178, 183, 202, 205, 216, 219, 224, 243, 244, 247, 251, 254, 259, 260–265, 320, 321, 327, 335, 340, 347, 374, 381
HDF5, 16, 271, 273–281, 287, 289
Header files, 9, 16, 36, 73, 81, 82, 137, 139, 191, 245, 257, 292, 348, 349
Heterogeneous, v
High-level interface, 352
HIP, 359
Hybrid parallelism, 370

I
Implementation files, 28, 73, 81, 82
Incomplete LU factorization with zero fill-in, 362
Inconsistency denial, 378
Inconsistency robustness, 378
Indexed datatypes, 172
Indirect solver
Inheritance, 66–69
Input/output, 47, 57, 58, 151
INSERT_VALUES flag, 354
Installation rules, 37
Iterations, 54, 55, 60, 61, 87, 95, 97, 147, 199, 200–203, 244, 247, 260–262, 355–357, 362
Iterative methods, 355, 362
Iterator, 85, 87–93, 244, 247, 356

K
Kokkos, 243–249, 251–267, 301, 359, 265, 270–272
Krylov methods, 356
KSP context, 355
KSPCreate, 355, 359, 362
KSPDestroy, 355, 359, 361, 362
KSPGetPC, 357, 362, 363
KSP object, 354–356, 361
KSPRICHARDSON, 356
KSPSetFromOptions, 355, 359, 361, 362

KSPSetOperators, 355
KSPSetTolerances, 355–357
KSPSolve, 355, 359, 361, 362

L
Lambda functions, 47, 74, 79–81, 93, 246, 247, 265, 376, 377
LAPACK, 333, 335, 343–347, 350, 359, 366, 369
Laptops, 107, 109, 113, 114, 118, 140, 310
Latency, 113, 381
Left preconditioning, 355
Linear algebra, 333, 335, 343, 363, 365, 370
Linear solvers, 333, 354–358, 361
Linear system, 340, 343–345, 352, 354–358, 361–363
Linking, 9, 27, 28, 40, 135, 137, 191, 192, 279, 292, 336
Linux, 3, 5, 8, 9, 12, 16–18, 22, 28, 39, 82, 115, 116, 119, 129, 136, 192, 273, 291, 303, 307, 309, 349
List, 3, 9, 13, 14, 17, 23, 24, 29, 37, 48, 53, 66, 70, 81, 86, 90, 116, 142, 169, 182, 196, 197, 210, 217, 289, 293, 295, 296, 304, 305, 307, 308, 321, 367, 368
Load balancing, 381
LocalOrdinal, 370
Loop dependence, 95
Loops, 5, 18, 55–57, 61, 74, 89, 90, 95–99, 127, 128, 199–204, 207, 208, 212, 218, 220, 240, 245–249, 252, 256, 257, 259, 260, 263, 282, 294, 295, 296, 316, 319, 342, 343, 356
Low-level interface, 352
LU factorization, 340–342, 362, 363

M
Macros, 4, 28, 35, 247, 248, 375
Mailbox, 374, 375, 381, 382
Makefile, 29–31
Managing SSH connections, 6
Map, 87, 88, 99, 209, 259, 287, 317, 370–372, 375, 381
Mat class, 352
MatAssemblyBegin, 353, 361
MatAssemblyEnd, 353, 361
MatCreate, 352, 353
MatGetLocalSubMatrix, 354
MatGetRow, 354
Matrix solver, 363
MatSetValues, 352, 353, 361
MATSOLVERMUMPS, 362, 363
Memcheck, 307–310, 312
Memory, 52, 53, 59–64, 66, 73–78, 85, 86, 95, 98, 103, 109–114, 118, 119, 123–125, 133, 135, 136, 146, 155, 159, 161, 169, 174, 175, 178, 181, 183–186, 189, 191, 205, 214, 216–219, 221, 223–230, 234, 237–240, 242, 246, 248, 249, 251–256, 263, 264, 266, 267, 275–277, 279, 287, 292, 303, 306–310, 313, 317, 328–330, 336, 339, 342, 349, 350, 352, 358, 362, 363, 365, 366, 370
Memory ownership, 73–75
Memory spaces, 59, 125, 175, 189, 191, 214, 223, 240, 248, 249, 253–256, 263, 266, 276, 277
Merge, 21, 24–26
Message Passing Interface (MPI), 37, 38, 133, 135–143, 145–161, 163–175, 177, 178, 180–186, 191, 192, 194, 217, 245, 279, 309–312, 333, 349–355, 358, 359, 361, 365, 368–373
Messages, 15, 20, 24, 35, 104, 125, 126, 133, 135, 140, 142, 143, 145–152, 168, 175, 181, 352, 356, 358, 375–377, 380–382
Method, 5, 21, 38, 39, 65, 66, 68, 69, 78, 101, 133, 136, 246, 257, 258, 259, 263, 264, 267, 274, 277, 278, 283–287, 292, 293, 303, 313, 326, 328, 333, 352, 355, 356, 358, 361, 362, 368, 375–377
Middleman, 381
Mirrors, 255, 256
Modifying sequence, 90, 91
Module, 27, 35, 36, 39, 40, 113, 118
Motherboard, 113–115, 118, 121–125, 168
MPI_Finalize, 139, 141, 147, 151, 154, 157, 158, 160, 167, 170–172, 174, 184, 349, 350, 352, 369, 372
MPI_Init, 139–141, 147, 151, 154, 157, 158, 160, 167, 170–172, 174, 184, 349, 350, 352, 368, 369, 372
Multiple Instruction streams, Multiple Data streams (MIMD), 102, 109, 112
Multiple Instruction streams, Single Data stream (MISD), 102, 111
MUMPs, 362, 363

N
NetCDF, 271, 281–289
Network, 101, 114, 124–128, 135, 143, 149, 150, 167, 183, 271, 281, 288, 380, 381
Network transparent, 380
Newton-based line search, 357
Newton's method, 356
Node, 31, 102, 116, 125–129, 133, 135, 138, 150, 181, 184, 308, 365, 380
Nonlinear solvers, 333, 356–358
Non-modifying sequence, 90, 91
Numerical libraries, 365
NVIDIA, 123, 124, 213–220, 224, 226, 231, 240, 247, 301, 325, 328, 330, 340, 343

O
Objects, 28, 30, 35, 36, 45, 46, 64–67, 69, 71, 72, 74, 75, 79, 80, 86, 88, 93, 136, 137, 150, 168, 175, 192, 208, 224, 247, 273–275, 277–279, 283, 285, 287, 291–295, 297, 298, 352–359, 361–363, 365–372, 376
One-sided communication, 136, 177, 183–185
OpenMP, 136, 138, 189, 191–197, 199–203, 205–208, 210–212, 243, 246–248, 310, 349, 359
Ordinary Differential Equations (ODE), 358

P
Package manager, 9, 39, 115, 136, 307
Parallel programming, 43, 95, 101, 107, 179
Parallel sections, 192, 194, 195, 199, 200, 211
ParameterList, 366–368
Partial Differential Equation (PDE), 351, 352, 358, 359, 362
Patterns, 4, 16, 30, 35, 43, 81, 95, 98, 101, 114, 152, 153, 161, 163, 166, 167, 170, 171, 235, 236, 244, 251, 253, 254, 265, 362, 370, 377, 379
Performant, 180, 273, 335
Peripheral Component Interconnect Express (PCIe), 113, 114, 121, 122, 124–126, 213, 241
Peripherals, 113, 114, 121
Persistent storage (non-volatile storage), 113, 118, 119
PETSC_DECIDE, 353, 354, 359, 362
PETSC_DEFAULT flag, 356
PetscCall, 352, 353, 355, 357, 359–362
PetscErrorCode, 352, 359, 360
PetscFinalize, 352, 353, 355, 357–359
PetscInitialize, 352, 353, 355, 357–359, 362
PetscOptionsGetValue, 352
PETSc solvers, 354
Pointers, 23, 25, 26, 53, 59, 61–64, 73–80, 103, 146, 149, 151, 154, 169, 170, 174, 175, 182, 224, 225, 240–242, 253, 278, 286, 292, 294–296, 306, 308, 366, 367
Point-to-point communication, 128, 143, 145, 147, 149, 151, 153, 154, 161, 181
Poisson equation, 359, 361
Polymorphism, 351
Portable Extensible Toolkit for Scientific computing (PETSc), 333, 351–363, 370
Preconditioner, 355, 357, 362, 363
Primitive datatypes, 169
Profiling, 86, 202, 243, 301, 313, 314, 316, 317, 321, 325, 328–330
Programming models, 213, 214, 216, 217, 243, 373, 382
Program structure, 51, 139, 237
Pull, 20, 21, 25, 156
Push, 20, 21, 23, 24, 59, 87, 278

Q
Queue, 86, 88, 237–239, 375, 381

R
Race condition, 204–208
Random Access Memory (RAM), 59, 111, 113, 114, 118, 119, 122–125, 129, 133, 214, 240, 248, 264
Randomized Numerical Linear Algebra (RNLA), 363
Rebase, 24–26
Reducing, 153, 154, 207, 214, 243, 326, 371
Repository, 6, 19–26, 138, 244, 291, 309, 359, 365
Return code, 141, 142
Richardson, 356
Right preconditioning, 355, 356
Relative tolerance (rtol), 356
Run-time estimation, 316

S

Single Program Multiple Data (SPMD), 368
Scalar, 225, 293, 295, 297, 335, 337, 338, 353, 359, 360, 370
Scaling, 112, 126, 127, 177–181, 336, 343
Scattering, 153, 156, 160
Scheduler, 5, 129, 138, 201, 265–267, 329, 381
Scoped_actor, 375–378
Scratch pads, 263, 264
Scripting, 5, 17, 18, 298
Secure Shell (SSH), 5–8, 21, 22
Sequential containers (STL containers), 85, 86
Shallow copy
Shared-memory paradigm, 95, 189
Shared pointers (smart pointer), 73, 76–79, 292, 366
Shell, 4, 5, 11, 16–18, 61, 220, 311, 363
Shuffling, 234
Single Instruction stream, Multiple Data streams (SIMD), 102, 112, 123, 211, 262, 348
SISD (serial), 102, 109, 111
SNES context, 356, 357
SNESCreate, 356, 357
SNESDestroy, 356, 357
SNESNEWTONLS, 357
SNESSetFunction, 357
SNESSetJacobian, 357
SNESSetTrustRegionTolerances, 357
SNESSetType, 356, 357
SNESSolve, 356, 357
Software, 8, 9, 16, 19, 23, 27, 28, 31, 32, 38, 39, 48, 49, 78, 85, 101, 107, 125, 129, 136, 177, 178, 243, 244, 271, 273, 281, 289, 303, 312, 327, 333, 339, 365, 366, 373, 374
Software applications, v
Software management, vi
Sorting, 90, 92
Sparse matrix, 340, 344, 352, 370, 371
Splitting communicator, 165
SSH connections, 5–8
Stack, 28, 39, 59–61, 67, 74, 88, 168, 304–306, 310, 327, 333
Staggered grids, 354
Standard error, 14
Standard implementation, 47, 48
Standard input, 14, 51, 57, 58
Standard output, 14, 58
Standard Template Library (STL), 43, 47, 78, 79, 85–90, 93, 95, 98, 99, 101, 136, 189, 247
Streams, 14, 15, 57, 58, 102, 121, 207, 237–239, 329, 374, 382
Strided datatypes, 170–172, 174
Strong scaling, 178–181
Structs, 65, 68, 69, 146, 148, 149, 173, 174, 245, 262, 263, 265, 276, 285, 286, 297, 359
Subviews, 253, 257, 258
Supercomputing, 136, 177, 213
Supervision hierarchy, 378–380
Supervision (Svn), 19, 378–380
Synchronization, 19, 24, 26, 150–152, 161, 181, 186, 203, 204, 206, 207, 216, 217, 222, 223, 233, 234, 260

T

Tags, 142, 145, 146, 148–150, 165
Task based programming, 264, 374
Tasks, 13, 28, 101, 123, 192, 203, 211, 212, 264–267, 326, 343, 344, 373, 374, 381
Templates, 43, 47, 59, 70–72, 79, 82, 85, 91–93, 100, 154, 157, 158, 160, 243, 249, 251, 253, 263, 265, 266, 326, 337–339, 345–347, 370
Testing, 37, 40, 56, 138, 181
Teuchos, 333, 365–369, 371, 372
Teuchos::ArrayRPC, 366, 367
Teuchos::Comm, 367, 368
Teuchos::CommandLineProcessor, 367, 368
Teuchos::Prt, 366
Teuchos::RPC, 367
Teuchos::TimeMonitor, 368
Text editors, 8, 15, 16
Threads, 116, 193–197, 199–206, 208–212, 216, 217, 221–227, 234, 253, 254, 260–264, 309, 326, 327, 349, 366, 376, 381
Thread teams, 203, 260–263
Time-dependent nonlinear problems, 351
Time-steppers, 358
Topological communicator, 163, 166, 182
Tpetra, 333, 365, 370–372
Trilinos, 20, 21, 333, 365–369, 372
Trust region, 357
Type, 4, 15, 19, 28, 30, 32, 34–37, 45–47, 49, 51–54, 58, 64, 66, 67, 69, 70–75, 76, 78–82, 85, 89, 91–93, 97, 99,

113, 118, 119, 123, 140, 142, 146, 149, 154–156, 166, 169–175, 207, 209, 219, 222, 225–227, 229, 230, 251, 252, 260–267, 275–278, 282, 284–288, 293–298, 306–308, 310, 311, 317–319, 321, 327, 329, 336, 337, 339, 343, 344, 348–350, 352, 362, 363, 367, 370–372, 375–377
Type deduction, 47, 73

U
Unifying memory, 240
Unique pointers (smart pointer), 73–76, 78, 79, 292, 366
Unstructured grids, 294, 296, 297, 354

V
Variables, 7, 16, 18, 29, 30, 32–35, 37–39, 51, 52, 54–56, 59–67, 73, 74, 79, 81, 96, 104, 129, 141, 146, 150, 154, 155, 160, 175, 191, 192, 195–197, 200, 201, 206, 208, 210, 222, 223, 225–227, 229, 230, 234, 239, 240, 266, 278, 281–288, 294, 304–308, 310, 326, 328, 351, 352, 361, 367, 368, 372, 376
VecAssemblyBegin, 353, 360
VecAssemblyEnd, 353, 360
Vec class, 352
VecCreate, 352, 353
VecGetArray, 354, 360
VecRestoreArray, 354, 360
VecSetValues, 352, 353
Vector, 86, 87, 89, 89, 93, 100, 112, 114, 123, 154, 157, 158, 160, 164, 170–173, 175, 196, 197, 200, 203, 207, 210, 211, 222, 223, 224, 226, 235, 241, 242, 244, 252, 258, 259, 261, 262, 266, 277, 278, 282–284, 286, 288, 297, 319, 322, 323, 335, 336, 337, 342, 345, 347, 353–357, 366, 370–372
Vector-level parallelism, 261, 262
Version control, 19, 40
Views, 16, 45, 47, 98, 101, 113, 133, 169, 213, 214, 240, 249, 251–258, 262–264, 314, 318, 321
Virtual Functions, 67, 69
Visualization Toolkit (VTK), 291–298
von Neumann, 109–112, 118

W
Warps, 216, 217, 221, 233, 234, 236, 327
Watchpoints, 306, 307
Weak scaling, 180
Windows, 3, 8, 48, 183–185, 295, 296, 303, 349
Workstations, 9, 107, 109, 113, 114, 117, 118, 122, 123, 125, 140, 310
Work-stealing, 381

The manufacturer's authorised representative in the EU is Springer Nature Customer Service Centre GmbH, Europaplatz 3, 69115 Heidelberg, Germany. If you have any concerns regarding our products, please contact ProductSafety@springernature.com

Printed and bound by CPI Group (UK) Ltd, Croydon, CR0 4YY
26/03/2026
02078979-0002